Springer Undergraduate Texts
in Mathematics and Technology

Springer Undergraduate Texts in Mathematics and Technology (SUMAT) publishes textbooks aimed primarily at the undergraduate. Each text is designed principally for students who are considering careers either in the mathematical sciences or in technology-based areas such as engineering, finance, information technology and computer science, bioscience and medicine, optimization or industry. Texts aim to be accessible introductions to a wide range of core mathematical disciplines and their practical, real-world applications; and are fashioned both for course use and for independent study.

More information about this series at http://www.springer.com/series/7438

Øyvind Ryan

Linear Algebra, Signal Processing, and Wavelets - A Unified Approach

MATLAB Version

 Springer

Øyvind Ryan
Department of Mathematics
University of Oslo
Oslo, Norway

The author has received funding from the Norwegian Non-Fiction Writers and Translators Association.

Additional material to this book can be downloaded from http://extras.springer.com.

ISSN 1867-5506   ISSN 1867-5514   (electronic)
Springer Undergraduate Texts in Mathematics and Technology
ISBN 978-3-030-01811-5   ISBN 978-3-030-01812-2   (eBook)
https://doi.org/10.1007/978-3-030-01812-2

Library of Congress Control Number: 2018964077

Mathematics Subject Classification (2010): 15-01, 65-01, 15A99, 65T50, 65T60

This Springer imprint is published by the registered company Springer Nature Switzerland AG
The registered company address is: Gewerbestrasse 11, 6330 Cham, Switzerland

To Natalia, Katrina, and Alina.

Preface

The idea for this book began with a new course at the University of Oslo, called "Applications of Linear Algebra", which was offered for the first time in 2012. At the university we had come to realize that students who have had a first course in linear algebra, already have the prerequisites to learn about several important and interesting topics in signal processing and wavelet theory. Unfortunately, most textbooks on these subjects are written in a language poorly suited to a linear algebra background, making much literature only accessible to engineering and signal processing students, and excluding a large number of students. It also seems that it is not a common textbook strategy to introduce signal processing and wavelets together from scratch, even though the two can very much motivate one another. Why not write a self-contained textbook, where linear algebra is the main foundation? This notion is what motivated the preparation of this book.

The author must admit that he is not particularly fond of the majority of signal processing literature, which often hides ideas and results which essentially boil down to basic linear algebra, using a signal processing nomenclature that is difficult for many mathematicians to understand. Some examples are:

- Matrix notation is often absent. Instead, linear operations, such as the DFT, are often expressed by component formulas. Matrix multiplication is absent, although basic operations such as convolution can be interpreted as this. As a result, important matrix factorizations are not included.
- Many operations that represent changes of coordinates (such as the DFT, DCT and DWT) are not represented as such. As a result, no tools, notation or results from linear algebra are used for these operations.
- Eigenvalues and eigenvectors are not mentioned, even when these concepts could shed light on the theory: It is often not mentioned that the Fourier basis vectors are eigenvectors for filters, with the frequency response being the corresponding eigenvalues. The property that *convolution in time corresponds to multiplication in frequency*, an important notion in signal processing, can be summarized as follows in linear algebra: for matrices with the same eigenvectors, the eigenvalues of the product are the product of the eigenvalues.
- Function spaces are rarely put into the context of vector or inner product spaces, even if Fourier series can be seen as a least squares approximation from such spaces.

Further, it is rarely mentioned that the formulas for the Fourier coefficients (the Fourier integrals) follow from the orthogonal decomposition theorem.

Others have also recognized the need to write new textbooks that employ more linear algebra. As an example, book [70] goes further in using matrix notation than many signal processing textbooks. Still, more can be done in this direction, so that students with only a basic linear algebra background will feel more at home.

This book provides an introduction to Fourier analysis and signal processing (in the first part of the book) and wavelets (in the second), assuming readers have completed an introductory course in linear algebra. Without this background knowledge, the book will be of limited value. An appendix has been included so that students can review the most important linear algebra concepts, but a full course on these topics is preferred in order to follow the book. We use book [32] as the primary reference work for linear algebra, meaning that many concrete definitions and concepts from that book are cited by concrete identifiers (i.e. page number, theorem number, etc.). This is a very popular choice for a first course in linear algebra, but there are many other excellent textbooks available. Leon [33] is a good example, which, just as [32], includes many exercises related to MATLAB. Several other good introductions to linear algebra claim to be applied in some way, such as [53], [75] and [69]. For a book that goes a bit further into the theory than these, please consult [49] or [37].

The book can be used at the undergraduate level, and can be taken directly after a first course in linear algebra. It is also possible to use the material at the graduate level. A number of theory aspects from basic linear algebra are further developed:

- Complex vector spaces and inner products are considered (many introductory linear algebra textbooks concentrate only on real vector spaces and inner product spaces).
- Inner product spaces that are function spaces are used extensively. Many introductory linear algebra textbooks consider such spaces, but often fail to offer extensive practice with them.
- More intuition on changes of coordinates is developed, in particular for the DFT, DCT and DWT.

The book itself can be viewed as an extension to a basic linear algebra textbook, and may in the future become additional chapters in such a book.

The style of the book is very different from most textbooks on signal processing, since the language of linear algebra is used consistently. Much of the material and many proofs have been redesigned from their sources to fit into a linear algebra context. The material on wavelets has also been redesigned for this purpose, but also because it is not that common to introduce wavelets at the undergraduate level. The book also attempts to break up much theory with explanatory code. This approach, too, is rarely found in the literature, where code is often moved to appendices, to separate it completely from the theory. The author believes that such separation is unfortunate in many cases, as it can lead students to spend all their time on the theory alone. Due to these considerations, this book has been more than 6 years in the making.

Since the book does not offer a comprehensive treatment of all basic signal processing concepts or nomenclature, some engineering students may feel excluded. To make the book more accessible for such students, a section is included to help reconcile the signal processing perspective and linear algebra perspective. The summaries throughout the book also address the connections and differences between the two perspectives. The section below on how to use the book in engineering courses provides further details.

The book attempts to go further than many books with the name "applications" in their title, in that it has a clear computational perspective. The theory motivates algorithms and code, for which many programming and best coding practice issues need to be addressed. The book focuses on integrating applications, such as modern standards for the compression of sound and images (MPEG, JPEG, JPEG2000). A public and open source github repository for Fourier analysis and wavelets accompanies the book, where all code from the book can be found (in particular the FFT and DWT libraries), as well as accompanying test code, documentation, and sample files. Throughout the book this repository will simply be referred to as "the library". The address of the repository is https://github.com/oyvindry/applinalgcode. The library contains more code than shown in the book, and this code will be continuously updated and improved. Note also that the scope of the library is larger than that of the book. As an example, a paper on more general boundary modes for wavelets is being completed in parallel with the book. Also, since most algorithms in the book are well suited for parallel computing, future versions of the library and the book may support GPU-based programming.

The library differs from many toolboxes in that the individual steps in a given implementation are firmly anchored in labeled formulas in the book. The focus on the computational perspective has been inspired by the project "Computing in Science Education" at the University of Oslo, an initiative to integrate computations into the basic science curriculum at the university from the very first semester.

Programming

It is assumed that the student has already been introduced to a programming language or computational tool. Preferably, he or she will have already taken a full course in programming first, since the book does not include an introduction to primitives such as loops, conditional statements, lists, function definitions, file handling, or plotting. At the University of Oslo, most students take such a Python-based course during the first semester, focusing on such primitives. This course uses book [31], which provides an excellent introduction to Python programming for beginning students.

This book comes in two versions: one based on MATLAB, and one on Python (where version 3 is the supported version). The version of the book you are reading uses MATLAB. There are several other recent examples of books adapted to different languages: See [41] and [42] for MATLAB and Python versions of a text on mechanics, and [34] and [35] for MATLAB and Python versions of a computational science textbook (the latter are also recommended introductions to programming in these languages for beginning students). If you search the internet for recommendations about what programming language to use in a basic linear algebra course, you may find statements such as "Python is too advanced for such a beginning course", or "You can get started much quicker with MATLAB". A good reply to such statements may be that the choice of programming language should most likely depend on how programming is already integrated into the courses at your university. Provided the integration at your university is thought-through and well-organized, the programming in the book should not prove to be too advanced for you, regardless of the choice of language.

To distinguish between the library and other toolboxes, most functions in the library have the ending "_impl". If you compare the code in the MATLAB and Python versions of the book, you will see that the code in the two languages is very similar:

- Function signatures and variable names are virtually the same, following the Python standard (words separated by "_" and using lowercase letters).
- Code indentation follows the Python standard, where it is an important part of the syntax.
- The heavily used Python package numpy does not use a prefix, since MATLAB does not prefix code (this means that other packages with conflicting names must use a prefix). For the same reason, the modules developed in the book also do not use a prefix.

There are also some differences in the MATLAB and Python versions, however.

- The Python code is split into modules, a very important structuring concept in Python. The book explains which modules the corresponding code is part of. In MATLAB there is no module concept, and many functions must be placed in files of the same name. As a result, there are more files in the MATLAB part of the library.
- In Python, it is customary to place test code in the same module as the functions being tested. Since MATLAB does not use modules, separate files are instead used for test code instead.
- MATLAB passes all input and return parameters by value, not by reference, as languages like Python do. This means that we can perform *in-place computation* in Python, i.e. the result can be written directly into the input buffer, avoiding memory copy. This can lead to much more efficient code, so the Python code performs in-place operations wherever possible. This affects the signatures of many functions: Several Python functions have no return values, since the result is written directly into the input buffer, while the MATLAB counterparts use a return value for the result. Examples of in-place algorithms developed in this book are the DFT, the DCT, filters, the DWT, and filter bank transforms.

There are also many other small differences between the two languages:

- Indices in vectors start with 1 in MATLAB, and 0 in Python (indexing starting with 0 is more natural for many of the algorithms discussed in the book).
- The distinction between different dimensions (1, 2, 3, and so on) is clearer in Python.
- Extracting subsets of an object corresponding to rows/columns with a given set of indices works more uniformly in MATLAB.

Many of these differences can be easily overcome, however, and there are many translation guides available online.[1]

MATLAB started as a command-based tool, and has evolved over the years to support many features of modern programming languages. Python on the other hand was designed as a general programming language from the outset. As a result, MATLAB is not as structured as Python, and there are several bad MATLAB programming habits around, which produce programs with illogical structures. We have sought to follow many Python conventions for MATLAB code, in an attempt to avoid many of these habits.

Python has support for classes, but their use in the library has been limited for two reasons. First of all, class hierarchies don't greatly simplify programming of the concepts in the book. Secondly, although MATLAB has some primitive support for classes, it remains unclear to the author how well this support actually works. By not using classes, the MATLAB and Python code in the library have virtually identical APIs.

---

[1] See for instance http://mathesaurus.sourceforge.net/matlab-numpy.html.

MATLAB has a built-in functionality for reading and writing sound and images, and another for playing sound and displaying images. To make the Python code similar to the MATLAB code, the library includes the **sound** and **images** modules, where the signatures are similar to their MATLAB counterparts. These functions simply call Python counterparts in such a way that the interface is the same.

Although the library contains everything developed in the book, the student is encouraged to follow the code development process used in the book, and to establish much of the code on his or her own. This can be an important part of the learning process. The code in the library has been written to be as simple and understandable as possible. To achieve this, some of the code is not as efficient as it could have been. With time the library may include more optimized versions of the algorithms presented in the book.

Structure of the Book

This book can be naturally divided into two parts. The first part provides an introduction to *harmonic analysis* (Chap. 1), *discrete Fourier analysis* (Chap. 2), and *filters* (Chap. 3). Chapter 1 starts with a general discussion of what sound is, and how to perform operations on digital sound. Then Fourier series is defined as a finite-dimensional model for sound, and the mathematics for computing and analyzing Fourier series are established. Chapter 2 moves on to digital sound. Sound is now modeled as vectors, and this establishes a parallel theory, in which Fourier series is replaced by the Discrete Fourier Transform. Three important topics discussed in this context are the Fast Fourier Transform, the Discrete Cosine Transform, and the sampling theorem. In Chap. 3 we look at an important type of operations on digital sound called filters, and it is revealed that these are exactly those linear transformations diagonalized by the Discrete Fourier Transform.

The second part of the book, starting with Chap. 4, gives an introduction to *wavelets* and *subband coding*. Chapter 4 starts with a motivation for introducing wavelets. While part one considers the frequency representation of sound, wavelets can adapt to the fact that frequency content may change with time. After motivating the first wavelets and setting up a general framework in Chap. 4, Chap. 5 establishes the connection between wavelets and filters, so that the theory from the first part applies. A more general framework than wavelets is established, which is the basis for what is called subband coding. Subband coding is revisited in Chap. 7 as well. In Chap. 6 we establish a theory that is used to construct useful wavelets. Chapter 8 is a small departure, addressing images and basis image processing, before Chap. 9 sets up the theory for wavelets in a two-dimensional framework, allowing us to experiment with them on images.

Each chapter includes a list of minimum requirements, which is intended to help students prepare for exams.

Notation and Differences in Notation from Signal Processing

We will follow the linear algebra notation you can find in classical linear algebra textbooks. Appendix A and the nomenclature provide a detailed account of the notation and conventions therein. The linear algebra notation and conventions differ from signal processing in several ways:

- In signal processing, the vector, sequence or function we analyze is often called the *signal*.
- In signal processing, vectors/sequences are usually called *Discrete-time signals*. The corresponding notation in signal processing would be x for the signal, and $x[n]$ for its components, and different signals with different base names are often named like $x_1[n]$, $x_2[n]$. One also often writes (capital) X for the DFT of the vector x.
- In signal processing, functions are usually called *Continuous-time signals*, and one often uses capital letters to denote continuous-time Fourier transforms, i.e. one often writes $X(\omega)$ for $\hat{x}(\omega)$.
- In mathematics, i is normally used for the imaginary complex number that satisfies $i^2 = -1$. In engineering literature, the name j is more frequently use.
- Matrices are scaled if necessary to make them unitary, in particular the DFT. This scaling is usually avoided in signal processing.

Regarding the first two points about using the capital letter of the input for the output, we will mostly avoid this by using dedicated letters for the input (typically $\boldsymbol{x}$), and for the output (typically $\boldsymbol{y}$ for the output of a DFT, and $\boldsymbol{z}$ for the output of a filter).

Resources for Students

Solutions to selected exercises in the book are available as electronic supplementary material for those who purchase the book. It is recommended to use these solutions with care. Much of the learning outcome depends on trial and error in solving exercises, and one should therefore not take the shortcut directly to the solutions: Although students may understand the solution to an exercise, they may not learn the thinking process needed to arrive at that solution. Jupyter notebooks for the programming-based examples and exercises (with solutions) in the book can also be found online at http://folk.uio.no/oyvindry/applinalg/. There is one notebook per chapter. Please consult the file `readme.txt` therein, where prerequisites for using those notebooks are covered. The notebooks will be updated continuously, and also upon request. The examples in the notebooks provide code that reproduce many figures in the book. The figures in the printed version of the book have been generated using Matplotlib, which has become a standard for plotting in Python. The sample code provided in the book may not run on its own, as it may rely on importing certain packages, or defining certain variables. These imports and definitions can be found in the notebooks, however.

Some of the code in the book produce sounds, of which many have file names associated with them. Those files can be found at http://folk.uio.no/oyvindry/applinalg/sounds/.

Students who program for this book are recommended to place their programs in the root folder obtained from the github page of the library, and use this as their working directory. Also, they should add the folders `matlab` and `python` to the path, and access sound files in the `sounds` folder, and image files in the `images` folders. Following these guidelines will provide a uniform access to those files, which also all resources for the book follow. Aside from this, no additional installation procedures are needed.

In several exercises the student is asked to implement his or her own versions of functions already residing in the library. He or she can then choose any function name, and using the same name as the library counterpart is also admissible. The order of the directories in the path is then important.

When working with the code in the library, it is useful to consult the documentation of the functions. This documentation can be seen when using auto-completion. Useful code examples from the library can also be found in the test code. This code can be found in files starting with `test_`, such as `test_fft.m`, `test_dwt.m`, and so on.

Resources for Instructors

To help instructors adopt the book in courses, the product page of the book on springer.com provides them with some supplementary material. This includes solutions to all exercises in the book.

More supplementary material for instructors may be added upon request, such as templates for slides for the various chapters. These can be further tailored by the instructor.

Advice for Course Adoption

The material in this book may be used either as part of a course on (applied) mathematics, or as part of an engineering course. Parts of it may also be used to supplement a graduate course. Below are some general recommendations on how a course may adopt the book, based on lessons learned at the University of Oslo.

Adopting the Book in Math Courses

For early bachelor level courses (e.g. for the 4'th semester), some parts of the book can be omitted to limit its scope. The following are only loosely linked to the rest, and are therefore good candidates for such omissions:

- Section 1.6 on the convergence of Fourier series,
- Sections 2.4, 2.5, and 3.5 on the Discrete Cosine Transform,
- Chapter 6 on the general construction of wavelets, and
- Sections 5.4 and 7.3 on multi-band signal processing.

These parts are also mathematically more advanced than the rest, and may therefore be more suitable for later bachelor courses. The first two also require analytic methods, which are not prerequisites for the remainder of the book. Another option is to drop Chaps. 8 and 9, as they are the only ones concentrating on higher dimensional data.

There are many important exercises in the book, covering many important topics not discussed in the text. Some of these are rather difficult, and some are long and more suitable for projects. Accordingly, another option is not to skip much of the material above, but to instead limit the number of exercises, and leave the focus on exercises and their solutions for subsequent courses. Many of the exercises are related to programming, and hopefully most universities provide the student with sufficient programming skills to solve these. If not, it is possible to limit the exercises related to programming without affecting the overall understanding of the theory.

Note also that most of the material in Chaps. 2 and 3 can be gone through independently of Chap. 1, and that Chap. 4 can be read independently from the first part of the book.

As previously mentioned, a course on linear algebra should be one major prerequisite. The appendix on linear algebra lists these prerequisites, but is not a substitute for them. The appendix can nevertheless be useful for planning the course, and instructors are free to add repetition exercises for the topics listed there.

Adopting the Book in Engineering Courses

If a full linear algebra course is part of the degree program, adoption can be similar to the approach described above. If not, one should spend time on going through the appendix on linear algebra, and add related exercises from a linear algebra textbook.

In an engineering program the student is likely to have had another introductory course on signal processing. He or she will have then already been exposed to many of the ideas in Chaps. 2 and 3, so that these chapters can be gone through quickly. It might be advisable to start with signal processing concepts that the students already know, and continue with how these concepts are presented differently in the book. As two examples:

- One could start with filters defined in the signal processing manner, and then move on to circulant Toeplitz matrices,
- One could start with the DFT, the frequency response, and the statement "convolution in time equals multiplication in frequency", and then explain how this is related to diagonalization.

In an engineering program some students may have already been exposed to multi-band signal processing. This would eliminate the need for Sects. 5.4 and 7.3. Chapter 6 on the general construction of wavelets may then be a useful add-on.

In Chap. 3 a section has been added on how the treatment of the topics in the book differs from that in signal processing: Comments of a similar nature can be found in the summaries throughout the book, in particular on multi-band signal processing in Chaps. 5 and 7. We have also attempted to explain how engineers work with these concepts, and in which fields they are applied. These comments will hopefully make the book more useful for those with a signal processing background.

Acknowledgments

The author would like to thank Professor Knut Mørken for his input on early versions of this book, and for suggesting that I use my special background to write literature that connects linear algebra to the topics of the book. Special thanks also go to Andreas Våvang Solbrå and Vegard Antun for their valuable contributions to the notes, both in reading early and later versions of them, and for teaching and following the course. Thanks also to the students who have taken the course, who have provided valuable feedback on how to present the topics in an understandable way. I would also like to thank all participants in the CSE project at the University of Oslo for their continuous inspiration.

The book has been typeset using a sophisticated program called DocOnce, which enabled the author to use a single source for both the MATLAB and Python versions, as well as other formats. Particular thanks go to Professor Hans Petter Langtangen, who, in addition to developing DocOnce and using it to write inspiring textbooks, also encouraged me to use it for making highly flexible course material. To the author's knowledge, this is the first mathematics book to be typeset with DocOnce.

In the author's view, this book represents only one step in what can be done to link linear algebra with other fields. He hopes to do much more in this direction in the years to come, and that this book can inspire others to do the same.

Oslo, Norway   Øyvind Ryan
December 2018

Contents

List of Examples and Exercises

Chapter 1
Sound and Fourier Series

A major part of the information we receive and perceive every day is in the form of audio. Most sounds are transferred directly from the source to our ears, like when we have a face to face conversation with someone or listen to the sounds in a forest or a street. However, a considerable part of the sounds are generated by loudspeakers in various kinds of audio machines like cell phones, digital audio players, home cinemas, radios, television sets and so on. The sounds produced by these machines are either generated from information stored inside, or electromagnetic waves are picked up by an antenna, processed, and then converted to sound.

What we perceive as sound corresponds to the physical phenomenon of slight variations in air pressure near our ears. Air pressure is measured by the SI-unit Pa (Pascal) which is equivalent to N/m^2 (force/area). In other words, 1 Pa corresponds to the force of one Newton exerted on an area of 1 square meter. Larger variations mean louder sounds, while faster variations correspond to sounds with a higher pitch.

Observation 1. Continuous Sound.

A sound can be represented as a function, corresponding to air pressure measured over time. When represented as a function, sound is often referred to as continuous sound.

Continuous sounds are defined for all time instances. On computers and various kinds of media players, however, the sound is *digital*, i.e. it is represented by a large number of function values, stored in a suitable number format. This makes it easier to manipulate and process on a computer.

Observation 2. Digital Sound.

A digital sound is a sequence $\boldsymbol{x} = \{x_i\}_{i=0}^{N-1}$ *that corresponds to measurements of a continuous sound f, recorded at a fixed rate of f_s measurements per second, i.e.*

$$x_k = f(k/f_s), \quad for \ k = 0, \ 1, \ \ldots, \ N-1.$$

f_s is called the *sampling frequency* or *sample rate*. The components in digital sound are called *samples*, and the time between successive samples is called the *sampling period*, often denoted T_s. Measuring the sound is also referred to as sampling the sound.

© Springer Nature Switzerland AG 2019
Ø. Ryan, *Linear Algebra, Signal Processing, and Wavelets - A Unified Approach*,
Springer Undergraduate Texts in Mathematics and Technology,
https://doi.org/10.1007/978-3-030-01812-2_1

Just as for the indices in a digital sound, vector and matrix indices will mostly start at 0 in this book. This indexing convention is not standard in most mathematics, where indices usually start at 1. It is, however, standard in many programming languages, and also in signal processing.

The quality of digital sound is often measured by the *bit rate* (number of bits per second), i.e. the product of the sampling rate and the number of bits (binary digits) used to store each sample. Both the sample rate and the number format influence the quality of the resulting sound. These are encapsulated in *digital sound formats*, some of which we describe below.

Telephony

For telephony it is common to sample the sound 8000 times per second and represent each sample value as a 13-bit integer. These integers are then converted to a kind of 8-bit floating-point format with a 4-bit significand. Telephony therefore generates a bit rate of $8 \times 8000 = 64,000$ bits per second, i.e. 64 kb/s.

The CD-Format

In the classical CD-format the audio signal is sampled 44,100 times per second and the samples stored as 16-bit integers. The value 44,100 for the sampling rate is not coincidental, and we will return to this shortly. 16-bit integers work well for music with a reasonably uniform dynamic range, but is problematic when the range varies. Suppose for example that a piece of music has a very loud passage. In this passage the samples will typically make use of almost the full range of integer values, from $-2^{15} - 1$ to 2^{15}. When the music enters a more quiet passage the sample values will necessarily become much smaller and perhaps only vary in the range -1000 to 1000, say. Since $2^{10} = 1024$ this means that in the quiet passage the music would only be represented with 10-bit samples. This problem can be avoided by using a floating-point format instead, but very few audio formats appear to do this.

The bit rate for CD-quality stereo sound is $44,100 \times 2 \times 16 \, \text{bits/s} = 1411.2 \, \text{kb/s}$. This quality measure is particularly popular for lossy audio formats where the uncompressed audio usually is the same (CD-quality). However, it should be remembered that even two audio files in the same file format and with the same bit rate may be of very different quality because the encoding programs may be of different quality.

Below we will read files in the `wav`-format. This format was developed by Microsoft and IBM, and is one of the most common file formats for CD-quality audio. It uses a 32-bit integer to specify the file size at the beginning of the file, which means that a WAV-file cannot be larger than 4 GB.

Newer Formats

Newer formats with higher quality are available. Music is distributed in various formats on DVDs (DVD-video, DVD-audio, Super Audio CD) with sampling rates up to 192,000 and up to 24 bits per sample. These formats also support surround sound (up to seven channels in contrast to the two stereo channels on a CD).

In this first chapter we will start by briefly discussing the basic properties of sound: *loudness* (the size of the variations), and *frequency* (the number of variations per second). We will then experiment with digital sound, and address to what extent sounds can be decomposed as a sum of different frequencies. Finally we will look at important operations on sound, called *filters*, which preserve frequencies.

The code examples in this chapter assume that your working directory is the root folder of the library, and that the `matlab` folder in the library has been added to the path.

1.1 Sound and Digital Sound: Loudness and Frequency

An example of a simple sound is shown in the left plot of Fig. 1.1, where air pressure is plotted against time. The initial air pressure has the value 101,325 Pa, which is the normal air pressure at sea level. Then the pressure varies more and more until it oscillates regularly between 101,323 Pa and 101,327 Pa. In the area where the air pressure is constant, no sound will be heard, but as the variations increase in size, the sound becomes louder and louder until about time $t = 0.03$ where the size of the oscillations becomes constant.

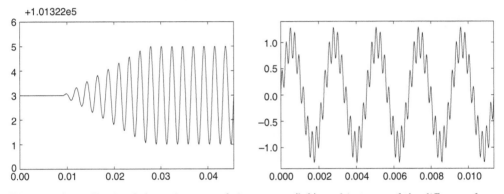

Fig. 1.1 An audio signal shown in terms of air pressure (left), and in terms of the difference from the ambient air pressure (right)

When discussing sound, one is usually only interested in the variations in air pressure, so the ambient air pressure (101,325 Pa) is subtracted from the measurements. This has been done in the right plot of Fig. 1.1, which shows another sound which displays a slow, cos-like, variation in air pressure, with some smaller and faster variations imposed on this. This combination of several kinds of systematic oscillations in air pressure is typical for general sounds.

Everyday sounds typically correspond to variations in air pressure of about 0.00002–2 Pa (0.00002 Pa corresponds to a just audible sound), while a jet engine may cause variations as large as 200 Pa. Short exposure to variations of about 20 Pa may in fact lead to hearing damage. The volcanic eruption at Krakatoa, Indonesia, in 1883, produced a sound wave with variations as large as almost 100,000 Pa, and the explosion could be heard 5000 km away. Since the range of the oscillations is so big, it is common to measure the loudness of a sound on a logarithmic scale:

Fact 3. Sound Pressure and Decibels.

It is common to relate a given sound pressure to the smallest sound pressure that can be perceived, as a level on a decibel *scale,*

$$L_p = 10 \log_{10} \left(\frac{p^2}{p_{\text{ref}}^2} \right) = 20 \log_{10} \left(\frac{p}{p_{\text{ref}}} \right).$$

Here p is the measured sound pressure while p_{ref} *is the sound pressure of a just perceivable sound, usually considered to be 0.00002 Pa.*

The square of the sound pressure appears in the definition of L_p since this represents the *power* of the sound which is relevant for what we perceive as loudness.

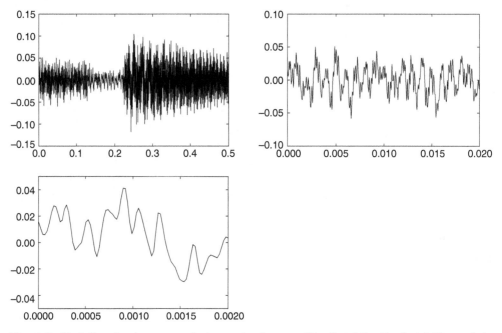

Fig. 1.2 Variations in air pressure during parts of a song. The first 0.5 s, the first 0.02 s, and the first 0.002 s

The sounds in Fig. 1.1 are synthetic in that they were constructed from mathematical formulas. The sounds in Fig. 1.2 on the other hand show the variation in air pressure for a song, where there is no mathematical formula involved. In the first half second there are so many oscillations that it is impossible to see the details, but if we zoom in on the first 0.002 s there seems to be a continuous function behind all the ink. In reality the air pressure varies more than this, even over this short time period, but the measuring equipment may not be able to pick up those variations, and it is also doubtful whether we would be able to perceive such rapid variations.

1.1.1 The Frequency of a Sound

The other important characteristic in sound is frequency, i.e. the speed of the variations. To make this concept more precise, let us start with a couple of definitions.

Definition 4. *Periodic Functions.*
A real function f is said to be periodic with period T if $f(t+T) = f(t)$ for all real numbers t.

Note that all the values of a periodic function f with period T are known if $f(t)$ is known for all t in the interval $[0,T)$. The following will be our prototype for periodic functions:

Observation 5. Frequency.
If ν is a real number, the function $f(t) = \sin(2\pi\nu t)$ is periodic with period $T = 1/\nu$. When t varies in the interval $[0,1]$, this function covers a total of ν periods. This is expressed by saying that f has frequency ν. *Frequency is measured in Hz (Hertz) which is the same as s^{-1} (the time t is measured in seconds). The function $\sin(2\pi\nu t)$ is also called a* pure tone.

Clearly $\sin(2\pi\nu t)$ and $\cos(2\pi\nu t)$ have the same frequency, and they are simply shifted versions of one another (since $\cos(2\pi\nu t) = \sin(2\pi\nu t + \pi/2)$). Both, as well as linear combinations of them, are called pure tones with frequency ν. Due to this, the complex functions $e^{\pm 2\pi i\nu t} = \cos(2\pi\nu t) \pm i\cos(2\pi\nu t)$ will also be called pure tones. They will also turn out to be useful in the following.

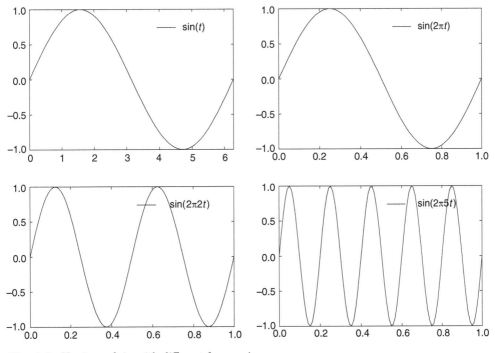

Fig. 1.3 Versions of sin with different frequencies

Figure 1.3 shows the pure tones $\sin t$, $\sin(2\pi t)$, $\sin(2\pi 2t)$, and $\sin(2\pi 5t)$. The corresponding frequencies are $1/(2\pi)$, 1, 2 and 5, and the corresponding periods are 2π, 1, $1/2$ and $1/5$. The last three plots are shown over $[0, 1]$, so that we can find the frequency by counting the number of periods which are covered.

If we are to perceive variations in air pressure as sound, they must fall within a certain range. It turns out that, for a human with good hearing to perceive a sound, the number of variations per second must be in the range 20–20,000.

1.1.2 Working with Digital Sound on a Computer

Before we can do anything at all with digital sound, we need to know how we can read and write such data from and to files, and also how to play the data on the computer. These commands are as follows.

```
[x, fs] = audioread(filename)            % Read from file
playerobj = audioplayer(x, fs)
play(playerobj)                          % Play the entire sound
playblocking(playerobj)                  % Play the entire sound...
                                           Block while playing
playblocking(playerobj, [start stop])    % Play the part of the sound...
                                           between sample start and stop
audiowrite(filename, x, fs)              % Write to file
```

The mysterious `playerobj` object simply encapsulates a digital sound and its sampling rate. `playblocking` plays the sound and blocks further execution until it has finished playing. We will have use for this functionality when we play several sounds in succession.

`play` basically sends the array of sound samples `x` and sample rate `fs` to the sound card, which uses some method for reconstructing the sound to an analog sound signal. This analog signal is then sent to the loudspeakers, and we hear the sound.

Sound samples can have different data types. We will always assume that they are of type `double`. The computer requires that they have values between -1 and 1 (0 corresponding to no variation in air pressure from ambience, and -1 and 1 the largest variations in air pressure). For sound samples outside $[-1, 1]$, the behavior is undefined when playing the sound.

Example 1.1: Listen to Different Channels

The audio sample file we will use is located in the folder `sounds`:

```
[x, f_s] = audioread('sounds/castanets.wav');
```

It has two sound channels. In such cases `x` is actually a matrix with two columns, each column representing a sound channel. To listen to each channel we can run the following code.

```
playerobj=audioplayer(x(:, 1), f_s);
playblocking(playerobj);


playerobj=audioplayer(x(:, 2), f_s);
playblocking(playerobj);
```

You may not hear a difference between the two channels. There may still be differences, however, which only are notable when the channels are sent to different loudspeakers.

We will in the following apply different operations to sound. We will then mostly apply these operations to the sound channels simultaneously.

Example 1.2: Playing a Sound Backwards

At times a popular game has been to play music backwards to try to find secret messages. In the old days of analog music this was not so easy, but with digital sound it is quite simple; we just reverse the samples. Thus, if $x = (x_i)_{i=0}^{N-1}$ are the samples of a digital sound, the samples $y = (y_i)_{i=0}^{N-1}$ of the reverse sound are

$$y_i = x_{N-i-1}.$$

When we reverse the sound samples, we have to reverse the elements in both sound channels. For our audio sample file this can be performed as follows.

```
z = x(end:(-1):1, :);
playerobj = audioplayer(z, f_s);
playblocking(playerobj);
```

Performing this on our sample file you obtain the sound file castanetsreverse.wav.

Example 1.3: Playing Pure Tones

You can also create and listen to sound samples on your own, without reading them from file. To create the samples of a pure tone (with only one channel) we can write

```
t = linspace(0, antsec, f_s*antsec);
x = sin(2*pi*f*t);
```

Here f is the frequency and antsec the length in seconds. A pure tone with frequency 440 Hz is found in the file puretone440.wav, and a pure tone with frequency 1500 Hz is found in the file puretone1500.wav.

Example 1.4: The Square Wave

There are many other ways in which a function can oscillate regularly. An example is the *square wave*, defined by

$$f_s(t) = \begin{cases} 1, & \text{if } 0 \leq t < T/2; \\ -1, & \text{if } T/2 \leq t < T. \end{cases}$$

Given a period T, it is 1 on the first half of each period, and -1 on the other. In the left part of Fig. 1.4 we have plotted the square wave with the same period we used for the pure tone.

To listen to the square wave, first create the samples for one period.

```
samplesperperiod=round(f_s/f); % The number of samples for one period
oneperiod = [ones(1,round(samplesperperiod/2)) ...
            -ones(1,round(samplesperperiod/2))];
```

Then we repeat one period to obtain a sound with the desired length:

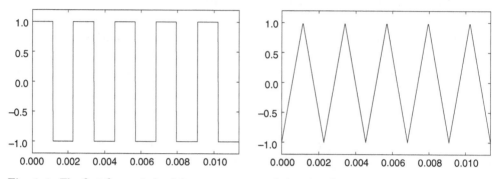

Fig. 1.4 The first five periods of the square wave and the triangle wave

```
x=repmat(oneperiod,1,antsec*f); % Repeat one period
playerobj=audioplayer(x, f_s);
playblocking(playerobj);
```

You can listen to the square wave in the file square440.wav. We hear a sound which seems to have the same "base frequency" as $\sin(2\pi 440t)$, but it is less pleasant to listen to: There seems to be some "sharp corners" in the sound.

Example 1.5: The Triangle Wave

Given a period T we define the *triangle wave* to increase linearly from -1 to 1 on the first half of each period, and decrease linearly from 1 to -1 on the second half of each period, i.e.

$$f_t(t) = \begin{cases} 4t/T - 1, & \text{if } 0 \le t < T/2; \\ 3 - 4t/T, & \text{if } T/2 \le t < T. \end{cases}$$

In the right part of Fig. 1.4 we have plotted the triangle wave with the same period we used for the pure tone. In Exercise 1.10 you will be asked to reproduce this plot, as well as construct and play the corresponding sound, which also can be found in the file triangle440.wav. Again you will note that the triangle wave has the same "base frequency" as $\sin(2\pi 440t)$, and is less pleasant to listen to than a pure tone. However, one can argue that it is somewhat more pleasant than a square wave.

In the next sections we will address why many sounds may be approximated well by adding many pure tones together. In particular, this will apply for the square wave and the triangle wave above, and we will also have something to say about why they sound so different.

Exercise 1.6: The Krakatoa Explosion

Compute the loudness of the Krakatoa explosion on the decibel scale, assuming that the variation in air pressure peaked at 100,000 Pa.

Exercise 1.7: The Sum of Two Pure Tones

a) Consider a sum of two pure tones, $f(t) = a\sin(2\pi\nu_1 t) + b\sin(2\pi\nu_2 t)$. For which values of a, b, ν_1, ν_2 is f periodic? What is the period of f when it is periodic?

b) Find two constant a and b so that the function $f(t) = a\sin(2\pi 440 t) + b\sin(2\pi 4400 t)$ resembles the right plot of Fig. 1.1 as closely as possible. Generate the samples of this sound, and listen to it.

Exercise 1.8: Playing with Different Sample Rates

If we provide another sample rate to the `play` functions, the sound card will assume a different time distance between neighboring samples. Play and listen to the audio sample file again, but with three different sample rates: $2f_s$, f_s, and $f_s/2$, where f_s is the sample rate returned by `audioread`.

Exercise 1.9: Playing Sound with Added Noise

Removing noise from recorded sound can be very challenging, but adding noise is simple. There are many kinds of noise, but one kind is easily obtained by adding random numbers to the samples of a sound. For this we can use the function `rand` as follows.

```
z = x + c*(2*rand(size(x))-1);
```

This adds noise to all channels. The function for returning random numbers returns numbers between 0 and 1, and above we have adjusted these so that they are between -1 and 1 instead, as for other sound which can be played by the computer. c is a constant (usually smaller than 1) that dampens the noise.

Write code which adds noise to the audio sample file, and listen to the result for damping constants c=0.4 and c=0.1. Remember to scale the sound values after you have added noise, since they may be outside $[-1, 1]$.

Exercise 1.10: Playing the Triangle Wave

Repeat what you did in Example 1.4, but now for the triangle wave of Example 1.5. Start by generating the samples for one period, then plot five periods, before you generate the sound over a period of 3 s and play it. Verify that you generate the same sound as in Example 1.5.

Exercise 1.11: Playing the Notes in an Octave

In music theory, an octave is a set of pure tones at frequencies $f_0, \ldots, f_{11}, f_{12}$ so that the ratio of neighboring tones are the same, and so that f_{12} is double the frequency of f_0, i.e. so that

$$\frac{f_1}{f_0} = \frac{f_2}{f_1} = \cdots = \frac{f_{12}}{f_{11}} = 2^{1/12}.$$

Make a program which plays all the pure tones in an octave, and listen to it with $f_0 = 440\,\text{Hz}$.

Exercise 1.12: The Karplus-Strong Algorithm for Making Guitar-Like Sounds

Given initial values $x_0, \ldots, x_p$, the difference equation

$$x_{n+p+1} - \frac{1}{2}(x_{n+1} + x_n) = 0$$

of order $p+1$ is known to create guitar like sounds. Show that all x_n lie in $[-1, 1]$ when the initial values do, and write a function

```
karplus_strong(x_init, f_s)
```

which takes the initial values $(x_0, x_1, \ldots, x_p)$ as input, and plays the resulting sound for $10\,\text{s}$ with sample rate f_s. Experiment with randomly generated initial values between -1 and 1, as well as different sample rates. What can you say about the frequencies in the resulting sound?

1.2 Fourier Series: Basic Concepts

We will now discuss the idea of decomposing a sound into a linear combination of pure tones. A coefficient in such a decomposition then gives the content at a given frequency. This will pave the way for constructing useful operations on sound, such as changing certain frequencies: Some frequencies may not be important for our perception of the sound, so that slightly changing these may not affect how we perceive them. We will first restrict to functions which are periodic with period T, so that they are uniquely defined by their values on $[0, T]$. Much of our analysis will apply for square integrable functions:

Definition 6. *Continuous and Square Integrable Functions.*
   The set of continuous, real functions defined on an interval $[0, T]$ is denoted $C[0, T]$.
   A real function f defined on $[0, T]$ is said to be *square integrable* if f^2 is Riemann-integrable, i.e., if the Riemann integral of f^2 on $[0, T]$ exists,

$$\int_0^T f(t)^2 \, dt < \infty.$$

The set of all square integrable functions on $[0, T]$ is denoted $L^2[0, T]$.

The sets of continuous and square integrable functions can be equipped with an inner product, a generalization of the scalar product for vectors.

Theorem 7. Inner Product Spaces.

Both $L^2[0, T]$ and $C[0, T]$ are real vector spaces. Moreover, if the two functions f and g lie in $L^2[0, T]$ (or in $C[0, T]$), then the product fg is Riemann-integrable (or in $C[0, T]$). Both spaces are also real inner product spaces[1] with inner product[2] defined by

$$\langle f, g \rangle = \frac{1}{T} \int_0^T f(t)g(t) \, dt, \tag{1.1}$$

and associated norm

$$\|f\| = \sqrt{\frac{1}{T} \int_0^T f(t)^2 dt}. \tag{1.2}$$

Proof. Since

$$|f + g|^2 \leq (2 \max(|f|, |g|))^2 \leq 4(|f|^2 + |g|^2).$$

$f + g$ is square integrable whenever f and g are. It follows that $L^2[0, T]$ is a vector space. The properties of an inner product space follow directly from the properties of Riemann-integrable functions. Also, since $|fg| \leq |f|^2 + |g|^2$, it follows that $\langle f, g \rangle < \infty$ whenever f and g are square integrable. It follows immediately that fg is Riemann-integrable whenever f and g are square integrable. $\square$

The mysterious factor $1/T$ is included so that the constant function $f(t) = 1$ has norm 1, i.e., its role is as a normalizing factor.

Definition 6 states how general we will allow our sounds to be. From linear algebra we know how to determine approximations using inner products, such as the one in Theorem 7. Recall that the projection[3] of a function f onto a subspace W (w.r.t. an inner product $\langle \cdot, \cdot \rangle$) is the function $g \in W$ which minimizes the *error* $\|f - g\|$. The error is also called *least squares error*, and the projection a best approximation of f from W. The projection is characterized by the fact that the error $f - g$ is orthogonal to W, i.e.

$$\langle f - g, h \rangle = 0, \quad \text{for all } h \in W.$$

If we have an orthogonal basis $\phi = \{\phi_i\}_{i=1}^m$ for W, the *orthogonal decomposition theorem* states that the best approximation from W is

$$g = \sum_{i=1}^m \frac{\langle f, \phi_i \rangle}{\langle \phi_i, \phi_i \rangle} \phi_i. \tag{1.3}$$

What we would like is a sequence of spaces

$$V_1 \subset V_2 \subset \cdots \subset V_n \subset \cdots$$

---

[1] See Section 6.1 in [32] for a review of inner products and orthogonality.

[2] See Section 6.7 in [32] for a review of function spaces as inner product spaces.

[3] See Section 6.3 in [32] for a review of projections and least squares approximations.

of increasing dimensions so that most sounds can be approximated arbitrarily well by choosing n large enough, and use the orthogonal decomposition theorem to compute the approximations. It turns out that pure tones can be used for this purpose.

Definition 8. *Fourier Series.*

Let $V_{N,T}$ be the subspace of $C[0,T]$ spanned by

$$\mathcal{D}_{N,T} = \{1, \cos(2\pi t/T), \cos(2\pi 2t/T), \cdots, \cos(2\pi Nt/T),$$
$$\sin(2\pi t/T), \sin(2\pi 2t/T), \cdots, \sin(2\pi Nt/T)\}. \qquad (1.4)$$

The space $V_{N,T}$ is called the N'*th order Fourier space*. The best approximation of f from $V_{N,T}$ with respect to the inner product (1.1) is denoted f_N, and is called the N'th-order *Fourier series* of f.

We see that $\mathcal{D}_{N,T}$ consists of the pure tones at frequencies $1/T, 2/T, \ldots, N/T$. Fourier series is similar to Taylor series, where instead polynomials are used for approximation, but we will see that there is a major difference in how the two approximations are computed. The theory of approximation of functions with Fourier series is referred to as *Fourier analysis*, and is a central tool in practical fields like image- and signal processing, as well as an important field of research in pure mathematics. The approximation $f_N \in V_{N,T}$ can serve as a compressed version of f if many of the coefficients can be set to 0 without the error becoming too big.

Note that all the functions in the set $\mathcal{D}_{N,T}$ are periodic with period T, but most have an even shorter period ($\cos(2\pi nt/T)$ also has period T/n). In general, the term *fundamental frequency* is used to denote the frequency corresponding to the shortest period.

The next theorem explains that $\mathcal{D}_{N,T}$ actually is an orthogonal basis for $V_{N,T}$, so that we can use the orthogonal decomposition theorem to obtain the coefficients in this basis.

Theorem 9. Fourier Coefficients.

The set $\mathcal{D}_{N,T}$ is an orthogonal basis for $V_{N,T}$. In particular, the dimension of $V_{N,T}$ is $2N+1$. If f is a function in $L^2[0,T]$, we denote by $a_0, \ldots, a_N$ and $b_1, \ldots, b_N$ the coordinates of f_N in the basis $\mathcal{D}_{N,T}$, i.e.

$$f_N(t) = a_0 + \sum_{n=1}^{N} \left(a_n \cos(2\pi nt/T) + b_n \sin(2\pi nt/T)\right). \qquad (1.5)$$

The $a_0, \ldots, a_N$ and $b_1, \ldots, b_N$ are called the (real) Fourier coefficients of f, and they are given by

$$a_0 = \langle f, 1 \rangle = \frac{1}{T} \int_0^T f(t)\, dt, \qquad (1.6)$$

$$a_n = 2\langle f, \cos(2\pi nt/T) \rangle = \frac{2}{T} \int_0^T f(t) \cos(2\pi nt/T)\, dt \quad \text{for } n \geq 1, \qquad (1.7)$$

$$b_n = 2\langle f, \sin(2\pi nt/T) \rangle = \frac{2}{T} \int_0^T f(t) \sin(2\pi nt/T)\, dt \quad \text{for } n \geq 1. \qquad (1.8)$$

Proof. Assume first that $m \neq n$. We compute the inner product

$$\langle \cos(2\pi mt/T), \cos(2\pi nt/T) \rangle$$

$$= \frac{1}{T} \int_0^T \cos(2\pi mt/T) \cos(2\pi nt/T) dt$$

$$= \frac{1}{2T} \int_0^T \left(\cos(2\pi mt/T + 2\pi nt/T) + \cos(2\pi mt/T - 2\pi nt/T) \right)$$

$$= \frac{1}{2T} \left[\frac{T}{2\pi(m+n)} \sin(2\pi(m+n)t/T) + \frac{T}{2\pi(m-n)} \sin(2\pi(m-n)t/T) \right]_0^T$$

$$= 0.$$

Here we have added the two identities $\cos(x \pm y) = \cos x \cos y \mp \sin x \sin y$ together to obtain an expression for $\cos(2\pi mt/T) \cos(2\pi nt/T) dt$ in terms of $\cos(2\pi mt/T + 2\pi nt/T)$ and $\cos(2\pi mt/T - 2\pi nt/T)$. By testing all other combinations of sin and cos also, we obtain the orthogonality of all functions in $\mathcal{D}_{N,T}$. We also obtain that

$$\langle \cos(2\pi mt/T), \cos(2\pi mt/T) \rangle = \frac{1}{2}$$

$$\langle \sin(2\pi mt/T), \sin(2\pi mt/T) \rangle = \frac{1}{2}$$

$$\langle 1, 1 \rangle = 1,$$

From the orthogonal decomposition theorem (1.3) it follows that

$$f_N(t) = \frac{\langle f, 1 \rangle}{\langle 1, 1 \rangle} 1 + \sum_{n=1}^N \frac{\langle f, \cos(2\pi nt/T) \rangle}{\langle \cos(2\pi nt/T), \cos(2\pi nt/T) \rangle} \cos(2\pi nt/T)$$

$$+ \sum_{n=1}^N \frac{\langle f, \sin(2\pi nt/T) \rangle}{\langle \sin(2\pi nt/T), \sin(2\pi nt/T) \rangle} \sin(2\pi nt/T)$$

$$= \frac{\frac{1}{T} \int_0^T f(t) dt}{1} + \sum_{n=1}^N \frac{\frac{1}{T} \int_0^T f(t) \cos(2\pi nt/T) dt}{\frac{1}{2}} \cos(2\pi nt/T)$$

$$+ \sum_{n=1}^N \frac{\frac{1}{T} \int_0^T f(t) \sin(2\pi nt/T) dt}{\frac{1}{2}} \sin(2\pi nt/T)$$

$$= \frac{1}{T} \int_0^T f(t) dt + \sum_{n=1}^N \left(\frac{2}{T} \int_0^T f(t) \cos(2\pi nt/T) dt \right) \cos(2\pi nt/T)$$

$$+ \sum_{n=1}^N \left(\frac{2}{T} \int_0^T f(t) \sin(2\pi nt/T) dt \right) \sin(2\pi nt/T).$$

Equations (1.6)–(1.8) now follow by comparison with Eq. (1.5). $\square$

From the orthogonality and the inner products of the Fourier basis functions it immediately follows that

$$\|f_N\|^2 = a_0^2 + \frac{1}{2} \sum_{n=1}^N (a_n^2 + b_n^2)$$

Since f is a function in time, and the a_n, b_n represent contributions from different frequencies, the Fourier series can be thought of as a change of coordinates, from what often is called the *time domain*, to the *frequency domain* (or *Fourier domain*). We will call the basis $\mathcal{D}_{N,T}$ the *N 'th order Fourier basis* for $V_{N,T}$. We note that $\mathcal{D}_{N,T}$ is not an orthonormal basis, only orthogonal.

In the signal processing literature, Eq. (1.5) is known as *the synthesis equation*, since the original function f is synthesized as a sum of the Fourier basis functions. Equations (1.6)–(1.8) are also called *analysis equations*.

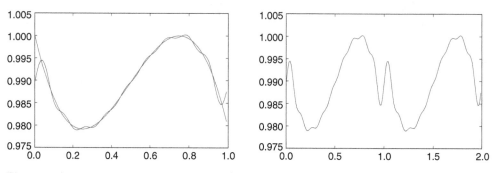

Fig. 1.5 A cubic polynomial on the interval $[0, 1]$, together with its Fourier series of order 9 (left). The Fourier series on a larger interval is shown right

Figure 1.5 shows the cubic polynomial $f(x) = -\frac{1}{3}x^3 + \frac{1}{2}x^2 - \frac{3}{16}x + 1$ defined on $[0, 1]$, together with its 9'th order Fourier series. The Fourier series is periodic with period 1, and we see that the approximation becomes poor at the ends of the interval, since the cubic polynomial does not satisfy $f(0) = f(1)$. At 0 and 1 it seems that the Fourier series hits the average of $f(0)$ and $f(1)$. We will return to why this may be so later. The Fourier series plotted on a larger interval is shown in the right plot of Fig. 1.5. Here the periodicity of the Fourier series is more visible.

Let us compute the Fourier series of some interesting functions.

Example 1.13: Fourier Coefficients of the Square Wave

Let us compute the Fourier coefficients of the square wave, as defined in Example 1.4. If we first use Eq. (1.6) we obtain

$$a_0 = \frac{1}{T} \int_0^T f_s(t)dt = \frac{1}{T} \int_0^{T/2} dt - \frac{1}{T} \int_{T/2}^T dt = 0.$$

Using Eq. (1.7) we get

$$a_n = \frac{2}{T} \int_0^T f_s(t) \cos(2\pi nt/T)dt$$

$$= \frac{2}{T} \int_0^{T/2} \cos(2\pi nt/T)dt - \frac{2}{T} \int_{T/2}^T \cos(2\pi nt/T)dt$$

$$= \frac{2}{T} \left[\frac{T}{2\pi n} \sin(2\pi n t/T) \right]_0^{T/2} - \frac{2}{T} \left[\frac{T}{2\pi n} \sin(2\pi n t/T) \right]_{T/2}^{T}$$

$$= \frac{2}{T} \frac{T}{2\pi n} ((\sin(n\pi) - \sin 0) - (\sin(2n\pi) - \sin(n\pi))) = 0.$$

Finally, using Eq. (1.8) we obtain

$$b_n = \frac{2}{T} \int_0^T f_s(t) \sin(2\pi n t/T) dt$$

$$= \frac{2}{T} \int_0^{T/2} \sin(2\pi n t/T) dt - \frac{2}{T} \int_{T/2}^T \sin(2\pi n t/T) dt$$

$$= \frac{2}{T} \left[-\frac{T}{2\pi n} \cos(2\pi n t/T) \right]_0^{T/2} + \frac{2}{T} \left[\frac{T}{2\pi n} \cos(2\pi n t/T) \right]_{T/2}^{T}$$

$$= \frac{2}{T} \frac{T}{2\pi n} ((-\cos(n\pi) + \cos 0) + (\cos(2n\pi) - \cos(n\pi)))$$

$$= \frac{2(1 - \cos(n\pi))}{n\pi}$$

$$= \begin{cases} 0, & \text{if } n \text{ is even;} \\ 4/(n\pi), & \text{if } n \text{ is odd.} \end{cases}$$

In other words, only the b_n-coefficients with n odd in the Fourier series are nonzero. This means that the Fourier series of the square wave is

$$\frac{4}{\pi} \sin(2\pi t/T) + \frac{4}{3\pi} \sin(2\pi 3t/T) + \frac{4}{5\pi} \sin(2\pi 5t/T) + \frac{4}{7\pi} \sin(2\pi 7t/T) + \cdots .$$

With $N = 20$ there are 10 terms in the sum. The corresponding Fourier series can be plotted over one period with the following code.

```
N = 20;
T = 1/440;
t = linspace(0, T, samplesperperiod);
x = zeros(1,length(t));
for k=1:2:19
    x = x + (4/(k*pi))*sin(2*pi*k*t/T);
end
figure()
plot(t, x, 'k-')
```

The left plot in Fig. 1.6 shows the result. In the right plot the values of the first 100 Fourier coefficients b_n are shown, to see that they actually converge to zero. This is clearly necessary in order for the Fourier series to converge.

Even though f oscillates regularly between -1 and 1 with period T, the discontinuities mean that it is far from the simple $\sin(2\pi t/T)$ which corresponds to a pure tone of frequency $1/T$. Clearly $b_1 \sin(2\pi t/T)$ is the dominant term in the Fourier series. This is not surprising since the square wave has the same period as this term, and this explains

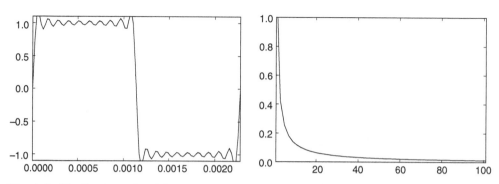

Fig. 1.6 The Fourier series of the square wave with $N = 20$, and the values for the first 100 Fourier coefficients b_n

why we heard the same pitch as the pure tone when we listened to the square wave in Example 1.4. Additional terms in the Fourier also contribute, and as we include more of these, we gradually approach the square wave. The square wave Fourier coefficients decrease as $1/n$, and this pollution makes the sound less pleasant. Note also that, there is a sharp jump in the Fourier series near the discontinuity at $T/2$. It turns out that, even as the number of terms in the Fourier series increases, this sharp jump does not decrease in amplitude. This kind of behavior near a discontinuity is also called a *Gibbs phenomenon*.

Let us listen to the Fourier series approximations of the square wave. For $N = 1$ the sound can be found in the file square440s1.wav. This sounds exactly like the pure tone with frequency $440\,\mathrm{Hz}$, as noted above. For $N = 5$ the sound can be found in the file square440s5.wav, and for $N = 9$ it can be found in the file square440s9.wav. The latter sounds are more like the square wave itself. As we increase N we can hear how the introduction of more frequencies gradually pollutes the sound more.

Example 1.14: Fourier Coefficients of the Triangle Wave

Let us also compute the Fourier coefficients of the triangle wave, as defined in Example 1.5. We now have

$$a_0 = \frac{1}{T} \int_0^{T/2} \frac{4}{T} \left(t - \frac{T}{4} \right) dt + \frac{1}{T} \int_{T/2}^{T} \frac{4}{T} \left(\frac{3T}{4} - t \right) dt.$$

Instead of computing this directly, it is quicker to see geometrically that the graph of f_t has as much area above as below the x-axis, so that this integral must be zero. Similarly, since f_t is symmetric about the midpoint $T/2$, and $\sin(2\pi nt/T)$ is antisymmetric about $T/2$, we have that $f_t(t)\sin(2\pi nt/T)$ also is antisymmetric about $T/2$, so that

$$\int_0^{T/2} f_t(t)\sin(2\pi nt/T)dt = - \int_{T/2}^{T} f_t(t)\sin(2\pi nt/T)dt.$$

This means that, for $n \geq 1$,

$$b_n = \frac{2}{T} \int_0^{T/2} f_t(t)\sin(2\pi nt/T)dt + \frac{2}{T} \int_{T/2}^{T} f_t(t)\sin(2\pi nt/T)dt = 0.$$

For the final coefficients, since both f and $\cos(2\pi nt/T)$ are symmetric about $T/2$, we get for $n \geq 1$,

$$
\begin{aligned}
a_n &= \frac{2}{T} \int_0^{T/2} f_t(t) \cos(2\pi nt/T)dt + \frac{2}{T} \int_{T/2}^{T} f_t(t) \cos(2\pi nt/T)dt \\
&= \frac{4}{T} \int_0^{T/2} f_t(t) \cos(2\pi nt/T)dt = \frac{4}{T} \int_0^{T/2} \frac{4}{T}\left(t - \frac{T}{4}\right) \cos(2\pi nt/T)dt \\
&= \frac{16}{T^2} \int_0^{T/2} t \cos(2\pi nt/T)dt - \frac{4}{T} \int_0^{T/2} \cos(2\pi nt/T)dt \\
&= \frac{4}{n^2\pi^2}(\cos(n\pi) - 1) \\
&= \begin{cases} 0, & \text{if } n \text{ is even;} \\ -8/(n^2\pi^2), & \text{if } n \text{ is odd.} \end{cases}
\end{aligned}
$$

where we have dropped the final tedious calculations (use integration by parts). From this it is clear that the Fourier series of the triangle wave is

$$
-\frac{8}{\pi^2} \cos(2\pi t/T) - \frac{8}{3^2\pi^2} \cos(2\pi 3t/T) - \frac{8}{5^2\pi^2} \cos(2\pi 5t/T) - \frac{8}{7^2\pi^2} \cos(2\pi 7t/T) + \cdots .
$$

In Fig. 1.7 we have repeated the plots used for the square wave, for the triangle wave. The figure indicates that the Fourier series converges faster.

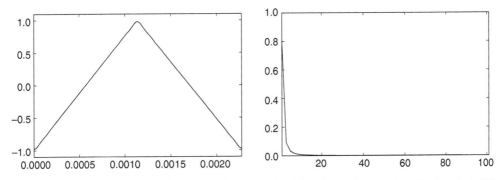

Fig. 1.7 The Fourier series of the triangle wave with $N = 20$, and the values for the first 100 Fourier coefficients $|a_n|$

Let us also listen to different Fourier series approximations of the triangle wave. For $N = 1$ it can be found in the file triangle440s1.wav. Again, this sounds exactly like the pure tone with frequency 440 Hz. For $N = 5$ the Fourier series approximation can be found in the file triangle440s5.wav, and for $N = 9$ it can be found in the file triangle440s9.wav. Again the latter sounds are more like the triangle wave itself, and as we increase N we can hear that more frequencies pollute the sound. However, since the triangle wave Fourier coefficients decrease as $1/n^2$ rather than $1/n$, the sound is somewhat less unpleasant. The faster convergence can also be heard.

Example 1.15: Fourier Coefficients of a Simple Function

There is an important lesson to be learned from the previous examples: Even if the signal is nice and periodic, it may not have a nice representation in terms of trigonometric functions. Thus, the Fourier basis may not be the best for expressing other functions. Unfortunately, many more such cases can be found, as we will now explain. Let us consider a periodic function which is 1 on $[0, T_0]$, but 0 is on $[T_0, T]$. This is a signal with short duration when T_0 is small compared to T. We compute that $y_0 = T_0/T$, and

$$a_n = \frac{2}{T} \int_0^{T_0} \cos(2\pi nt/T)dt = \frac{1}{\pi n} \left[\sin(2\pi nt/T)\right]_0^{T_0} = \frac{\sin(2\pi nT_0/T)}{\pi n}$$

for $n \geq 1$. Similar computations hold for b_n. We see that $|a_n|$ is of the order $1/(\pi n)$, and that infinitely many n contribute, This function may be thought of as a simple building block, corresponding to a small time segment. However, in terms of the Fourier basis it is not so simple. The time segment building block may be useful for restricting a function to smaller time segments, and later on we will see that it still can be useful.

1.2.1 Fourier Series for Symmetric and Antisymmetric Functions

In Example 1.13 we saw that the Fourier coefficients b_n vanished, resulting in a sine-series for the Fourier series of the square wave. Similarly, in Example 1.14 we saw that a_n vanished, resulting in a cosine-series for the triangle wave. This is not a coincident, and is captured by the following result.

Theorem 10. Symmetry and Antisymmetry.
If f is antisymmetric about 0 (that is, if $f(-t) = -f(t)$ for all t), then $a_n = 0$, so the Fourier series is actually a sine-series. If f is symmetric about 0 (which means that $f(-t) = f(t)$ for all t), then $b_n = 0$, so the Fourier series is actually a cosine-series.

The point is that the square wave is antisymmetric about 0, and the triangle wave is symmetric about 0.

Proof. Note first that we can write

$$a_n = \frac{2}{T} \int_{-T/2}^{T/2} f(t)\cos(2\pi nt/T)dt \qquad b_n = \frac{2}{T} \int_{-T/2}^{T/2} f(t)\sin(2\pi nt/T)dt,$$

i.e. we can change the integration bounds from $[0, T]$ to $[-T/2, T/2]$. This follows from the fact that all $f(t)$, $\cos(2\pi nt/T)$ and $\sin(2\pi nt/T)$ are periodic with period T.
Suppose first that f is symmetric. We obtain

$$b_n = \frac{2}{T} \int_{-T/2}^{T/2} f(t)\sin(2\pi nt/T)dt$$

$$= \frac{2}{T} \int_{-T/2}^{0} f(t)\sin(2\pi nt/T)dt + \frac{2}{T} \int_{0}^{T/2} f(t)\sin(2\pi nt/T)dt$$

$$= \frac{2}{T} \int_{-T/2}^{0} f(t)\sin(2\pi nt/T)dt - \frac{2}{T} \int_{0}^{-T/2} f(-t)\sin(-2\pi nt/T)dt$$

$$= \frac{2}{T} \int_{-T/2}^{0} f(t)\sin(2\pi nt/T)dt - \frac{2}{T} \int_{-T/2}^{0} f(t)\sin(2\pi nt/T)dt = 0.$$

where we have made the substitution $u = -t$, and used that sin is antisymmetric. The case when f is antisymmetric can be proved in the same way, and is left as an exercise. □

Exercise 1.16: Shifting the Fourier Basis Vectors

Show that $\sin(2\pi nt/T + a) \in V_{N,T}$ when $|n| \leq N$, regardless of the value of a.

Exercise 1.17: Listening to the Fourier Series of the Triangle Wave

a) Plot the Fourier series of the triangle wave.

b) Write code so that you can listen to the Fourier series of the triangle wave. How high must you choose N for the Fourier series to be indistinguishable from the triangle wave itself?

Exercise 1.18: Riemann-Integrable Functions Which Are Not Square Integrable

Find a function f which is Riemann-integrable on $[0, T]$, and so that $\int_0^T f(t)^2 dt$ is infinite.

Exercise 1.19: When Are Fourier Spaces Included in Each Other?

Given the two Fourier spaces V_{N_1,T_1}, V_{N_2,T_2}. Find necessary and sufficient conditions in order for $V_{N_1,T_1} \subset V_{N_2,T_2}$.

Exercise 1.20: Fourier Series of Antisymmetric Functions Are Sine Series

Prove the second part of Theorem 10, i.e. show that if f is antisymmetric about 0 (i.e. $f(-t) = -f(t)$ for all t), then $a_n = 0$, i.e. the Fourier series is actually a sine series.

Exercise 1.21: More Connections Between Symmetric-/Antisymmetric Functions and Sine-/Cosine Series

Show that

a) Any cosine series $a_0 + \sum_{n=1}^{N} a_n \cos(2\pi nt/T)$ is a symmetric function.

b) Any sine series $\sum_{n=1}^{N} b_n \sin(2\pi nt/T)$ is an antisymmetric function.

c) Any periodic function can be written as a sum of a symmetric—and an antisymmetric function by writing $f(t) = \frac{f(t)+f(-t)}{2} + \frac{f(t)-f(-t)}{2}$.

d) If $f_N(t) = a_0 + \sum_{n=1}^{N}(a_n \cos(2\pi nt/T) + b_n \sin(2\pi nt/T))$, then

$$\frac{f_N(t) + f_N(-t)}{2} = a_0 + \sum_{n=1}^{N} a_n \cos(2\pi nt/T)$$

$$\frac{f_N(t) - f_N(-t)}{2} = \sum_{n=1}^{N} b_n \sin(2\pi nt/T).$$

Exercise 1.22: Fourier Series of Low-Degree Polynomials

Find the Fourier series coefficients of the periodic functions with period T defined by $f(t) = t$, $f(t) = t^2$, and $f(t) = t^3$, on $[0, T]$.

Exercise 1.23: Fourier Series of Polynomials

Write down difference equations for finding the Fourier coefficients of $f(t) = t^k$ from those of $f(t) = t^{k-1}$, and write a program which uses this recursion to compute the Fourier coefficients of $f(t) = t^k$. Use the program to verify what you computed in Exercise 1.22.

Exercise 1.24: Fourier Series of a Given Polynomial

Use the previous exercise to find the Fourier series for $f(x) = -\frac{1}{3}x^3 + \frac{1}{2}x^2 - \frac{3}{16}x + 1$ on the interval $[0, 1]$. Plot the 9th order Fourier series for this function. You should obtain the plots from Fig. 1.5.

1.3 Complex Fourier Series

In Sect. 1.2 we saw how a function can be expanded in a series of sines and cosines. These functions are related to the complex exponential function via Euler's formula

$$e^{ix} = \cos x + i \sin x$$

where i is the imaginary unit with the property that $i^2 = -1$. Because the algebraic properties of the exponential function are much simpler than those of cos and sin, it is often an advantage to work with complex numbers, even though the given setting is real numbers. This is definitely the case in Fourier analysis. More precisely, we will make the substitutions

$$\cos(2\pi n t/T) = \frac{1}{2}\left(e^{2\pi i n t/T} + e^{-2\pi i n t/T}\right) \tag{1.9}$$

$$\sin(2\pi n t/T) = \frac{1}{2i}\left(e^{2\pi i n t/T} - e^{-2\pi i n t/T}\right) \tag{1.10}$$

in Definition 8. From these identities it is clear that the set of complex exponential functions $e^{2\pi i n t/T}$ also is a basis of periodic functions (with the same period) for $V_{N,T}$. We may therefore reformulate Definition 8 as follows:

Definition 11. *Complex Fourier Basis.*
   We define the set of functions

$$\mathcal{F}_{N,T} = \{e^{-2\pi i N t/T}, e^{-2\pi i (N-1)t/T}, \cdots, e^{-2\pi i t/T},$$
$$1, e^{2\pi i t/T}, \cdots, e^{2\pi i (N-1)t/T}, e^{2\pi i N t/T}\}, \tag{1.11}$$

and call this the order N *complex Fourier basis* for $V_{N,T}$.

   The function $e^{2\pi i n t/T}$ is also called a pure tone with frequency n/T, just as sines and cosines are. We would like to show that these functions also are orthogonal. To show this, we need to say more on the inner product we have defined by Eq. (1.1). A weakness with this definition is that we have assumed real functions f and g, so that it can not be used for the $e^{2\pi i n t/T}$. We will therefore extend the definition of the inner product to complex functions as follows:

$$\langle f, g \rangle = \frac{1}{T}\int_0^T f\bar{g}\, dt. \tag{1.12}$$

Complex inner products are formally defined in Appendix A, where we also motivate why the second term should be conjugated in complex inner products. It is straightforward to see that (1.12) satisfies the requirements of a complex inner product as defined in Appendix A. The associated norm is

$$\|f\| = \sqrt{\frac{1}{T}\int_0^T |f(t)|^2 dt}. \tag{1.13}$$

With the new definition (1.12) it is an exercise to see that the functions $e^{2\pi int/T}$ are orthonormal. Using the orthogonal decomposition theorem we can therefore write

$$f_N(t) = \sum_{n=-N}^{N} \frac{\langle f, e^{2\pi int/T} \rangle}{\langle e^{2\pi int/T}, e^{2\pi int/T} \rangle} e^{2\pi int/T} = \sum_{n=-N}^{N} \langle f, e^{2\pi int/T} \rangle e^{2\pi int/T}$$

$$= \sum_{n=-N}^{N} \left(\frac{1}{T} \int_0^T f(t) e^{-2\pi int/T} dt \right) e^{2\pi int/T}.$$

We summarize this in the following theorem, which is a version of Theorem 9 using the complex Fourier basis:

Theorem 12. Complex Fourier Coefficients.
We denote by $y_{-N}, \ldots, y_0, \ldots, y_N$ *the coordinates of* f_N *in the basis* $\mathcal{F}_{N,T}$*, i.e.*

$$f_N(t) = \sum_{n=-N}^{N} y_n e^{2\pi int/T}. \tag{1.14}$$

The y_n *are called the complex Fourier coefficients of* f*, and they are given by.*

$$y_n = \langle f, e^{2\pi int/T} \rangle = \frac{1}{T} \int_0^T f(t) e^{-2\pi int/T} dt. \tag{1.15}$$

Let us consider two immediate and important consequences of the orthonormal basis we have established. The first one follows directly from the orthonormality.

Theorem 13. Parseval's Theorem.
We have that

$$\|f_N\|^2 = \sum_{n=-N}^{N} |y_n|^2$$

Theorem 14. Bessel's Inequality.
For any $f \in L^2[0,T]$ *we have that* $\|f\|^2 \geq \|f_N\|^2$*. In particular, the sequence* $\|f_N\|^2 = \sum_{n=-N}^{N} |y_n|^2$ *is convergent, so that* $y_n \to 0$*.*

Proof. Since $f_N(t)$ is the projection of f onto $V_{N,T}$ we have that

$$\|f\|^2 = \|f - f_N\|^2 + \|f_N\|^2 \geq \|f_N\|^2,$$

□

In particular the Fourier coefficients of square integrable functions go to zero. The results does not say that $\sum_{n=-N}^{N} |y_n|^2 \to \|f\|^2$, which would imply that $\|f - f_N\| \to 0$. This is more difficult to analyze, and we will only prove a particular case of it in Sect. 1.6.
If we reorder the real and complex Fourier bases so that the two functions $\{\cos(2\pi nt/T), \sin(2\pi nt/T)\}$ and $\{e^{2\pi int/T}, e^{-2\pi int/T}\}$ have the same index in the

bases, Eqs. (1.9)–(1.10) give us that the change of coordinates matrix[4] from $\mathcal{D}_{N,T}$ to $\mathcal{F}_{N,T}$, denoted $P_{\mathcal{F}_{N,T}\leftarrow\mathcal{D}_{N,T}}$, is represented by repeating the matrix

$$\frac{1}{2}\begin{pmatrix} 1 & 1/i \\ 1 & -1/i \end{pmatrix}$$

along the diagonal (with an additional 1 for the constant function 1). In other words, since a_n, b_n are coefficients relative to the real basis and y_n, y_{-n} the corresponding coefficients relative to the complex basis, we have for $n > 0$,

$$\begin{pmatrix} y_n \\ y_{-n} \end{pmatrix} = \frac{1}{2}\begin{pmatrix} 1 & 1/i \\ 1 & -1/i \end{pmatrix}\begin{pmatrix} a_n \\ b_n \end{pmatrix}.$$

This can be summarized by the following theorem:

Theorem 15. Change of Coordinates Between Real and Complex Fourier Bases.
The complex Fourier coefficients y_n and the real Fourier coefficients a_n, b_n of a function f are related by

$$y_0 = a_0,$$
$$y_n = \frac{1}{2}(a_n - ib_n),$$
$$y_{-n} = \frac{1}{2}(a_n + ib_n),$$

for $n = 1, \ldots, N$.

Combining with Theorems 10, 15 can help us state properties of complex Fourier coefficients for symmetric- and antisymmetric functions. We look into this in Exercise 1.35.

Due to the somewhat nicer formulas for the complex Fourier coefficients, we will write most Fourier series in complex form in the following. Let us consider some examples where we compute complex Fourier series.

Example 1.25: Complex Fourier Coefficients of a Part of a Pure Tone

Let us consider the pure tone $f(t) = e^{2\pi it/T_2}$ with period T_2, but let us consider it only on the interval $[0, T]$ instead, where $T < T_2$, and extended to a periodic function with period T. The Fourier coefficients are

$$y_n = \frac{1}{T}\int_0^T e^{2\pi it/T_2}e^{-2\pi int/T}dt = \frac{1}{2\pi iT(1/T_2 - n/T)}\left[e^{2\pi it(1/T_2 - n/T)}\right]_0^T$$

$$= \frac{1}{2\pi i(T/T_2 - n)}\left(e^{2\pi iT/T_2} - 1\right).$$

---

[4] See Section 4.7 in [32], to review the mathematics behind change of coordinates.

Here it is only the term $1/(T/T_2 - n)$ which depends on n, so that y_n can only be large when n is close T/T_2. In Fig. 1.8 we have plotted $|y_n|$ for two different combinations of T, T_2.

In both plots it is seen that many Fourier coefficients contribute, but this is more visible when $T/T_2 = 0.5$. When $T/T_2 = 0.9$, most contribution is seen to be in the y_1-coefficient. This sounds reasonable, since f then is closest to the pure tone $f(t) = e^{2\pi it/T}$ of frequency $1/T$ (which in turn has $y_1 = 1$ and all other $y_n = 0$).

Apart from computing complex Fourier series, there is an important lesson to be learned from this example: In order for a periodic function to be approximated by other periodic functions, their period must somehow match.

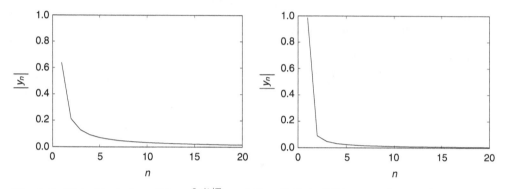

Fig. 1.8 Plot of $|y_n|$ when $f(t) = e^{2\pi it/T_2}$, and $T_2 > T$. Left: $T/T_2 = 0.5$. Right: $T/T_2 = 0.9$

Example 1.26: Complex Fourier Coefficients When There Are Different Frequencies on Different Parts

Most sounds change in content over time. Assume that f is equal to a pure tone of frequency n_1/T on $[0, T/2)$, and equal to a pure tone of frequency n_2/T on $[T/2, T)$, i.e.

$$f(t) = \begin{cases} e^{2\pi in_1 t/T} & \text{on } [0, T_2] \\ e^{2\pi in_2 t/T} & \text{on} [T_2, T) \end{cases}.$$

When $n \neq n_1, n_2$ we have that

$$y_n = \frac{1}{T}\left(\int_0^{T/2} e^{2\pi in_1 t/T} e^{-2\pi int/T} dt + \int_{T/2}^{T} e^{2\pi in_2 t/T} e^{-2\pi int/T} dt \right)$$

$$= \frac{1}{T}\left(\left[\frac{T}{2\pi i(n_1 - n)} e^{2\pi i(n_1-n)t/T} \right]_0^{T/2} + \left[\frac{T}{2\pi i(n_2 - n)} e^{2\pi i(n_2-n)t/T} \right]_{T/2}^{T} \right)$$

$$= \frac{e^{\pi i(n_1-n)} - 1}{2\pi i(n_1 - n)} + \frac{1 - e^{\pi i(n_2-n)}}{2\pi i(n_2 - n)}.$$

Let us restrict to the case when n_1 and n_2 are both even. We see that

$$
y_n = \begin{cases} \frac{1}{2} + \frac{1}{\pi i(n_2-n_1)} & n = n_1, n_2 \\ 0 & n \text{ even }, n \neq n_1, n_2 \\ \frac{n_1-n_2}{\pi i(n_1-n)(n_2-n)} & n \text{ odd} \end{cases}
$$

Here we have computed the cases $n = n_1$ and $n = n_2$ as above. In Fig. 1.9 we have plotted $|y_n|$ for two different combinations of n_1, n_2.

We see that, when n_1, n_2 are close, the Fourier coefficients are close to those of a pure tone of frequency n/T with $n \approx n_1, n_2$, but that also other frequencies contribute. When n_1, n_2 are further apart, we see that the Fourier coefficients are like the sum of the two base frequencies. Other frequencies contribute also here.

There is an important lesson to be learned from this as well: We should be aware of changes in a sound over time, and it may not be smart to use a frequency representation over a large interval when we know that there are simpler frequency representations on the smaller intervals.

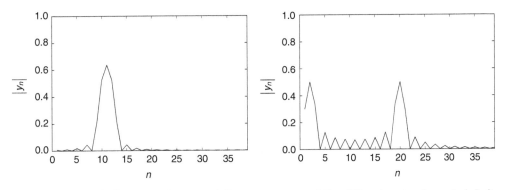

Fig. 1.9 Plot of $|y_n|$ when we have two different pure tones at the different parts of a period. Left: $n_1 = 10$, $n_2 = 12$. Right: $n_1 = 2$, $n_2 = 20$

Example 1.27: Complex Fourier Coefficients of $f(t) = \cos^3(2\pi t/T)$

In some cases it is not necessary to compute the Fourier integrals at all, in order to compute the Fourier series. Let us consider the function $f(t) = \cos^3(2\pi t/T)$, where T is the period of f. We can write

$$
\cos^3(2\pi t/T) = \left(\frac{1}{2}(e^{2\pi it/T} + e^{-2\pi it/T}) \right)^3
$$

$$
= \frac{1}{8}(e^{2\pi i3t/T} + 3e^{2\pi it/T} + 3e^{-2\pi it/T} + e^{-2\pi i3t/T})
$$

$$
= \frac{1}{8}e^{2\pi i3t/T} + \frac{3}{8}e^{2\pi it/T} + \frac{3}{8}e^{-2\pi it/T} + \frac{1}{8}e^{-2\pi i3t/T}.
$$

From this we see that the complex Fourier series is given by $y_1 = y_{-1} = \frac{3}{8}$, and that $y_3 = y_{-3} = \frac{1}{8}$. We see that the function lies in $V_{3,T}$, i.e. there are finitely many terms in the Fourier series. In general, if the function is some trigonometric function, we can often use trigonometric identities to find an expression for the Fourier series directly.

Exercise 1.28: Orthonormality of Complex Fourier Basis

Show that the complex functions $e^{2\pi i n t/T}$ are orthonormal.

Exercise 1.29: Complex Fourier Series of $f(t) = \sin^2(2\pi t/T)$

Compute the complex Fourier series of the function $f(t) = \sin^2(2\pi t/T)$.

Exercise 1.30: Complex Fourier Series of Polynomials

Repeat Exercise 1.22, computing the complex Fourier series instead of the real Fourier series.

Exercise 1.31: Complex Fourier Series and Pascal's Triangle

In this exercise we will find a connection with certain Fourier series and the rows in Pascal's triangle.

a) Show that both $\cos^n(t)$ and $\sin^n(t)$ are in $V_{N,2\pi}$ for $1 \leq n \leq N$.

b) Write down the N'th order complex Fourier series for $f_1(t) = \cos t$, $f_2(t) = \cos^2 t$, and $f_3(t) = \cos^3 t$.

c) In b) you should be able to see a connection between the Fourier coefficients and the three first rows in Pascal's triangle. Formulate and prove a general relationship between row n in Pascal's triangle and the Fourier coefficients of $f_n(t) = \cos^n t$.

Exercise 1.32: Complex Fourier Coefficients of the Square Wave

Compute the complex Fourier coefficients of the square wave using Eq. (1.15), i.e. repeat the calculations from Example 1.13 for the complex case. Use Theorem 15 to verify your result.

Exercise 1.33: Complex Fourier Coefficients of the Triangle Wave

Repeat Exercise 1.32 for the triangle wave.

Exercise 1.34: Complex Fourier Coefficients of Low-Degree Polynomials

Use Eq. (1.15) to compute the complex Fourier coefficients of the periodic functions with period T defined by, respectively, $f(t) = t$, $f(t) = t^2$, and $f(t) = t^3$, on $[0, T]$. Use Theorem 15 to verify your calculations from Exercise 1.22.

Exercise 1.35: Complex Fourier Coefficients for Symmetric and Antisymmetric Functions

In this exercise we will prove a version of Theorem 10 for complex Fourier coefficients.

a) If f is symmetric about 0, show that y_n is real, and that $y_{-n} = y_n$.

b) If f is antisymmetric about 0, show that the y_n are purely imaginary, $y_0 = 0$, and that $y_{-n} = -y_n$.

c) Show that $\sum_{n=-N}^{N} y_n e^{2\pi i n t/T}$ is symmetric when $y_{-n} = y_n$ for all n, and rewrite it as a cosine-series.

d) Show that $\sum_{n=-N}^{N} y_n e^{2\pi i n t/T}$ is antisymmetric when $y_0 = 0$ and $y_{-n} = -y_n$ for all n, and rewrite it as a sine-series.

1.4 Some Properties of Fourier Series

We continue by establishing some important properties of Fourier series, in particular the Fourier coefficients for some important functions. In these lists, we will use the notation $f \to y_n$ to indicate that y_n is the n'th (complex) Fourier coefficient of $f(t)$.

Theorem 16. Fourier Series Pairs.
   The functions 1, $e^{2\pi i n t/T}$, *and* $\chi_{-a,a}$ *have the Fourier coefficients*

$$e^{2\pi i n t/T} \to e_n = (\ldots, 0, \ldots, 0, 1, 0, \ldots, 0, \ldots)$$

$$\chi_{[-a,a]} \to \frac{\sin(2\pi n a/T)}{\pi n}.$$

The 1 *in* e_n *is at position* n *and the function* χ_A *is the characteristic function of the set* A, *defined by*

$$\chi_A(t) = \begin{cases} 1, & \text{if } t \in A; \\ 0, & \text{otherwise.} \end{cases} \tag{1.16}$$

The first pair is easily verified, so the proof is omitted. The case for $\chi_{-a,a}$ is very similar to the square wave, but easier to prove, and therefore also omitted.

Theorem 17. Fourier Series Properties.

The mapping $f \to y_n$ is linear: if $f \to x_n$, $g \to y_n$, then

$$af + bg \to ax_n + by_n$$

for all n. Moreover, if f is real and periodic with period T, the following properties hold:

1. *$y_n = \overline{y_{-n}}$ for all n.*
2. *If $f(t) = f(-t)$ (i.e. f is symmetric), then all y_n are real, so that b_n are zero and the Fourier series is a cosine series.*
3. *If $f(t) = -f(-t)$ (i.e. f is antisymmetric), then all y_n are purely imaginary, so that the a_n are zero and the Fourier series is a sine series.*
4. *If $g(t) = f(t-d)$ (i.e. g is the function f delayed by d) and $f \to y_n$, then $g \to e^{-2\pi i n d/T} y_n$.*
5. *If $g(t) = e^{2\pi i d t/T} f(t)$ with d an integer, and $f \to y_n$, then $g \to y_{n-d}$.*
6. *Let d be a number. If $f \to y_n$, then $f(d+t) = f(d-t)$ for all t if and only if the argument of y_n is $-2\pi n d/T$ for all n.*

Proof. The proof of linearity is left to the reader. Property 1 follows immediately by writing

$$y_n = \frac{1}{T}\int_0^T f(t)e^{-2\pi i n t/T}\,dt = \overline{\frac{1}{T}\int_0^T f(t)e^{2\pi i n t/T}\,dt}$$
$$= \overline{\frac{1}{T}\int_0^T f(t)e^{-2\pi i (-n)t/T}\,dt} = \overline{y_{-n}}.$$

Also, if $f(t) = f(-t)$, we have that

$$y_n = \frac{1}{T}\int_0^T f(t)e^{-2\pi i n t/T}\,dt = \frac{1}{T}\int_0^T f(-t)e^{-2\pi i n t/T}\,dt$$
$$= -\frac{1}{T}\int_0^{-T} f(t)e^{2\pi i n t/T}\,dt = \frac{1}{T}\int_0^T f(t)e^{2\pi i n t/T}\,dt = \overline{y_n}.$$

The first statement in Property 2 follows from this. The second statement follows directly by noting that

$$y_n e^{2\pi i n t/T} + y_{-n}e^{-2\pi i n t/T} = y_n(e^{2\pi i n t/T} + e^{-2\pi i n t/T}) = 2y_n \cos(2\pi n t/T),$$

or by invoking Theorem 10. Property 3 is proved in a similar way. To prove property 4, the Fourier coefficients of $g(t) = f(t-d)$ are

$$\frac{1}{T}\int_0^T g(t)e^{-2\pi i n t/T}\,dt = \frac{1}{T}\int_0^T f(t-d)e^{-2\pi i n t/T}\,dt$$
$$= \frac{1}{T}\int_0^T f(t)e^{-2\pi i n (t+d)/T}\,dt$$
$$= e^{-2\pi i n d/T}\frac{1}{T}\int_0^T f(t)e^{-2\pi i n t/T}\,dt = e^{-2\pi i n d/T}y_n.$$

For property 5, the Fourier coefficients of $g(t) = e^{2\pi i d t/T} f(t)$ are

$$\frac{1}{T} \int_0^T g(t) e^{-2\pi i n t/T} dt = \frac{1}{T} \int_0^T e^{2\pi i d t/T} f(t) e^{-2\pi i n t/T} dt$$

$$= \frac{1}{T} \int_0^T f(t) e^{-2\pi i (n-d) t/T} dt = y_{n-d}.$$

If $f(d+t) = f(d-t)$ for all t, we define the function $g(t) = f(t+d)$ which is symmetric about 0, so that it has real Fourier coefficients. But then the Fourier coefficients of $f(t) = g(t-d)$ are $e^{-2\pi i n d/T}$ times the (real) Fourier coefficients of g by property 4. It follows that y_n, the Fourier coefficients of f, has argument $-2\pi n d/T$. The proof in the other direction follows by noting that any function where the Fourier coefficients are real must be symmetric about 0, once the Fourier series is known to converge. This proves property 6. □

Let us analyze these properties, to see that they match the notion we already have for frequencies and sound. The first property says that the positive and negative frequencies in a (real) sound essentially are the same. The fourth property says that, if we delay a sound, the frequency content also is essentially the same. This also matches our intuition on sound, since we think of the frequency representation as something which is independent from when it is played. The fifth property says that, if we multiply a sound with a pure tone, the frequency representation is shifted (delayed), according to the value of the frequency. This is something we see in early models for the transmission of audio, where an audio signal is transmitted after having been multiplied with what is called a *carrier wave*. The carrier wave can be a pure tone. The result is a signal where the frequencies have been shifted with the frequency of the carrier wave. The point of shifting the frequency like this is to use a frequency range where one knows that other signals do not interfere. The last property looks a bit mysterious. We will not have use for this property before the next chapter.

From Theorem 17 we see that there are several cases of duality between a function and its Fourier series:

- Delaying a function corresponds to multiplying the Fourier coefficients with a complex exponential. Vice versa, multiplying a function with a complex exponential corresponds to delaying the Fourier coefficients.
- Symmetry/antisymmetry for a function corresponds to the Fourier coefficients being real/purely imaginary. Vice versa, a function which is real has Fourier coefficients which are conjugate symmetric.

Actually, one can show that these dualities are even stronger if we had considered Fourier series of complex functions instead of real functions. We will not go into this.

1.4.1 Rate of Convergence for Fourier Series

We now know enough to say a few things about the rate of convergence of Fourier series, even if it may not be that the Fourier series converges to the function itself (in Sect. 1.6 we will find sufficient conditions for this). We have already seen examples which illustrate different convergence rates: The square wave seemed to have very slow convergence near the discontinuities, while the triangle wave did not have this same problem. To analyze this further, the following simple lemma will be useful.

Lemma 18. Differentiation of Fourier Series.

Assume that f is differentiable. Then $(f_N)'(t) = (f')_N(t)$. In other words, the derivative of the Fourier series equals the Fourier series of the derivative.

Proof. We first compute

$$\langle f, e^{2\pi int/T} \rangle = \frac{1}{T} \int_0^T f(t) e^{-2\pi int/T} dt$$

$$= \frac{1}{T} \left(\left[-\frac{T}{2\pi in} f(t) e^{-2\pi int/T} \right]_0^T + \frac{T}{2\pi in} \int_0^T f'(t) e^{-2\pi int/T} dt \right)$$

$$= \frac{T}{2\pi in} \frac{1}{T} \int_0^T f'(t) e^{-2\pi int/T} dt = \frac{T}{2\pi in} \langle f', e^{2\pi int/T} \rangle.$$

where we used integration by parts, and that $-\frac{T}{2\pi in} f(t) e^{-2\pi int/T}$ is periodic with period T. It follows that $\langle f, e^{2\pi int/T} \rangle = \frac{T}{2\pi in} \langle f', e^{2\pi int/T} \rangle$. From this we get that

$$(f_N)'(t) = \left(\sum_{n=-N}^{N} \langle f, e^{2\pi int/T} \rangle e^{2\pi int/T} \right)' = \frac{2\pi in}{T} \sum_{n=-N}^{N} \langle f, e^{2\pi int/T} \rangle e^{2\pi int/T}$$

$$= \sum_{n=-N}^{N} \langle f', e^{2\pi int/T} \rangle e^{2\pi int/T} = (f')_N(t).$$

where we substituted the connection between the inner products we just found. □

1.4.2 Differentiating Fourier Series

The connection between the Fourier series of and its derivative can be used to simplify the computation of other Fourier series. As an example, let us see how we can simplify our previous computation for the triangle wave. It is straightforward to see the relationship $f_t'(t) = \frac{4}{T} f_s(t)$ from the plots of f_s and f_t. From this relationship, and from the expression for the Fourier series of the square wave from Example 1.13, it follows that

$$((f_t)')_N(t) = \frac{4}{T} \left(\frac{4}{\pi} \sin(2\pi t/T) + \frac{4}{3\pi} \sin(2\pi 3t/T) + \frac{4}{5\pi} \sin(2\pi 5t/T) + \cdots \right).$$

If we integrate this we obtain

$$(f_t)_N(t) = -\frac{8}{\pi^2} \left(\cos(2\pi t/T) + \frac{1}{3^2} \cos(2\pi 3t/T) + \frac{1}{5^2} \cos(2\pi 5t/T) + \cdots \right) + C.$$

What remains is to find the integration constant C. This is simplest found if we set $t = T/4$, since then all cosine terms are 0. Clearly then $C = 0$, and we arrive at the same expression as in Example 1.14. This approach clearly had less computations involved. There is a minor point here which we have not addressed: the triangle wave is not differentiable at two points, as required by Lemma 18. It is, however, not too difficult to see that this result also holds in cases where we have a finite number of points of nondifferentiability.

We have the following corollary to Lemma 18:

Corollary 19. Connection Between the Fourier Coefficients of $f(t)$ and $f'(t)$.

If the complex Fourier coefficients of f are y_n and f is differentiable, then the Fourier coefficients of $f'(t)$ are $\frac{2\pi in}{T} y_n$.

Turning this around, the Fourier coefficients of $f(t)$ are $T/(2\pi in)$ times those of $f'(t)$. If f is s times differentiable, we can repeat this argument to conclude that the Fourier coefficients of $f(t)$ are $(T/(2\pi in))^s$ times those of $f^{(s)}(t)$. In other words, the Fourier coefficients of a function which is many times differentiable decay to zero very fast.

Observation 20. Convergence Speed of Differentiable Functions.

The Fourier series converges quickly when the function is many times differentiable.

This result applies for our comparison between the square- and triangle waves, since the square wave is discontinuous, while the triangle wave is continuous with a discontinuous first derivative. The functions considered in Examples 1.25 and 1.26 were also not continuous, which implies slow convergence for these Fourier series also.

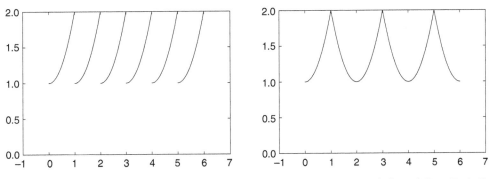

Fig. 1.10 Two different extensions of f to a periodic function on the whole real line. Periodic extension (left) and symmetric extension (right)

Often a function is defined on an interval $[0, T]$. We can of course consider the T-periodic extension of this, but if $f(0) \neq f(T)$, the periodic extension will have a discontinuity at T (and is also ill-defined at T), even if f is continuous on $[0, T]$. This is shown in the left plot in Fig. 1.10. As a consequence, we expect a slowly converging Fourier series. We can therefore ask ourselves the following question:

Idea 21. Continuous Extension.

Assume that f is continuous on $[0, T]$. Can we construct a periodic function $\breve{f}$, which agrees with f on $[0, T]$, and which is both continuous and periodic?

The period of $\breve{f}$ must clearly be larger than T. If this is possible the Fourier series of $\breve{f}$ could produce better approximations to f. It turns out that the following extension strategy does the job:

Definition 22. *Symmetric Extension of a Function.*
Let f be a function defined on $[0, T]$. By the *symmetric extension* of f, denoted $\breve{f}$, we mean the function defined on $[0, 2T]$ by

$$\breve{f}(t) = \begin{cases} f(t), & \text{if } 0 \le t \le T; \\ f(2T - t), & \text{if } T < t \le 2T. \end{cases}$$

A symmetric extension is shown in the right plot in Fig. 1.10. Clearly the following hold:

Theorem 23. Continuous Extension.
If f is continuous on $[0, T]$, then $\breve{f}$ is continuous on $[0, 2T]$, and $\breve{f}(0) = \breve{f}(2T)$. If we extend $\breve{f}$ to a periodic function on the whole real line (which we also will denote by $\breve{f}$), this function is continuous, agrees with f on $[0, T]$, and is symmetric.

In particular the Fourier series of $\breve{f}$ is a cosine series. $\breve{f}$ is symmetric since, for $0 \le t \le T$,

$$\breve{f}(-t) = \breve{f}(2T - t) = f(2T - (2T - t)) = f(t) = \breve{f}(t).$$

In summary, we now have two possibilities for approximating a function f defined on $[0, T]$:

- By the Fourier series of f,
- By the Fourier series of $\breve{f}$ restricted to $[0, T]$.

Example 1.36: Periodic and Symmetric Extension

Let f be the function with period T defined by $f(t) = 2t/T - 1$ for $0 \le t < T$. In each period the function increases linearly from -1 to 1. Because f is discontinuous at the boundaries, we would expect the Fourier series to converge slowly. The Fourier series is a sine-series since f is antisymmetric, and we can compute b_n as

$$b_n = \frac{2}{T} \int_0^T \frac{2}{T} \left(t - \frac{T}{2} \right) \sin(2\pi n t/T) dt = \frac{4}{T^2} \int_0^T \left(t - \frac{T}{2} \right) \sin(2\pi n t/T) dt$$

$$= \frac{4}{T^2} \int_0^T t \sin(2\pi n t/T) dt - \frac{2}{T} \int_0^T \sin(2\pi n t/T) dt = -\frac{2}{\pi n},$$

so that

$$f_N(t) = -\sum_{n=1}^N \frac{2}{n\pi} \sin(2\pi n t/T),$$

which indeed converges slowly to 0. Let us now instead consider the symmetric extension of f. Clearly this is the triangle wave with period $2T$, and the Fourier series of this was

$$(\breve{f})_N(t) = -\sum_{n \le N,\, n \text{ odd}} \frac{8}{n^2 \pi^2} \cos(2\pi n t/(2T)).$$

The second series clearly converges faster than the first, since its Fourier coefficients are $a_n = -8/(n^2\pi^2)$ (with n odd), while the Fourier coefficients in the first series are $b_n = -2/(n\pi)$.

If we use $T = 1/440$, the symmetric extension has period $1/220$, which gives a triangle wave where the first term in the Fourier series has frequency $220\,\text{Hz}$. Listening to this we should hear something resembling a $220\,\text{Hz}$ pure tone, since the first term in the Fourier series is the most dominating in the triangle wave. Listening to the periodic extension we should hear a different sound. The first term in the Fourier series has frequency $440\,\text{Hz}$, but this drowns a bit in the contribution of the other terms in the Fourier series, due to the slow convergence of the Fourier series.

The Fourier series with 7 terms for the periodic- and symmetric extensions of f are shown in Fig. 1.11.

It is clear from the plots that the Fourier series for f is not a very good approximation, while for the symmetric extension, we cannot differ between the Fourier series and the function.

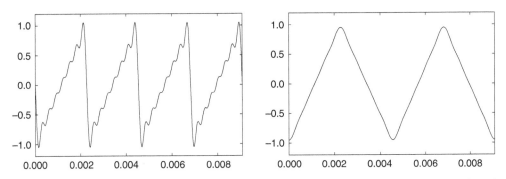

Fig. 1.11   The Fourier series with $N = 7$ terms of the periodic- (left) and symmetric (right) extensions of the function in Example 1.36

Exercise 1.37: Fourier Series of a Delayed Square Wave

Define the function f with period T on $[-T/2, T/2)$ by

$$f(t) = \begin{cases} 1, & \text{if } -T/4 \le t < T/4; \\ -1, & \text{if } T/4 \le |t| < T/2. \end{cases}$$

f is just the square wave, delayed with $d = -T/4$. Compute the Fourier coefficients of f directly, and use Property 4 in Theorem 17 to verify your result.

Exercise 1.38: Find a Function from Its Fourier Series

Find a function f which has the complex Fourier series

$$\sum_{n \text{ odd}} \frac{4}{\pi(n+4)} e^{2\pi i n t/T}.$$

Hint

Attempt to use one of the properties in Theorem 17 on the Fourier series of the square wave.

Exercise 1.39: Relation Between the Complex Fourier Coefficients of f and the Cosine-Coefficients of $\breve{f}$

Show that the complex Fourier coefficients y_n of f, and the cosine-coefficients a_n of $\breve{f}$ are related by $a_{2n} = y_n + y_{-n}$. This result is not enough to obtain the entire Fourier series of $\breve{f}$, but at least it gives us half of it.

1.5 Filters on Periodic Functions

It is easy to use Fourier series to analyze or improve sound: Noise in sound often corresponds to the presence of some high frequencies with large coefficients. By removing these, we remove the noise. For example, setting all Fourier coefficients except the first one to zero, changes the unpleasant square wave to a pure tone. Such operations on continuous-time are called *filters* and are defined as follows, regardless of whether the input is periodic or not.

Definition 24. *Filters.*
A linear transformation s is called a *filter* if, for any frequency $e^{2\pi i \nu t}$, we have that

$$s(e^{2\pi i \nu t}) = \lambda_s(\nu) e^{2\pi i \nu t}$$

for some function λ_s. We say that filters preserve frequencies. λ_s is also called the *frequency response* of s, and $\lambda_s(\nu)$ describes how s treats frequency ν.

Any pure tone is an eigenvector of a filter, with the frequency response providing the eigenvalues. We will later study filters where the input is *discrete-time*, and then use the terms *analog* and *digital* to distinguish between continuous-time- and discrete-time filters, respectively.

It is clear that $s(f) \in V_{N,T}$ when $f \in V_{N,T}$ and s is a filter. Thus, a filter can be restricted to the Fourier spaces, and below we analyze filters using this restriction. Let us first remark some properties which apply regardless of discrete-time or continuous-time, and regardless of whether the input is periodic, or a Fourier series. We need the following definition:

Definition 25. *Time-Invariance.*
Assume that s is a linear transformation. Let the function x be input to s, with $y = s(x)$ the output function. Let also z and w be delays of x and y with d time units, i.e. $z(t) = x(t - d)$ and $w(t) = y(t - d)$. s is said to be *time-invariant* if, for any d and x, $s(z) = w$, i.e. s sends the delayed input to the delayed output.

It turns out that time-invariance is one of the fundamental properties of filters:

Proposition 26. Properties of Filters.
The following hold:

1. *The composition of two filters is again a filter,*
2. *any two filters s_1 and s_2 commute, i.e. $s_1 s_2 = s_2 s_1$,*
3. *$\lambda_{s_1 s_2}(\nu) = \lambda_{s_1}(\nu)\lambda_{s_2}(\nu)$ for any filters s_1 and s_2,*
4. *filters are time-invariant.*

Due to the result above, filters are also called *LTI filters*, LTI standing for Linear, Time-Invariant. Clearly $s_1 + s_2$ is a filter whenever s_1 and s_2 are. The set of all filters is thus a vector space which is closed under multiplication. Such a space is called an *algebra*. Since all such filters commute, it is called a *commutative algebra*.

Proof. If s_1 and s_2 have frequency responses $\lambda_{s_1}(\nu)$ and $\lambda_{s_2}(\nu)$, respectively, we obtain

$$s_1(s_2(e^{2\pi i\nu t})) = \lambda_{s_2}(\nu)s_1(e^{2\pi i\nu t}) = \lambda_{s_1}(\nu)\lambda_{s_2}(\nu)e^{2\pi i\nu t}.$$

From this it follows that the composition of s_1 and s_2 also is a filter, that the filters commute, and that $\lambda_{s_1 s_2}(\nu) = \lambda_{s_1}(\nu)\lambda_{s_2}(\nu)$. Finally,

$$s(e^{2\pi i\nu(t-d)}) = e^{-2\pi i\nu d}s(e^{2\pi i\nu t}) = e^{-2\pi i\nu d}\lambda_s(\nu)e^{2\pi i\nu t} = \lambda_s(\nu)e^{2\pi i\nu(t-d)},$$

which shows that filters are time-invariant on any single frequency. Due to linearity, filters are also time-invariant on any sum of frequencies as well. □

Since the Fourier spaces have a mapping between time and frequency, it is straightforward to find expressions for filters restricted to the Fourier spaces. If $f(t) = \sum_{n=-N}^{N} y_n e^{2\pi int/T}$, since $y_n = \frac{1}{T}\int_0^T f(s)e^{-2\pi ins/T}ds$ we obtain

$$
\begin{aligned}
s(f) = s\left(\sum_{n=-N}^{N} y_n e^{2\pi int/T}\right) &= \sum_{n=-N}^{N} \lambda_s(n/T)y_n e^{2\pi int/T}\\
&= \frac{1}{T}\int_0^T \left(\sum_{n=-N}^{N} \lambda_s(n/T)e^{2\pi in(t-s)/T}\right)f(s)ds\\
&= \frac{1}{T}\int_0^T g(t-s)f(s)ds
\end{aligned}
$$

where $g(t) = \sum_{n=-N}^{N} \lambda_s(n/T)e^{2\pi int/T}$, i.e. g is a function with Fourier coefficients being values on the frequency response.

When f and g are T-periodic functions, the expression $\frac{1}{T}\int_0^T g(t-s)f(s)ds$ is called the *circular convolution* of g and f, and written $g \circledast f$. The normalization factor $\frac{1}{T}$ is included for convenience. Since both f and g are T-periodic, it is straightforward to see that a circular convolution also can be written as $\frac{1}{T}\int_0^T g(s)f(t-s)ds$ (make the substitution $u = t - s$).

Theorem 27. Filters and Circular Convolution.

A filter s restricted to $V_{N,T}$ can be written as a circular convolution with $g(t) = \sum_{n=-N}^{N} \lambda_s(n/T)e^{2\pi i nt/T} \in V_{N,T}$. Conversely, any circular convolution with a $g \in V_{N,T}$ gives rise to a filter s with frequency response

$$\lambda_s(\nu) = \frac{1}{T} \int_0^T g(u)e^{-2\pi i \nu u} du.$$

g is also called the convolution kernel *of the filter.*

Combining two filters also can be thought of as a circular convolution. To see why, if g_1 and g_2 are the convolution kernels of the filters s_1 and s_2, we obtain

$$s_1 s_2(f) = g_1 \circledast (g_2 \circledast f) = (g_1 \circledast g_2) \circledast f,$$

where we used that $\circledast$ is associative, which is straightforward to prove. The equation $\lambda_{s_1 s_2}(\nu) = \lambda_{s_1}(\nu)\lambda_{s_2}(\nu)$ from Proposition 26 is therefore often summarized as *convolution in time corresponds to multiplication in frequency*. We will return to this notion as well as circular convolution later.

Proof. We only need to comment on the last statement. But

$$g(t) = \sum_{n=-N}^{N} \lambda_s(n/T)e^{2\pi i nt/T}$$

implies that $\lambda_s(n/T) = \frac{1}{T} \int_0^T g(t)e^{-2\pi i nt/T} dt$ by the uniqueness of the Fourier coefficients. Replacing n/T with ν gives the desired result. □

Circular convolution of functions gives meaning also for functions which do not reside in some Fourier space, as long as the integral exists. Many important filters can be constructed by using a g in some Fourier space, however, as we will see in the next section. In particular, filters with $g \in V_{N,T}$ can be used to extract good approximations to f from $V_{N,T}$. One of those is obtained by choosing the filter so that $\lambda_s(n/T) = 1$ for $-N \le n \le N$, i.e.

$$\lambda_s(\nu) = \begin{cases} 1 & \text{if } |\nu| \le N/T \\ 0 & \text{if } |\nu| > N/T \end{cases}. \tag{1.17}$$

This is called a *low-pass filter*, since it annihilates high frequencies (i.e. $\nu > |N/T|$). The corresponding convolution kernel is

$$g(t) = \sum_{n=-N}^{N} e^{2\pi i nt/T} = 1 + 2\sum_{n=1}^{N} \cos(2\pi nt/T),$$

and is also called the *Dirichlet kernel*, denoted $D_N(t)$. The output of the filter is seen to be

$$\frac{1}{T} \int_0^T D_N(t-s)f(s)ds = \frac{1}{T} \sum_{n=-N}^{N} \int_0^T f(s)e^{2\pi i n(t-s)/T} ds$$

$$= \sum_{n=-N}^{N} \frac{1}{T} \int_0^T f(s)e^{-2\pi i ns/T} ds\, e^{2\pi i nt}$$

$$= \sum_{n=-N}^{N} y_n e^{2\pi i nt/T} = f_N(t).$$

Filtering with the Dirichlet kernel thus produces the Fourier series, and proving $\lim_{N\to\infty} f_N(t) = f(t)$ is thus the same as proving $\lim_{N\to\infty} \frac{1}{T}\int_0^T D_N(u)f(t-u)du = f(t)$. We will see in the next section that establishing the second limit here is a useful strategy for proving the pointwise convergence of Fourier series.

Exercise 1.40: Symmetric Filters Preserve Sine- and Cosine-Series

An analog filter where $\lambda_s(\nu) = \lambda_s(-\nu)$ is also called a *symmetric filter*.

a) Prove that, if the input to a symmetric filter is a Fourier series which is a cosine- or sine series, then the output also is a cosine- or sine series.

b) Show that $s(f) = \int_{-a}^a g(s)f(t-s)ds$ is a symmetric filter whenever g is symmetric around 0, and zero outside $[-a,a]$.

Exercise 1.41: Circular Convolution and Fourier Series

Let $f,g \in V_{N,T}$, and let $h = f \circledast g$. Assume that f, g, and h have Fourier coefficients x_n, y_n, and z_n respectively. Show that $z_n = x_n y_n$.

1.6 Convergence of Fourier Series*

A major topic in Fourier analysis is finding conditions which secure convergence of the Fourier series of a function f. This turns out to be very difficult in general, and depends highly on the type of convergence we consider. This section will cover some important results on this, and is a bit more technical than the remainder of the book. We will consider both

- *pointwise convergence*, i.e. that $f_N(t) \to f(t)$ for all t, and
- *convergence in* $\|\cdot\|$, i.e. that $\|f_N - f\| \to 0$.

The latter unfortunately does not imply the first, which is harder to prove. Although a general theorem about the pointwise convergence of the Fourier series for square integrable functions exists, this result is way too hard to prove here. Instead we will restrict ourselves to the following class of functions, for which it is possible to state a proof for both modes of convergence. This class also contains most functions we encounter in the book, such as the square wave and the triangle wave:

Definition 28. *Piecewise Continuous Functions.*
   A T-periodic function is said to be *piecewise continuous* if there exists a finite set of points

$$0 \le a_0 < a_1 < \cdots < a_{n-1} < a_n < T$$

so that

1. f is continuous on each interval between adjacent such points,
2. the one-sided limits $f(a_i^+) := \lim_{t \to a_i^+} f(t)$ and $f(a_i^-) := \lim_{t \to a_i^-} f(t)$ exist, and
3. $f(a_i) = \frac{f(a_i^+) + f(a_i^-)}{2}$ (i.e. the value at a "jump" is the average of the one-sided limits).

For piecewise continuous functions, convergence in $\| \cdot \|$ for the Fourier series will follow from the following theorem.

Theorem 29. Approximating Piecewise Continuous Functions.

Let f be piecewise continuous. For each $N \geq 1$ we can find an $S_N \in V_{N,T}$, so that $\lim_{N \to \infty} S_N(t) = f(t)$ for all t. Also, $S_N \to f$ uniformly as $N \to \infty$ on any interval $[a, b]$ where f is continuous.

The functions S_N are found in a constructive way in the proof of this theorem, but note that these are not the same as the Fourier series f_N! Therefore, the theorem says nothing about the convergence of the Fourier series itself.

Proof. In the proof we will use the concept of a *summability kernel*, which is a sequence of functions k_N defined on $[0, T]$ satisfying

1. $\frac{1}{T} \int_0^T k_N(t) dt = 1$,
2. $\frac{1}{T} \int_0^T |k_N(t)| dt \leq C$ for all N, and for some constant C,
3. For all $0 < \delta < T/2$, $\lim_{N \to \infty} \frac{1}{T} \int_\delta^{T-\delta} |k_N(t)| dt = 0$.

In Exercise 1.44 you are aided through the construction of one important summability kernel, denoted F_N and called the *Fejer kernel*. The Fejer kernel has the following additional properties

- $F_N \in V_{N,T}$,
- its periodic extension satisfies $F_N(t) = F_N(-t)$,
- $0 \leq F_N(t) \leq \frac{T^2}{4(N+1)t^2}$.

Denote by s_N the filter with kernel F_N, i.e.

$$s_N(f) = \frac{1}{T} \int_0^T F_N(u) f(t - u) du,$$

Denote also the function $s_N(f)$ by S_N. Since $F_N \in V_{N,T}$, we know from the last section that $S_N \in V_{N,T}$ also. If we now use that $F_N(t) = F_N(-t)$, and make the substitution $v = -u$, it easily follows that $S_N(t) = \frac{1}{T} \int_0^T F_N(u) f(t+u) du$ as well. We can thus write

$$S_N(t) = \frac{1}{2T} \int_0^T F_N(u)(f(t + u) + f(t - u)) du.$$

Since also $f(t) = \frac{1}{T} \int_0^T F_N(u) f(t) du$, it follows that

$$S_N(t) - f(t) = \frac{1}{2T} \int_0^T F_N(u)(f(t + u) + f(t - u) - 2f(t)) du. \qquad (1.18)$$

When $S_N(t) - f(t)$ is written on this form, one can see that the integrand is continuous at $u = 0$ as a function of u: This is obvious if f is continuous at t. If on the other hand

f is not continuous at t, we have by assumption that $f(t + u) + f(t - u) \to 2f(t)$ as $u \to 0$. Given $\epsilon > 0$, we can therefore find a $\delta > 0$ so that

$$|f(t + u) + f(t - u) - 2f(t)| < \epsilon$$

for all $|u| < \delta$. Now, split the integral (1.18) in three: $\int_0^T = \int_{-T/2}^{-\delta} + \int_{-\delta}^{\delta} + \int_{\delta}^{T/2}$. For the second of these we have

$$\frac{1}{2T} \int_{-\delta}^{\delta} |F_N(u)(f(t + u) + f(t - u) - 2f(t))| du$$

$$\leq \frac{1}{2T} \int_{-\delta}^{\delta} \epsilon F_n(u) du \leq \frac{1}{2T} \int_{-T/2}^{T/2} \epsilon F_n(u) du = \frac{\epsilon}{2}.$$

For the third of these we have

$$\frac{1}{2T} \int_{\delta}^{T/2} |F_N(u)(f(t + u) + f(t - u) - 2f(t))| du$$

$$\leq \frac{1}{2T} \int_{\delta}^{T/2} 4\|f\|_{\infty} \frac{T^2}{4(N + 1)u^2} du \leq \frac{\|f\|_{\infty} T}{2(N + 1)\delta^2},$$

where $\|f\|_{\infty} = \max_{x \in [0,T]} |f(x)|$. A similar calculation can be done for the first integral. Clearly then we can choose N so big that the sum of the first and third integrals are less than $\epsilon/2$, and we then get that $|S_N(t) - f(t)| < \epsilon$. This shows that $S_N(t) \to f(t)$ as $N \to \infty$ for any t. For the final statement, if $[a, b]$ is an interval where f is continuous, choose the δ above so small that $[a - \delta, b + \delta]$ still contains no discontinuities. Since continuous functions are uniformly continuous on compact intervals, it is not too hard to see that the convergence of f to S_N on $[a, b]$ is uniform. This completes the proof. □

Since $S_N(t) = \frac{1}{T} \int_0^T F_N(u) f(t - u) du \in V_{N,T}$, and f_N is a best approximation from $V_{N,T}$, we have that $\|f_N - f\| \leq \|S_N - f\|$. If f is continuous, the result says that $\|f - S_N\|_{\infty} \to 0$, which implies that $\|f - S_N\| \to 0$, so that $\|f - f_N\| \to 0$. Therefore, for f continuous, both $\|f - S_n\|_{\infty} \to 0$ and $\|f - f_N\| \to 0$ hold, so that we have established both modes of convergence. If f has a discontinuity, it is obvious that $\|f - S_N\|_{\infty} \to 0$ can not hold, since S_N is continuous. $\|f - f_N\| \to 0$ still holds, however. The reason is that any function with only a finite number of discontinuities can be approximated arbitrarily well with continuous functions w.r.t. $\| \cdot \|$. The proof of this is left as an exercise.

Both the square- and triangle waves are piecewise continuous (at least if we redefined the value of the square wave at the discontinuity). Therefore both their Fourier series converge to f in $\| \cdot \|$. Since the triangle wave is continuous, S_N also converges uniformly to f_t.

The result above states that S_N converges pointwise to f—it does not say that f_N converges pointwise to f. This suggests that S_N may be better suited to approximate f. In Fig. 1.12 we have plotted $S_N(t)$ and $f_N(t)$ for the square wave. Clearly the approximations are very different.

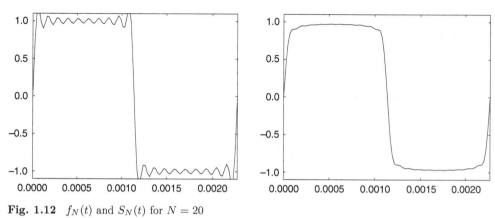

Fig. 1.12 $f_N(t)$ and $S_N(t)$ for $N = 20$

The pointwise convergence of $f_N(t)$ is more difficult to analyze, so we will make some additional assumptions.

Theorem 30. Pointwise Convergence of Fourier Series.

Assume that f is piecewise continuous and that the one-sided limits

$$D_+f(t) = \lim_{h \to 0^+} \frac{f(t+h) - f(t^+)}{h} \qquad D_-f(t) = \lim_{h \to 0^-} \frac{f(t+h) - f(t^-)}{h}$$

exist. Then $\lim_{N \to \infty} f_N(t) = f(t)$.

Proof. We will use the Dirichlet kernel, defined in Sect. 1.5. It turns out that this kernel satisfies only one of the properties of a summability kernel (see Exercise 1.43), but this will turn out to be enough for our purposes, due to the additional assumption on the one-sided limits for the derivative. We have already proved that

$$f_N(t) = \int_0^T D_N(u)f(t-u)du.$$

A formula similar to (1.18) can be easily proved using the same substitution $v = -u$:

$$f_N(t) - f(t) = \frac{1}{2T} \int_0^T (f(t+u) + f(t-u) - 2f(t))D_N(u)du. \qquad (1.19)$$

Substituting the expression for the Dirichlet kernel obtained in Exercise 1.43, the integrand can be written as

$$f(t+u) - f(t^+) + f(t-u) - f(t^-))D_N(u)$$
$$= \left(\frac{f(t+u) - f(t^+)}{\sin(\pi u/T)} + \frac{f(t-u) - f(t^-)}{\sin(\pi u/T)} \right) \sin(\pi(2N+1)u/T)$$
$$= h(u)\sin(\pi(2N+1)u/T).$$

We have that

$$\frac{f(t+u) - f(t^+)}{\sin(\pi u/T)} = \frac{f(t+u) - f(t^+)}{\pi u/T} \frac{\pi u/T}{\sin(\pi u/T)}$$

$$\rightarrow \begin{cases} \frac{T}{\pi} D_+ f(t) & \text{when } u \rightarrow 0^+, \\ \frac{T}{\pi} D_- f(t) & \text{when } u \rightarrow 0^-, \end{cases},$$

and similarly for $\frac{f(t-u) - f(t^-)}{\sin(\pi u/T)}$. It follows that the function h defined above is a piecewise continuous function in u. The proof will be done if we can show that $\int_0^T h(u) \sin(\pi(2N+1)u/T)dt \rightarrow 0$ as $N \rightarrow \infty$ for any piecewise continuous h. Since

$$\sin(\pi(2N+1)u/T) = \sin(2\pi Nu/T)\cos(\pi u/T) + \cos(2\pi Nu/T)\sin(\pi u/T),$$

and since $h(u)\cos(\pi u/T)$ and $h(u)\sin(\pi u/T)$ also are piecewise continuous, it is enough to show that $\int h(u)\sin(2\pi Nu/T)du \rightarrow 0$ and $\int h(u)\cos(2\pi Nu/T)du \rightarrow 0$. These are simply the order N Fourier coefficients of h. Since h is in particular square integrable, it follows from Bessel's inequality (Theorem 14) that the Fourier coefficients of h go to zero, and the proof is done. □

The requirement on the one-sided limits of the derivative above can be can be replaced by less strict conditions. This gives rise to what is known as *Dini's test*. One can also replace with the less strict requirement that f has a finite number of local minima and maxima. This is referred to as *Dirichlet's theorem*, after Dirichlet who proved it in 1829. There also exist much more general conditions that secure pointwise convergence of the Fourier series. The most general results require deep mathematical theory to prove.

Both the square wave and the triangle wave have one-sided limits for the derivative. Therefore both their Fourier series converge to f pointwise.

1.6.1 Interpretation of the Filters Corresponding to the Fejer and Dirichlet Kernels

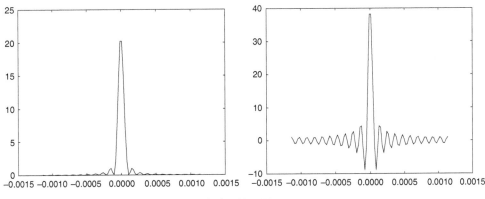

Fig. 1.13 The Fejer and Dirichlet kernels for $N = 20$

The Fejer and Dirichlet kernels are shown in Fig. 1.13. It is shown in Exercise 1.44 that $S_N(t) = \frac{1}{N+1} \sum_{n=0}^{N} f_n(t)$, where again S_N was the function $s_n(f)$, with s_n the filter corresponding to the Fejer kernel. This is also called the *Cesaro mean* of the Fourier series. From the same exercise it will be clear that the frequency response is

$$\lambda_s(\nu) = \begin{cases} 1 - \frac{T|\nu|}{N+1} & \text{if } |\nu| \leq N/T \\ 0 & \text{if } |\nu| > N/T \end{cases}, \tag{1.20}$$

On $[-N/T, N/T]$, this first increases linearly to 1, then decreases linearly back to 0. Outside $[-N/T, N/T]$ we get zero. The frequency responses for the two filters are shown in Fig. 1.14.

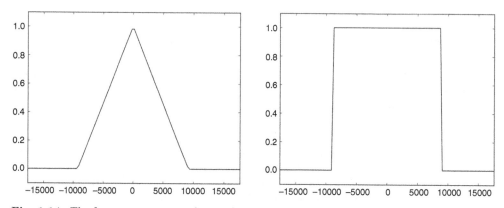

Fig. 1.14 The frequency responses for the filters with Fejer and Dirichlet kernels

Both are low-pass filters. The lowest frequency $\nu = 0$ is treated in the same way by both filters, but the higher frequencies are differed: The Dirichlet kernel keeps them exactly, while the Fejer kernel attenuates them, i.e. does not include all the frequency content at the higher frequencies. That filtering with the Fejer kernel gave something with better convergence properties (i.e. $S_N(t)$) can be interpreted as follows: We should be careful when we include the contribution from the higher frequencies, as this may affect the convergence.

Exercise 1.42: Approximation in Norm with Continuous Functions

Assume that f is a function defined on $[a, b]$ with only a finite number of discontinuities, and with all one-sided limits existing. Show that there exists a continuous function g so that $\|f - g\| < \epsilon$.

Exercise 1.43: The Dirichlet Kernel

Recall that the Dirichlet kernel was defined as

$$D_N(t) = \sum_{n=-N}^{N} e^{2\pi int/T}$$

a) Show that $D_N(t) = \frac{\sin(\pi(2N+1)t/T)}{\sin(\pi t/T)}$.

b) Prove that $D_N(t)$ satisfies one of the properties of a summability kernel, but fails on the other ones.

Exercise 1.44: The Fejer Summability Kernel

The Fejer kernel is defined as

$$F_N(t) = \sum_{n=-N}^{N} \left(1 - \frac{|n|}{N+1}\right) e^{2\pi int/N}.$$

Clearly $F_N \in V_{N,T}$, and of degree N.

a) Show that $F_N(t) = \frac{1}{N+1} \left(\frac{\sin(\pi(N+1)t/T)}{\sin(\pi t/T)}\right)^2$, and conclude from this that $0 \le F_N(t) \le \frac{T^2}{4(N+1)t^2}$.

Hint

Use that $\frac{2}{\pi}|u| \le |\sin u|$ when $u \in [-\pi/2, \pi/2]$.

b) Show that $F_N(t)$ satisfies the three properties of a summability kernel.

c) Show that $\frac{1}{T} \int_0^T F_N(u)f(t-u)du = \frac{1}{N+1} \sum_{n=0}^{N} f_n(t)$.

Hint

Show that $F_N(t) = \frac{1}{N+1} \sum_{n=0}^{N} D_n(t)$, and use Exercise 1.43b).

1.7 The MP3 Standard

Digital audio first became commonly available when the CD was introduced in the early 1980s. As the storage capacity and processing speeds of computers increased, it became possible to transfer audio files to computers and both play and manipulate the data, as we have described. Large amounts of data are needed for audio, so that advanced compression techniques are desirable. Lossless coding techniques like Huffman and Lempel-Ziv coding [65, 64] were known and with these the file size could be reduced by 50%. However, by also allowing the data to be altered a little bit it turned out that it was possible to reduce the file size down to about 10%, without much loss in quality. The MP3 audio format takes advantage of this.

MP3, or more precisely *MPEG-1 Audio Layer 3*, is part of an audio-visual standard called MPEG. MPEG has evolved over the years, from MPEG-1 to MPEG-2, and then to MPEG-4. The data on a DVD disc can be stored with either MPEG-1 or MPEG-2, while the data on a bluray disc can be stored with either MPEG-2 or MPEG-4. MP3 became an international standard in 1991, and virtually all audio software and music players support this format. MP3 is just a sound format. It leaves a substantial amount of freedom in the encoder, so that different encoders can exploit properties of sound in various ways, in order to alter the sound in removing inaudible components therein. As a consequence there are many different MP3 encoders available, of varying quality. In particular, an encoder which works well for higher bit rates (high quality sound) may not work so well for lower bit rates.

With MP3, the sound is split into *frequency bands*, each band corresponding to a particular frequency range. In the simplest model, 32 frequency bands are used. A frequency analysis of the sound, based on what is called a *psycho-acoustic model*, is the basis for further transformation of these bands. The psycho-acoustic model computes the significance of each band for the human perception of the sound. When we hear a sound, there is a mechanical stimulation of the ear drum, and the amount of stimulus is directly related to the size of the sample values of the digital sound. The movement of the ear drum is then converted to electric impulses that travel to the brain where they are perceived as sound. The perception process uses a transformation of the sound so that a steady oscillation in air pressure is perceived as a sound with a fixed frequency. In this process certain kinds of perturbations of the sound are hardly noticed by the brain, and this is exploited in lossy audio compression.

When the psycho-acoustic model is applied, *scale factors* and *masking thresholds* are computed for each frequency band. The masking thresholds have to do with a phenomenon called *masking*, meaning that a loud sound will make a simultaneous low sound inaudible. For compression this means that if certain frequencies of a signal are very prominent, most of the other frequencies can be removed, even when they are quite large. If the sounds are below the masking threshold, it is simply omitted by the encoder. Masking effects are just one example of what are called psycho-acoustic effects, and all such effects can be taken into account in a psycho-acoustic model. Another obvious such effect is that the human auditory system can only perceive frequencies in the range 20–20,000 Hz. An obvious way to do compression is therefore to remove frequencies outside this range, although there are indications that these frequencies may influence the listening experience inaudibly.

The scale factors tell the encoder about the precision to be used for each frequency band: If the model decides that one band is very important for our perception of the sound, it assigns a big scale factor to it, so that more effort is put into encoding it by the encoder (i.e. it uses more bits to encode this band).

Using appropriate scale factors and masking thresholds provide compression, since bits used to encode the sound are spent on parts important for our perception. Developing a useful psycho-acoustic model requires detailed knowledge of human perception of sound. Different MP3 encoders use different such models, so they may produce very different results.

The information remaining after frequency analysis and using a psycho-acoustic model is coded efficiently with (a variant of) Huffman coding. MP3 supports bit rates from 32 to 320 kb/s and the sampling rates 32, 44.1, and 48 kHz. Variable bit rates are also supported (i.e. the bit rate varies in different parts of the file). An MP3 encoder also stores metadata about the sound, such as title, album, and artist name.

MP3 too has evolved in the same way as MPEG, from MP1 to MP2, and to MP3, each one more sophisticated than the other, providing better compression. MP3 is not the latest development of audio coding in the MPEG family: its successor is AAC (Advanced Audio Coding), which can achieve better quality than MP3 at the same bit rate.

There exist a wide variety of documents on the MP3 standard. In [52] an overview (written in a signal processing friendly language) is given. Rossing [62] contains much more about acoustics in general, and in particular on psycho-acoustic models.

1.8 Summary

We defined digital sound, which can be obtained by sampling continuous sound. We also demonstrated simple operations on digital sound such as adding noise, playing at different rates etc.

We discussed the basic question of what is sound is, and concluded that it could be modeled as a sum of frequency components. If the function was periodic we could define its Fourier series, which can be thought of as an approximation scheme for periodic functions using finite-dimensional spaces of trigonometric functions. We established the basic properties of Fourier series, and some duality relationships between functions and their Fourier series. We also computed the Fourier series of the square- and triangle waves, and we saw how symmetric extensions could speed up convergence of a Fourier series. Finally analog filters were defined, and we showed their usefulness in analyzing the convergence of Fourier series.

Filters are one of the most important objects of study in signal processing, and much more material in this book will be devoted to them. Chapter 3 will in particular be devoted to digital filters. The MP3 standard uses filters to split a sound into frequency bands.

In Fourier analysis, one has the *Continuous-time Fourier transform*, or CTFT, which is the parallel to Fourier series for non-periodic functions. We will return to this in Chap. 6. In the present context, the CTFT actually corresponds to the frequency response of an analog filter. Many books stick to the term convolution rather than the term filter.

Many excellent textbooks on Fourier analysis exist. Many go very deep, and are thus suited for graduate level and higher. Tao [73] gives a short introduction suitable in an undergraduate course. Deitmar [15] is a more comprehensive introduction, suitable at late undergraduate or early graduate level. Katznelson [30] is a classical Fourier analysis textbook.

What You Should Have Learned in This Chapter

- Computer operations for reading, writing, and listening to sound.
- Construct sounds such as pure tones and the square wave, from mathematical formulas.
- Definition of the Fourier spaces, and the orthogonality of the Fourier basis.
- Fourier series approximations and formulas for the Fourier coefficients.
- For symmetric/antisymmetric functions, Fourier series are actually cosine/sine series.
- Certain properties of Fourier series, for instance how delay of a function or multiplication with a complex exponential affect the Fourier coefficients.
- The Fourier series motivation behind the symmetric extension of a function.

Chapter 2
Digital Sound and Discrete Fourier Analysis

In Chap. 1 we saw how a periodic function can be decomposed into a linear combination of sines and cosines, or equivalently, a linear combination of complex exponential functions. This kind of decomposition is, however, not very convenient from a computational point of view. The coefficients are given by integrals that in most cases cannot be evaluated exactly, so some kind of numerical integration technique needs to be applied. Transformation to the *frequency domain*, where meaningful operations on sound easily can be constructed, amounts to a linear transformation called the *Discrete Fourier transform*. We will start by defining this, and see how it can be implemented efficiently.

2.1 Discrete Fourier Analysis and the Discrete Fourier Transform

In this section we will parallel the developments we did for Fourier series, assuming instead that vectors (rather than functions) are involved. As with Fourier series we will assume that the vector is periodic. This means that we can represent it with the values from only the first period. In the following we will only work with these values, but we will remind ourselves from time to time that the values actually come from a periodic vector. As for functions, we will call the periodic vector the *periodic extension* of the finite vector. To illustrate this, we have in Fig. 2.1 shown a vector $\boldsymbol{x}$ and its periodic extension $\boldsymbol{x}$. Note that, if f is a function with period T, then the vector $\boldsymbol{x}$ with components $x_k = f(kT/N)$ will be a periodic vector with period N.

© Springer Nature Switzerland AG 2019   47
Ø. Ryan, *Linear Algebra, Signal Processing, and Wavelets - A Unified Approach*,
Springer Undergraduate Texts in Mathematics and Technology,
https://doi.org/10.1007/978-3-030-01812-2_2

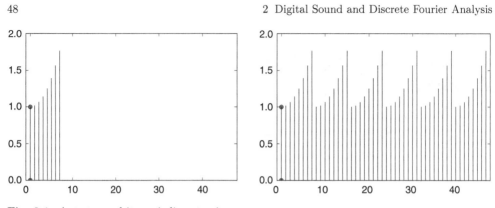

Fig. 2.1 A vector and its periodic extension

At the outset our vectors will be real, but since we will use complex exponentials we must be able to work with complex vectors also. We therefore first need to define the standard inner product and norm for complex vectors.

Definition 1. *Euclidean Inner Product.*

For complex vectors of length N the Euclidean inner product is given by

$$\langle \boldsymbol{x}, \boldsymbol{y} \rangle = \boldsymbol{y}^H \boldsymbol{x} = \sum_{k=0}^{N-1} x_k \overline{y_k}. \tag{2.1}$$

The associated norm is

$$\|\boldsymbol{x}\| = \sqrt{\sum_{k=0}^{N-1} |x_k|^2}. \tag{2.2}$$

In the previous chapter we saw that, using a Fourier series, a function with period T could be approximated by linear combinations of the functions (the pure tones) $\{e^{2\pi i n t/T}\}_{n=0}^{N}$. This can be generalized to vectors (digital sounds), with pure tones replaced with the following.

Definition 2. *Discrete Fourier Analysis.*

In Discrete Fourier analysis, a vector $\boldsymbol{x} = (x_0, \ldots, x_{N-1})$ is represented as a linear combination of the N vectors

$$\boldsymbol{\phi}_n = \frac{1}{\sqrt{N}} \left(1, e^{2\pi i n/N}, e^{2\pi i 2n/N}, \ldots, e^{2\pi i kn/N}, \ldots, e^{2\pi i n(N-1)/N} \right).$$

These vectors are called the normalized complex exponentials, or the pure digital tones of order N. n is also called the *frequency index*. The whole collection $\mathcal{F}_N = \{\boldsymbol{\phi}_n\}_{n=0}^{N-1}$ is called the N-point *Fourier basis*.

Note that pure digital tones can be considered as samples of a pure tone, taken uniformly over one period: $f(t) = e^{2\pi i n t/T}/\sqrt{N}$ is a pure tone with frequency n/T, and its samples are

$$f(kT/N) = \frac{e^{2\pi i n(kT/N)/T}}{\sqrt{N}} = \frac{e^{2\pi i nk/N}}{\sqrt{N}},$$

which is component k in $\boldsymbol{\phi}_n$. When mapping a pure tone to a digital pure tone, the index n corresponds to frequency $\nu = n/T$, and N the number of samples taken over one period. Since $Tf_s = N$, where f_s is the sampling frequency, we have the following connection between frequency and frequency index:

$$\nu = \frac{nf_s}{N} \text{ and } n = \frac{\nu N}{f_s} \tag{2.3}$$

The following lemma shows that the Fourier basis is orthonormal, so that it indeed is a basis.

Lemma 3. Complex Exponentials form an Orthonormal Basis.
The normalized complex exponentials $\{\boldsymbol{\phi_n}\}_{n=0}^{N-1}$ of order N form an orthonormal basis in $\mathbb{R}^N$.

The proof given below is a direct proof of this fact. We remark that a constructive and more insightful proof follows from the next chapter, where it is shown that the vectors can be realized as eigenvectors of symmetric matrices with distinct eigenvalues. Orthogonality is then guaranteed by a well known result in linear algebra. This kind of proof also says more about how natural these vectors actually are.

Proof. Let n_1 and n_2 be two distinct integers in the range $[0, N-1]$. The inner product of $\boldsymbol{\phi}_{n_1}$ and $\boldsymbol{\phi}_{n_2}$ is then given by

$$\begin{aligned}
\langle \boldsymbol{\phi}_{n_1}, \boldsymbol{\phi}_{n_2} \rangle &= \frac{1}{N} \langle e^{2\pi i n_1 k/N}, e^{2\pi i n_2 k/N} \rangle \\
&= \frac{1}{N} \sum_{k=0}^{N-1} e^{2\pi i n_1 k/N} e^{-2\pi i n_2 k/N} \\
&= \frac{1}{N} \sum_{k=0}^{N-1} e^{2\pi i (n_1 - n_2) k/N} = \frac{1}{N} \frac{1 - e^{2\pi i (n_1 - n_2)}}{1 - e^{2\pi i (n_1 - n_2)/N}} = 0.
\end{aligned}$$

In particular, this orthogonality means that the complex exponentials form a basis. Clearly also $\langle \boldsymbol{\phi}_n, \boldsymbol{\phi}_n \rangle = 1$, so that the N-point Fourier basis is in fact an orthonormal basis. $\square$

Note that the normalizing factor $\frac{1}{\sqrt{N}}$ was not present for pure tones in the previous chapter. Also, the normalizing factor $\frac{1}{T}$ from the last chapter is not part of the definition of the inner product in this chapter. This slightly different notation for functions and vectors will not cause confusion in what follows.

The focus in Discrete Fourier analysis is to change coordinates from the standard basis to the Fourier basis, perform some operations on the coordinates, and then change coordinates back to the standard basis. Such operations are of crucial importance, and in this section we study such change of coordinates. We start with the following definition.

Definition 4. *Discrete Fourier Transform.*
We will denote the change of coordinates matrix from the standard basis of $\mathbb{R}^N$ to the Fourier basis $\mathcal{F}_N$ by F_N. We will also call this the (N-point) *Fourier matrix*.

The matrix $\sqrt{N} F_N$ is also called the (N-point) *discrete Fourier transform*, or DFT. If $\boldsymbol{x}$ is a vector in R^N, then $\boldsymbol{y} = \text{DFT}\boldsymbol{x}$ are called the DFT coefficients of $\boldsymbol{x}$. (the DFT coefficients are thus the coordinates in $\mathcal{F}_N$, scaled with $\sqrt{N}$). DFT$\boldsymbol{x}$ is sometimes written as $\hat{\boldsymbol{x}}$.

Note that we define the Fourier matrix and the DFT as two different matrices, the one being a scaled version of the other. The reason for this is that there are different traditions in different fields. In pure mathematics, the Fourier matrix is mostly used since it is, as we will see, a unitary matrix. In signal processing, the scaled version provided by the DFT is mostly used. We will normally write x for the given vector in $\mathbb{R}^N$, and y for its DFT. In applied fields, the Fourier basis vectors are also called *synthesis vectors*, since they can be used to "synthesize" the vector x, with weights provided by the coordinates in the Fourier basis, i.e.

$$x = y_0\phi_0 + y_1\phi_1 + \cdots + y_{N-1}\phi_{N-1}. \tag{2.4}$$

Equation (2.4) is also called the synthesis equation.

Using Appendix A we can now find an expression for the matrix F_N as follows. Since F_N^{-1} is the change of coordinates from the Fourier basis to the standard basis, the ϕ_i are columns in F_N^{-1}, These are orthonormal by Lemma 3, so that $F_N = ((F_N)^{-1})^{-1} = ((F_N)^{-1})^H$, where $A^H = (\overline{A})^T$ is the conjugate transpose of A. The rows of F_N are thus the Fourier basis vectors conjugated, and also $(F_N)^{-1} = (F_N)^H$, i.e. F_N is unitary, the complex parallel to orthogonal. We thus have the following result.

Theorem 5. Fourier Matrix Is Unitary.

The Fourier matrix F_N is the unitary $N \times N$-matrix with entries given by

$$(F_N)_{nk} = \frac{1}{\sqrt{N}}e^{-2\pi ink/N},$$

for $0 \leq n, k \leq N - 1$.

Since the Fourier matrix is easily inverted, the DFT is also easily inverted. Note that, since $(F_N)^T = F_N$, we have that $(F_N)^{-1} = \overline{F_N}$. Let us make the following definition.

Definition 6. *IDFT.*

The matrix $\overline{F_N}/\sqrt{N}$ is the inverse of the DFT matrix $\sqrt{N}F_N$. We call this inverse matrix the *inverse discrete Fourier transform*, or IDFT.

The DFT and IDFT are also called the *forward DFT* and *reverse DFT*. That $y = \text{DFT}x$ and $x = \text{IDFT}y$ can also be expressed in component form as

$$y_n = \sum_{k=0}^{N-1} x_k e^{-2\pi ink/N} \qquad x_k = \frac{1}{N}\sum_{n=0}^{N-1} y_n e^{2\pi ink/N} \tag{2.5}$$

In applied fields such as signal processing, it is more common to state the DFT and IDFT in these component forms, rather than in the matrix forms $y = \text{DFT}x$ and $x = \text{IDFT}y$. Let us now see how these formulas work out in practice by considering some examples.

Example 2.1: DFT of a Cosine

Let x be the vector of length N defined by $x_k = \cos(2\pi 5k/N)$, and y the vector of length N defined by $y_k = \sin(2\pi 7k/N)$. Let us see how we can compute $F_N(2x + 3y)$.

By the definition of the Fourier matrix as a change of coordinates, $F_N(\phi_n) = e_n$. We therefore get

$$
\begin{aligned}
F_N\left(2\boldsymbol{x} + 3\boldsymbol{y}\right) &= F_N(2\cos(2\pi 5 \cdot /N) + 3\sin(2\pi 7 \cdot /N)) \\
&= F_N(2\frac{1}{2}(e^{2\pi i 5 \cdot /N} + e^{-2\pi i 5 \cdot /N}) + 3\frac{1}{2i}(e^{2\pi i 7 \cdot /N} - e^{-2\pi i 7 \cdot /N})) \\
&= F_N(\sqrt{N}\phi_5 + \sqrt{N}\phi_{N-5} - \frac{3i}{2}\sqrt{N}(\phi_7 - \phi_{N-7})) \\
&= \sqrt{N}(F_N(\phi_5) + F_N(\phi_{N-5}) - \frac{3i}{2}F_N\phi_7 + \frac{3i}{2}F_N\phi_{N-7}) \\
&= \sqrt{N}e_5 + \sqrt{N}e_{N-5} - \frac{3i}{2}\sqrt{N}e_7 + \frac{3i}{2}\sqrt{N}e_{N-7}.
\end{aligned}
$$

Example 2.2: DFT of a Square Wave

Let us attempt to apply the DFT to a vector $\boldsymbol{x}$ which is 1 on indices close to 0, and 0 elsewhere. Assume that

$$
x_{-L} = \cdots = x_{-1} = x_0 = x_1 = \cdots = x_L = 1,
$$

while all other values are 0. This is similar to a square wave, with some modifications: First of all we assume symmetry around 0, while the square wave of Example 1.4 assumes antisymmetry around 0. Secondly the values of the square wave are now 0 and 1, contrary to -1 and 1 before. Finally, we have a different proportion of where the two values are assumed. Nevertheless, we will also refer to this as a square wave.

Since indices with the DFT are between 0 an $N - 1$, and since $\boldsymbol{x}$ is assumed to have period N, the indices $[-L, L]$ where our vector is 1 translates to the indices $[0, L]$ and $[N - L, N - 1]$ (i.e., it is 1 on the first and last parts of the vector). Elsewhere our vector is zero. Since $\sum_{k=N-L}^{N-1} e^{-2\pi ink/N} = \sum_{k=-L}^{-1} e^{-2\pi ink/N}$ (since $e^{-2\pi ink/N}$ is periodic with period N), the DFT of $\boldsymbol{x}$ is

$$
\begin{aligned}
y_n &= \sum_{k=0}^{L} e^{-2\pi ink/N} + \sum_{k=N-L}^{N-1} e^{-2\pi ink/N} = \sum_{k=0}^{L} e^{-2\pi ink/N} + \sum_{k=-L}^{-1} e^{-2\pi ink/N} \\
&= \sum_{k=-L}^{L} e^{-2\pi ink/N} = e^{2\pi inL/N}\frac{1 - e^{-2\pi in(2L+1)/N}}{1 - e^{-2\pi in/N}} \\
&= e^{2\pi inL/N}e^{-\pi in(2L+1)/N}e^{\pi in/N}\frac{e^{\pi in(2L+1)/N} - e^{-\pi in(2L+1)/N}}{e^{\pi in/N} - e^{-\pi in/N}} \\
&= \frac{\sin(\pi n(2L + 1)/N)}{\sin(\pi n/N)}.
\end{aligned}
$$

This computation does in fact also give us the IDFT of the same vector, since the IDFT just requires a change of sign in all the exponents, in addition to the $1/N$ normalizing factor. From this example we see that, in order to represent $\boldsymbol{x}$ in terms of frequency components, all components are actually needed. The situation would have been easier if only a few frequencies were needed.

Example 2.3: Computing the DFT by Hand

In most cases it is difficult to compute a DFT by hand, due to the entries $e^{-2\pi ink/N}$ in the matrices. The DFT is therefore usually calculated on a computer only. However, in the case $N = 4$ the calculations are quite simple. In this case the Fourier matrix takes the form

$$\text{DFT}_4 = \begin{pmatrix} 1 & 1 & 1 & 1 \\ 1 & -i & -1 & i \\ 1 & -1 & 1 & -1 \\ 1 & i & -1 & -i \end{pmatrix}.$$

We now can compute the DFT of a vector like $(1, 2, 3, 4)^T$ simply as

$$\text{DFT}_4 \begin{pmatrix} 1 \\ 2 \\ 3 \\ 4 \end{pmatrix} = \begin{pmatrix} 1+2+3+4 \\ 1-2i-3+4i \\ 1-2+3-4 \\ 1+2i-3-4i \end{pmatrix} = \begin{pmatrix} 10 \\ -2+2i \\ -2 \\ -2-2i \end{pmatrix}.$$

In general, computing the DFT implies using floating point multiplication. Here there was no need for this, since DFT_4 has unit entries which are either real or purely imaginary.

Example 2.4: Direct Implementation of the DFT

The DFT can be implemented very simply and directly by the code

```
function y = dft_impl(x)
    N = size(x, 1);
    y = zeros(size(x));
    for n = 1:N
        D = exp(2*pi*1i*(n-1)*(0:(N-1))/N);
        y(n) = dot(D, x);
    end
```

n has been replaced by $n-1$ in this code since n runs from 1 to N (array indices must start at 1 in MATLAB). In Exercise 2.14 we will extend this to a general implementation we will use later. Note that we do not allocate the entire matrix F_N in this code, as this quickly leads to out of memory situations, even for N of moderate size. Instead we construct one row of F_N at a time, and use this to compute one entry in the output. The method dot can be used here, since each entry in matrix multiplication can be viewed as an inner product. It is likely that the dot function is more efficient than using a for-loop, since MATLAB may have an optimized version of this. Note that dot in MATLAB conjugates the first components, contrary to what we do in our definition of a complex inner product. This is why we have dropped a sign in the exponent here.

This can be rewritten to a direct implementation of the IDFT also. We will look at this in the exercises, where we also make the method more general, so that the DFT can be applied to a series of vectors at a time (it can then be applied to all the channels in a sound in one call). Multiplying a full $N \times N$ matrix by a vector requires roughly N^2 arithmetic operations. The DFT algorithm above will therefore take a long time when N becomes moderately large. It turns out that a much more efficient algorithm exists for computing the DFT, which we will study at the end of the chapter. MATLAB also has a built-in implementation of the DFT which uses such an efficient algorithm.

Example 2.5: Thresholding the DFT Coefficients

Since the DFT coefficients represent contributions at given frequencies, they are extremely useful for performing operations on sound, and also for compression. The function `forw_comp_rev_DFT` in the library can manipulate the DFT coefficients of the audio sample file in several ways, and plays the result. In particular it can set DFT coefficients below a given threshold to zero. The idea behind doing so is that frequency components with small values may contribute little to the perception of the sound, so that they can be discarded. This gives a sound with fewer frequencies, more suitable for compression. `forw_comp_rev_DFT` accepts the named parameter `threshold` for this purpose. This parameter will also force writing the percentage of the DFT coefficients which are zeroed out to the display. If you run `forw_comp_rev_DFT` with `threshold` equal to 20, the resulting sound can be found in the file castanetsthreshold002.wav, and the function says that about 68% of the DFT coefficients were set to zero. You can clearly hear the disturbance in the sound, but we have not lost that much. If we instead try `threshold` equal to 70, the resulting sound is the file castanetsthreshold01.wav, and about 94% of the DFT coefficients were set to zero. The quality is much poorer now, but one can still can recognize the sound. This suggests that most of the information is contained in frequencies with the highest values.

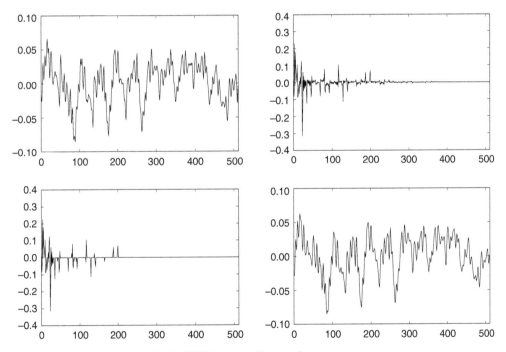

Fig. 2.2 Experimenting with the DFT on a small part of a song

In Fig. 2.2 we have illustrated this principle for 512 sound samples from a song. The samples of the sound and (the absolute value of) its DFT are shown at the top. At the bottom all values of the DFT with absolute value smaller than 0.02 are set to zero (52 values then remain), and the sound is reconstructed with the IDFT, and then shown in the right plot. The start and end signals look similar, even though the last can be represented with less than 10% of the frequencies from the first.

Example 2.6: Quantizing the DFT Coefficients

The thresholding from the previous example affects only frequencies with low contribution. This makes it too simple for practical use: A more neutral way would be to let each DFT coefficient occupy a certain number of bits. This is also called *quantization*, and is closer to what modern audio standards do. This is useful if the number of bits we use is less than what we have at the start. `forw_comp_rev_DFT` accepts a named parameter `n`, which secures that a DFT coefficient with bit representation

$$\ldots d_2 d_1 d_0 . d_{-1} d_{-2} d_{-3} \ldots$$

is truncated so that the bits $d_{n-1}, d_{n-2}, \ldots$ are discarded. High values of n thus mean more rounding, so that one should hear that the sound degrades with increasing n. The following three examples verify this:

- with $n = 3$ the result can be found in the file castantesquantizedn3.wav,
- with $n = 5$ the result can be found in the file castantesquantizedn5.wav,
- with $n = 7$ the result can be found in the file castantesquantizedn7.wav.

In practice this quantization procedure is also too simple, since the human auditory system is more sensitive to certain frequencies, and should thus allocate a higher number of bits for those. Modern audio standards take this into account.

2.1.1 Properties of the DFT

The DFT has properties which are very similar to those of Fourier series, as they were listed in Theorem 17. The following theorem sums them up:

Theorem 7. Properties of the DFT.
  Let $\boldsymbol{x}$ be a real vector of length N. The DFT has the following properties:

1. $(\widehat{\boldsymbol{x}})_{N-n} = \overline{(\widehat{\boldsymbol{x}})_n}$ *for* $0 \leq n \leq N - 1$.
2. *If* $x_k = x_{N-k}$ *for all n (so $\boldsymbol{x}$ is symmetric), then $\widehat{\boldsymbol{x}}$ is a real vector.*
3. *If* $x_k = -x_{N-k}$ *for all k (so $\boldsymbol{x}$ is antisymmetric), then $\widehat{\boldsymbol{x}}$ is a purely imaginary vector.*
4. *If d is an integer and $\boldsymbol{z}$ is the vector with components $z_k = x_{k-d}$ (the vector $\boldsymbol{x}$ with its elements delayed by d), then $(\widehat{\boldsymbol{z}})_n = e^{-2\pi i d n/N} (\widehat{\boldsymbol{x}})_n$.*
5. *If d is an integer and $\boldsymbol{z}$ is the vector with components $z_k = e^{2\pi i d k/N} x_k$, then $(\widehat{\boldsymbol{z}})_n = (\widehat{\boldsymbol{x}})_{n-d}$.*

Note that there is no loss in generality in assuming that the input vector $\boldsymbol{x}$ is real, since a complex DFT can be computed as two real DFT's, i.e. $DFT_N(\boldsymbol{x}_1 + i\boldsymbol{x}_2) = DFT_N \boldsymbol{x}_1 + i DFT_N \boldsymbol{x}_2$.

Proof. The methods used in the proof are very similar to those used in the proof of Theorem 17. From the definition of the DFT we have

$$(\widehat{x})_{N-n} = \sum_{k=0}^{N-1} e^{-2\pi i k(N-n)/N} x_k = \sum_{k=0}^{N-1} e^{2\pi i k n/N} x_k = \overline{\sum_{k=0}^{N-1} e^{-2\pi i k n/N} x_k} = \overline{(\widehat{x})_n}$$

which proves property 1. To prove property 2, we write

$$(\widehat{x})_n = \sum_{k=0}^{N-1} x_k e^{-2\pi i k n/N} = \sum_{k=0}^{N-1} x_{N-k} e^{-2\pi i k n/N} = \sum_{k=0}^{N-1} x_k e^{-2\pi i (N-k)n/N}$$

$$= \sum_{k=0}^{N-1} x_k e^{2\pi i k n/N} = \overline{\sum_{k=0}^{N-1} x_k e^{-2\pi i k n/N}} = \overline{(\widehat{x})_n}.$$

It follows that $\widehat{x} = \overline{\widehat{x}}$, so that $\widehat{x}$ is real. Property 3 follows similarly. To prove property 4 observe that

$$(\widehat{z})_n = \sum_{k=0}^{N-1} x_{k-d} e^{-2\pi i k n/N} = \sum_{k=0}^{N-1} x_k e^{-2\pi i (k+d)n/N}$$

$$= e^{-2\pi i d n/N} \sum_{k=0}^{N-1} x_k e^{-2\pi i k n/N} = e^{-2\pi i d n/N} (\widehat{x})_n.$$

For the proof of property 5 we note that the DFT of z is

$$(\widehat{z})_n = \sum_{k=0}^{N-1} e^{2\pi i d k/N} x_n e^{-2\pi i k n/N} = \sum_{k=0}^{N-1} x_n e^{-2\pi i (n-d)k/N} = (\widehat{x})_{n-d}.$$

This completes the proof. □

These properties have similar interpretations as the ones listed in Theorem 17 for Fourier series. Property 1 says that we need to store only about one half of the DFT coefficients, since the remaining coefficients can be obtained by conjugation. In particular, when N is even, we only need to store $y_0, y_1, \ldots, y_{N/2}$. This also means that, if we plot the (absolute value) of the DFT of a real vector, we will see a symmetry around the index $n = N/2$. The theorem generalizes all but the last property of Theorem 17. We will handle the generalization of this property later.

Example 2.7: The DFT of a Signal Modulated with a Complex Exponential

To see how we can use the fourth property of Theorem 7, consider a vector $x = (x_0, x_1, x_2, x_3, x_4, x_5, x_6, x_7)$ with length $N = 8$, and assume that x is so that $\mathrm{DFT}_8(x) = (1, 2, 3, 4, 5, 6, 7, 8)$. Consider the vector z with components $z_k = e^{2\pi i 2k/8} x_k$. Let us compute $\mathrm{DFT}_8(z)$. Since multiplication of x with $e^{2\pi i k d/N}$ delays the output $y = \mathrm{DFT}_N(x)$ with d elements, setting $d = 2$, the $\mathrm{DFT}_8(z)$ can be obtained by delaying $\mathrm{DFT}_8(x)$ by two elements, so that $\mathrm{DFT}_8(z) = (7, 8, 1, 2, 3, 4, 5, 6)$. It is straightforward to compute this directly also:

$$(\text{DFT}_N \boldsymbol{z})_n = \sum_{k=0}^{N-1} z_k e^{-2\pi i k n/N} = \sum_{k=0}^{N-1} e^{2\pi i 2k/N} x_k e^{-2\pi i k n/N}$$

$$= \sum_{k=0}^{N-1} x_k e^{-2\pi i k(n-2)/N} = (\text{DFT}_N(\boldsymbol{x}))_{n-2}.$$

Exercise 2.8: Computing the DFT by Hand

Compute $F_4 \boldsymbol{x}$ when $\boldsymbol{x} = (2, 3, 4, 5)$.

Exercise 2.9: Exact form of Low-Order DFT Matrix

As in Example 2.3, state the exact Cartesian form of the Fourier matrix for the cases $N = 6$, $N = 8$, and $N = 12$.

Exercise 2.10: DFT of a Delayed Vector

We have a real vector $\boldsymbol{x}$ with length N, and define the vector $\boldsymbol{z}$ by delaying all elements in $\boldsymbol{x}$ with 5 cyclically, i.e. $z_5 = x_0$, $z_6 = x_1, \ldots, z_{N-1} = x_{N-6}$, and $z_0 = x_{N-5}, \ldots, z_4 = x_{N-1}$. For a given n, if $|(F_N \boldsymbol{x})_n| = 2$, what is then $|(F_N \boldsymbol{z})_n|$?

Exercise 2.11: Using the Symmetry Property

Given a real vector $\boldsymbol{x}$ of length 8 where $(F_8(\boldsymbol{x}))_2 = 2 - i$, what is $(F_8(\boldsymbol{x}))_6$?

Exercise 2.12: DFT of $\cos^2(2\pi k/N)$

Let $\boldsymbol{x}$ be the vector of length N where $x_k = \cos^2(2\pi k/N)$. What is then $F_N \boldsymbol{x}$?

Exercise 2.13: DFT of $c^k \boldsymbol{x}$

Let $\boldsymbol{x}$ be the vector with entries $x_k = c^k$. Show that the DFT of $\boldsymbol{x}$ is given by the vector with components

$$y_n = \frac{1 - c^N}{1 - ce^{-2\pi i n/N}}$$

for $n = 0, \ldots, N - 1$.

Exercise 2.14: DFT Implementation

Extend the code for the function `dft_impl` in Example 2.4 so that

1. The function also takes a second parameter called `forward`. If this is true the DFT is applied. If it is false, the IDFT is applied. If this parameter is not present, then the forward transform should be assumed.
2. If the input x is two-dimensional (i.e. a matrix), the DFT/IDFT should be applied to each column of x. This ensures that, in the case of sound, the FFT is applied to each channel in the sound when the entire sound is used as input.

Also, write documentation for the code.

Exercise 2.15: Symmetry

Assume that N is even.

a) Prove that, if $x_{k+N/2} = x_k$ for all $0 \leq k < N/2$, then $y_n = 0$ when n is odd.

b) Prove that, if $x_{k+N/2} = -x_k$ for all $0 \leq k < N/2$, then $y_n = 0$ when n is even.

c) Prove the converse statements in a) and b).

d) Prove the following:

- $x_n = 0$ for all odd n if and only if $y_{k+N/2} = y_k$ for all $0 \leq k < N/2$.
- $x_n = 0$ for all even n if and only if $y_{k+N/2} = -y_k$ for all $0 \leq k < N/2$.

Exercise 2.16: DFT on Complex and Real Data

Let x_1, x_2 be real vectors, and set $x = x_1 + ix_2$. Use Theorem 7 to show that

$$(F_N(x_1))_k = \frac{1}{2}\left((F_N(x))_k + \overline{(F_N(x))_{N-k}}\right)$$
$$(F_N(x_2))_k = \frac{1}{2i}\left((F_N(x))_k - \overline{(F_N(x))_{N-k}}\right)$$

This shows that we can compute two DFT's on real data from one DFT on complex data, and $2N$ extra additions.

Exercise 2.17: Vandermonde Matrices

In linear algebra, an $N \times N$-matrix where column j is on the form $(x_j^0, x_j^1, \ldots, x_j^{N-1})$ for some number x_j is called a *Vandermonde matrix*. The number x_j is called a *generator* for (the j'th column) of the matrix.

a) Show that a Vandermonde matrix where all the generators are distinct and nonzero is non-singular.

Hint

Combine elementary row operations with a cofactor expansion along the last column.

b) Let A be the $s \times s$-matrix obtained by taking s distinct columns of DFT_N, and s successive columns of DFT_N. Show that A is non-singular.

Exercise 2.18: Fourier Basis Vectors with Zeros Inserted

Later we will encounter vectors created by zeroing out certain components in the Fourier basis vectors, and we will have use for the DFT of these.

a) Show that

$$(e^{2\pi i r \cdot 0/N}, 0, e^{2\pi i r \cdot 2/N}, 0, \ldots, e^{2\pi i r(N-2)/N}, 0) = \frac{1}{2}(e^{2\pi i r k/N} + e^{2\pi i(r+N/2)k/N})$$

$$(0, e^{2\pi i r \cdot 1/N}, 0, e^{2\pi i r \cdot 3/N}, \ldots, 0, e^{2\pi i r(N-1)/N}) = \frac{1}{2}(e^{2\pi i r k/N} - e^{2\pi i(r+N/2)k/N}).$$

Do this by computing the DFT of the left hand sides.

b) Assume that N (the length of the vector) is divisible by M, and let s be a fixed number. Show that the vector where only components on the form $Mk + s$ in $(e^{2\pi i r \cdot 0/N}, e^{2\pi i r \cdot 1/N}, \ldots, e^{2\pi i r(N-1)/N})$ are kept (the others zeroed out) can be written as

$$\frac{1}{M} \sum_{t=0}^{M-1} e^{-2\pi i s t/M} e^{2\pi i(r+tN/M)k/N}.$$

In other words, such vectors can be written as a sum of M Fourier basis vectors.

2.2 Connection Between the DFT and Fourier Series: Sampling and the Sampling Theorem

So far we have focused on the DFT as a tool to rewrite a vector in terms of the Fourier basis vectors. In practice, the given vector $\boldsymbol{x}$ will often be sampled from some real data given by a function $f(t)$. We may then compare the frequency content of $\boldsymbol{x}$ and f, and ask how they are related: What is the relationship between the Fourier coefficients of f and the DFT-coefficients of $\boldsymbol{x}$?

In order to study this, assume for simplicity that $f \in V_{M,T}$ for some M. This means that f equals its Fourier approximation f_M,

$$f(t) = f_M(t) = \sum_{n=-M}^{M} z_n e^{2\pi i n t/T}, \text{ where } z_n = \frac{1}{T} \int_0^T f(t) e^{-2\pi i n t/T} \, dt. \tag{2.6}$$

We here have changed our notation for the Fourier coefficients from y_n to z_n, in order not to confuse them with the DFT coefficients. We recall that in order to represent the frequency n/T fully, we need the corresponding exponentials with both positive and negative arguments, i.e., both $e^{2\pi i n t/T}$ and $e^{-2\pi i n t/T}$.

Fact 8. Frequency vs. Fourier Coefficients.

Suppose f is given by its Fourier series (2.6). Then the total frequency content for the frequency n/T is given by the two coefficients z_n and z_{-n}.

We have the following connection between the Fourier coefficients of f and the DFT of the samples of f.

Proposition 9. Relation Between Fourier Coefficients and DFT Coefficients.

Let $N > 2M$, $f \in V_{M,T}$, and let $\boldsymbol{x} = \{f(kT/N)\}_{k=0}^{N-1}$ be N uniform samples from f over $[0,T]$. The Fourier coefficients z_n of f can be computed from

$$(z_0, z_1, \ldots, z_M, \underbrace{0, \ldots, 0}_{N-(2M+1)}, z_{-M}, z_{-M+1}, \ldots, z_{-1}) = \frac{1}{N} DFT_N \boldsymbol{x}. \qquad (2.7)$$

In particular, the total contribution in f from frequency n/T, for $0 \le n \le M$, is given by y_n and y_{N-n}, where $\boldsymbol{y}$ is the DFT of $\boldsymbol{x}$.

Proof. Let $\boldsymbol{x}$ and $\boldsymbol{y}$ be as defined, so that

$$x_k = \frac{1}{N} \sum_{n=0}^{N-1} y_n e^{2\pi i n k/N}. \qquad (2.8)$$

Inserting the sample points $t = kT/N$ into the Fourier series, we must have that

$$x_k = f(kT/N) = \sum_{n=-M}^{M} z_n e^{2\pi i n k/N} = \sum_{n=-M}^{-1} z_n e^{2\pi i n k/N} + \sum_{n=0}^{M} z_n e^{2\pi i n k/N}$$

$$= \sum_{n=N-M}^{N-1} z_{n-N} e^{2\pi i (n-N)k/N} + \sum_{n=0}^{M} z_n e^{2\pi i n k/N}$$

$$= \sum_{n=0}^{M} z_n e^{2\pi i n k/N} + \sum_{n=N-M}^{N-1} z_{n-N} e^{2\pi i n k/N}.$$

This states that $\boldsymbol{x} = \text{NIDFT}_N(z_0, z_1, \ldots, z_M, \underbrace{0, \ldots, 0}_{N-(2M+1)}, z_{-M}, z_{-M+1}, \ldots, z_{-1})$. Equation (2.7) follows by applying the DFT to both sides. We also see that $z_n = y_n/N$ and $z_{-n} = y_{2M+1-n}/N = y_{N-n}/N$, when $\boldsymbol{y}$ is the DFT of $\boldsymbol{x}$. It now also follows immediately that the frequency content in f for the frequency n/T is given by y_n and y_{N-n}. This completes the proof. $\square$

In Proposition 9 we take N samples over $[0,T]$, i.e. we sample at rate $f_s = N/T$ samples per second. When $|n| \le M$, a pure tone with frequency $\nu = n/T$ is then seen to correspond to the DFT indices n and $N - n$. Since $T = N/f_s$, $\nu = n/T$ can also be written as $\nu = n f_s/N$. Moreover, the highest frequencies in Proposition 9 are those close to $\nu = M/T$, which correspond to DFT indices close to $N - M$ and M, which are the nonzero frequencies closest to $N/2$. DFT index $N/2$ corresponds to the frequency $N/(2T) = f_s/2$, which corresponds to the highest frequency we can reconstruct from samples for any M. Similarly, the lowest frequencies are those close to $\nu = 0$, which correspond to DFT indices close to 0 and N. Let us summarize this as follows.

Observation 10. Connection Between DFT Index and Frequency.

Assume that $\boldsymbol{x}$ are N samples of a sound taken at sampling rate f_s samples per second, and let $\boldsymbol{y}$ be the DFT of $\boldsymbol{x}$. Then the DFT indices n and $N - n$ give the frequency contribution at frequency $\nu = nf_s/N$. Moreover, the low frequencies in $\boldsymbol{x}$ correspond to the y_n with n near 0 and N, while the high frequencies in $\boldsymbol{x}$ correspond to the y_n with n near $N/2$.

The theorem says that any $f \in V_{M,T}$ can be reconstructed from its samples (since we can write down its Fourier series), as long as $N > 2M$. That $f \in V_{M,T}$ is important. In Fig. 2.3 the function $f(t) = \sin(2\pi 8t) \in V_{8,1}$ is shown. For this we need to choose N so that $N > 2M = 16$ samples. Here $N = 23$ samples were taken, so that reconstruction from the samples is possible.

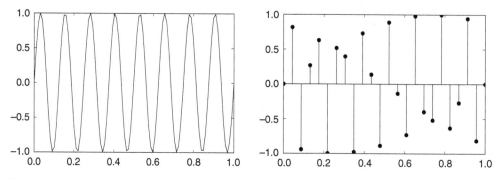

Fig. 2.3 An example on how the samples are picked from an underlying continuous time function (left), and the samples on their own (right)

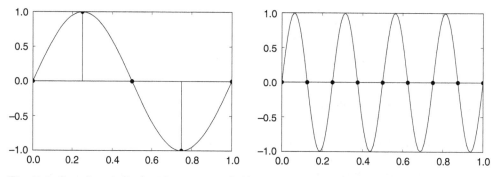

Fig. 2.4 Sampling $\sin(2\pi t)$ with two points (left), and sampling $\sin(2\pi 4t)$ with eight points (right)

That the condition $N < 2M$ is also necessary can easily be observed in Fig. 2.4. Right we have plotted $\sin(2\pi 4t) \in V_{4,1}$, with $N = 8$ sample points taken uniformly from $[0, 1]$. Here $M = 4$, so that we require $2M + 1 = 9$ sample points, according to Proposition 9. Clearly there is an infinite number of possible functions in $V_{M,T}$ passing through the sample points (which are all zero): Any $f(t) = c\sin(2\pi 4t)$ will do. Left we consider one period of $\sin(2\pi t)$. Since this is in $V_{M,T} = V_{1,1}$, reconstruction should be possible if we have $N \geq 2M + 1 = 3$ samples.

The special case $N = 2M + 1$ is interesting. No zeros are then inserted in the vector in Eq. (2.7). Since the DFT is one-to-one, this means that there is a one-to-one correspondence between sample values and functions in $V_{M,T}$ (i.e. Fourier series), i.e. we can always find a unique interpolant in $V_{M,T}$ from $N = 2M + 1$ samples. In Exercise 2.21 you will asked to write code where you start with a given function f, Take $N = 2M + 1$ samples, and plot the interpolant from $V_{M,T}$ against f. Increasing M should give an interpolant which is a better approximation to f, and if f itself resides in some $V_{M,T}$ for some M, we should obtain equality when we choose M big enough. We have in elementary calculus courses seen how to determine a polynomial of degree $N - 1$ that interpolates a set of N data points. In mathematics many other classes besides polynomials exist which are useful for interpolation. The Fourier basis is just one example.

Besides reconstructing a function from its samples, Proposition 9 also enables us to approximate a function from its samples, in case we have no bound on the highest frequency. Note that the unique function $\tilde{f}$ in $V_{M,T}$ going through the samples, is in general different from the best approximation f_M to f from $V_{M,T}$. Computing $\tilde{f}$ is easier, however, since one only needs to apply a DFT to the samples, while computing f_M requires us to compute the Fourier integrals.

The condition $N > 2M$ in Proposition 9 can also be written as $N/T > 2M/T$. The left side is now the sampling rate f_s, while the right side is the double of the highest frequency in f. The result can therefore also be restated as follows

Proposition 11. Reconstruction from Samples.

Any $f \in V_{M,T}$ can be reconstructed uniquely from a uniform set of samples $\{f(kT/N)\}_{k=0}^{N-1}$, as long as $f_s > 2|\nu|$, where ν denotes the highest frequency in f.

We also refer to $f_s = 2|\nu|$ as the *critical sampling rate*, since it is the minimum sampling rate we need in order to reconstruct f from its samples. If f_s is substantially larger than $2|\nu|$ we say that f is *oversampled*, since we have more samples than we really need. Similarly we say that f is *undersampled* if f_s is smaller than $2|\nu|$. Proposition 9 gives one formula for the reconstruction. In the literature another formula can be found, which we now will deduce. This alternative version of Theorem 9 is also called *the sampling theorem*. We start by substituting $N = T/T_s$ (i.e. $T = NT_s$, with T_s being the sampling period) in the Fourier series for f:

$$ f(kT_s) = \sum_{n=-M}^{M} z_n e^{2\pi i n k/N} \qquad -M \le k \le M. $$

Equation (2.7) said that the Fourier coefficients could be found from the samples from

$$ (z_0, z_1, \ldots, z_M, \underbrace{0, \ldots, 0}_{N-(2M+1)}, z_{-M}, z_{-M+1}, \ldots, z_{-1}) = \frac{1}{N} \mathrm{DFT}_N \boldsymbol{x}. $$

By delaying the n index with $-M$, this can also be written as

$$ z_n = \frac{1}{N} \sum_{k=0}^{N-1} f(kT_s) e^{-2\pi i n k/N} = \frac{1}{N} \sum_{k=-M}^{M} f(kT_s) e^{-2\pi i n k/N}, \quad -M \le n \le M. $$

Inserting this in the reconstruction formula we get

$$
\begin{aligned}
f(t) &= \frac{1}{N} \sum_{n=-M}^{M} \sum_{k=-M}^{M} f(kT_s) e^{-2\pi i n k/N} e^{2\pi i n t/T} \\
&= \sum_{k=-M}^{M} \frac{1}{N} \left(\sum_{n=-M}^{M} f(kT_s) e^{2\pi i n(t/T - k/N)} \right) \\
&= \sum_{k=-M}^{M} \frac{1}{N} e^{-2\pi i M(t/T-k/N)} \frac{1 - e^{2\pi i(2M+1)(t/T-k/N)}}{1 - e^{2\pi i(t/T-k/N)}} f(kT_s) \\
&= \sum_{k=-M}^{M} \frac{1}{N} \frac{\sin(\pi(t-kT_s)/T_s)}{\sin(\pi(t-kT_s)/T)} f(kT_s)
\end{aligned}
$$

Let us summarize our findings as follows:

Theorem 12. Sampling Theorem and the Ideal Interpolation Formula for Periodic Functions.

Let f be a periodic function with period T, and assume that f has no frequencies higher than ν Hz. Then f can be reconstructed exactly from its samples $f(-MT_s), \ldots, f(MT_s)$ (where T_s is the sampling period, $N = \frac{T}{T_s}$ is the number of samples per period, and $N = 2M + 1$) when the sampling rate $f_s = \frac{1}{T_s}$ is bigger than 2ν. Moreover, the reconstruction can be performed through the formula

$$
f(t) = \sum_{k=-M}^{M} f(kT_s) \frac{1}{N} \frac{\sin(\pi(t-kT_s)/T_s)}{\sin(\pi(t-kT_s)/T)}. \tag{2.9}
$$

Formula (2.9) is also called the *ideal interpolation formula* for periodic functions. The function $\frac{1}{N}\frac{\sin(\pi(t-kT_s)/T_s)}{\sin(\pi(t-kT_s)/T)}$ is also called an *interpolation kernel*. Note that f itself may not be equal to a finite Fourier series, and reconstruction is in general not possible then. The ideal interpolation formula can in such cases still be used, but the result we obtain may be different from $f(t)$.

In fact, the following more general result holds, which we will not prove. The result is also valid for functions which are not periodic, and is frequently stated in the literature:

Theorem 13. Sampling Theorem and the Ideal Interpolation Formula, General Version.

Assume that f has no frequencies higher than ν Hz. Then f can be reconstructed exactly from its samples $\ldots, f(-2T_s), f(-T_s), f(0), f(T_s), f(2T_s), \ldots$ when the sampling rate is bigger than 2ν. Moreover, the reconstruction can be performed through the formula

$$
f(t) = \sum_{k=-\infty}^{\infty} f(kT_s) \frac{\sin(\pi(t-kT_s)/T_s)}{\pi(t-kT_s)/T_s}. \tag{2.10}
$$

When f is periodic, it is possible to deduce this partly from the interpolation formula for periodic functions, see Exercise 2.22. When f is not periodic we require more tools from Fourier analysis, however.

Exercise 2.19: Code Example

Explain what the code below does, line by line:

```
[x, fs] = audioread('sounds/castanets.wav');
N = size(x, 1);
y = fft(x);
y((round(N/4)+1):(round(N/4)+N/2), :) = 0;
newx = abs(ifft(y));
newx = newx/max(max(newx));
playerobj = audioplayer(newx, fs);
playblocking(playerobj)
```

Comment in particular why we adjust the sound samples by dividing with the maximum value of the sound samples. What changes in the sound do you expect to hear?

Exercise 2.20: Mapping DFT Indices to Frequencies

In the code from the previous exercise it turns out that $f_s = 44{,}100\,\text{Hz}$, and that the number of sound samples is $N = 292{,}570$. Which frequencies in the sound file will be changed on the line where we zero out some of the DFT coefficients?

Exercise 2.21: Plotting the Interpolant

Implement code where you do the following:

- at the top you define the function $f(x) = \cos^6(x)$, and $M = 3$,
- compute the unique interpolant from $V_{M,T}$ (i.e. by taking $N = 2M+1$ samples over one period), as guaranteed by Proposition 9,
- plot the interpolant against f over one period.

Finally run the code also for $M = 4$, $M = 5$, and $M = 6$. Explain why the plots coincide for $M = 6$, but not for $M < 6$. Does increasing M above $M = 6$ have any effect on the plots?

Exercise 2.22: Proving the General Version of the Sampling Theorem for Functions in a Fourier Space

a) Show that the function

$$g(t) = \begin{cases} \frac{1}{\sin t} - \frac{1}{t} & , \text{if } t > 0 \\ 0 & , \text{if } t = 0 \end{cases}$$

is continuous and strictly increasing on $(-\pi, \pi)$.

Any function of period T can also be viewed as a function of period sT, for any integer s. Replacing N with sN, T with sT, and taking the difference between two corresponding terms in the ideal interpolation formula for periodic functions, and the one for non-periodic functions, we get

$$f(kT_s)\sin(\pi(t-kT_s)/T_s)\left(\frac{1}{sN\sin(\pi(t-kT_s)/(sT))}-\frac{1}{\pi(t-kT_s)/T_s}\right).$$

b) With

$$a_k=\frac{1}{\sin(\pi(t-kT_s)/(sT))}-\frac{1}{\pi(t-kT_s)/(sT)}=g(\pi(t-kT_s)/(sT)),$$

the difference between the ideal interpolation formula for periodic and non-periodic functions is

$$\frac{1}{sN}\sum_{k=-sM}^{sM}f(kT_s)\sin(\pi(t-kT_s)/T_s)a_k$$

Show that, for N odd, this can be made arbitrarily small by choosing s big enough.

Hint

Split the sum into N alternating series, so that there is a common $f(kT_s)$ in each series. The result from a) will be needed in order to bound each of those series.

Thus, we can prove the more general interpolation formula for a periodic f by considering f as a function with period sT, and letting $s\to\infty$.

2.3 The Fast Fourier Transform (FFT)

The main application of the DFT is computing frequency information in large data sets. Since this is so useful in many areas, it is of vital importance that the DFT can be computed with efficient algorithms. The straightforward implementation of the DFT with matrix multiplication is not efficient for large data sets. However, it turns out that the DFT matrix may be factored in a way that leads to much more efficient algorithms, and this is the topic of the present section. We will discuss the most widely used implementation of the DFT, usually referred to as the *Fast Fourier Transform* (FFT). The invention of the FFT made the DFT computationally feasible in many fields. The FFT is for instance used much in processing and compression of sound, images, and video.

The most basic FFT algorithm is easy to state and prove. It applies for a general complex input vector $\boldsymbol{x}$, with length N being an even number.

Theorem 14. FFT Algorithm When N Is Even.
Let $\boldsymbol{y}=DFT_N\boldsymbol{x}$ be the N-point DFT of $\boldsymbol{x}$, with N an even number, and let $D_{N/2}$ be the $(N/2)\times(N/2)$-diagonal matrix with entries $(D_{N/2})_{n,n}=e^{-2\pi in/N}$ for $0\le n<N/2$. Then we have that

$$(y_0,y_1,\ldots,y_{N/2-1})=DFT_{N/2}\boldsymbol{x}^{(e)}+D_{N/2}DFT_{N/2}\boldsymbol{x}^{(o)} \tag{2.11}$$

$$(y_{N/2},y_{N/2+1},\ldots,y_{N-1})=DFT_{N/2}\boldsymbol{x}^{(e)}-D_{N/2}DFT_{N/2}\boldsymbol{x}^{(o)} \tag{2.12}$$

where $\boldsymbol{x}^{(e)}, \boldsymbol{x}^{(o)} \in \mathbb{R}^{N/2}$ consist of the even- and odd-indexed entries of $\boldsymbol{x}$, respectively, i.e.

$$\boldsymbol{x}^{(e)} = (x_0, x_2, \ldots, x_{N-2}) \qquad\qquad \boldsymbol{x}^{(o)} = (x_1, x_3, \ldots, x_{N-1}).$$

The formulas (2.11)–(2.12) reduce the computation of an N-point DFT to two $N/2$-point DFT's, and this is the basic fact which speeds up computations considerably. Note that the same term $D_{N/2}\mathrm{DFT}_{N/2}\boldsymbol{x}^{(o)}$ appears in both formulas above. It is important that this is computed only once, and inserted in both equations.

Proof. Suppose first that $0 \le n \le N/2 - 1$. We start by splitting the sum in the expression for the DFT into even and odd indices,

$$
\begin{aligned}
y_n &= \sum_{k=0}^{N-1} x_k e^{-2\pi i n k/N} = \sum_{k=0}^{N/2-1} x_{2k} e^{-2\pi i n 2k/N} + \sum_{k=0}^{N/2-1} x_{2k+1} e^{-2\pi i n(2k+1)/N} \\
&= \sum_{k=0}^{N/2-1} x_{2k} e^{-2\pi i n k/(N/2)} + e^{-2\pi i n/N} \sum_{k=0}^{N/2-1} x_{2k+1} e^{-2\pi i n k/(N/2)} \\
&= \left(\mathrm{DFT}_{N/2}\boldsymbol{x}^{(e)}\right)_n + e^{-2\pi i n/N}\left(\mathrm{DFT}_{N/2}\boldsymbol{x}^{(o)}\right)_n,
\end{aligned}
$$

where we have substituted $\boldsymbol{x}^{(e)}$ and $\boldsymbol{x}^{(o)}$ as in the text of the theorem, and recognized the $N/2$-point DFT in two places. Assembling this for $0 \le n < N/2$ we obtain Eq. (2.11). For the second half of the DFT coefficients, i.e. $\{y_{N/2+n}\}_{0 \le n \le N/2-1}$, we similarly have

$$
\begin{aligned}
y_{N/2+n} &= \sum_{k=0}^{N-1} x_k e^{-2\pi i(N/2+n)k/N} = \sum_{k=0}^{N-1} x_k e^{-\pi i k} e^{-2\pi i n k/N} \\
&= \sum_{k=0}^{N/2-1} x_{2k} e^{-2\pi i n 2k/N} - \sum_{k=0}^{N/2-1} x_{2k+1} e^{-2\pi i n(2k+1)/N} \\
&= \sum_{k=0}^{N/2-1} x_{2k} e^{-2\pi i n k/(N/2)} - e^{-2\pi i n/N} \sum_{k=0}^{N/2-1} x_{2k+1} e^{-2\pi i n k/(N/2)} \\
&= \left(\mathrm{DFT}_{N/2}\boldsymbol{x}^{(e)}\right)_n - e^{-2\pi i n/N}\left(\mathrm{DFT}_{N/2}\boldsymbol{x}^{(o)}\right)_n.
\end{aligned}
$$

Equation (2.12) now follows similarly. $\square$

An algorithm for the IDFT can be deduced in exactly the same way: All we need is to change the sign in the exponents of the Fourier matrix, and divide by $1/N$. If we do this we get the following the IFFT algorithm. Recall that we use the notation $\overline{A}$ for the matrix where all the elements of A have been conjugated.

Theorem 15. IFFT Algorithm When N Is Even.
 Let N be an even number and let $\tilde{\boldsymbol{x}} = \overline{DFT_N}\boldsymbol{y}$. Then we have that

$$(\tilde{x}_0, \tilde{x}_1, \ldots, \tilde{x}_{N/2-1}) = \overline{DFT_{N/2}}\boldsymbol{y}^{(e)} + \overline{D_{N/2}DFT_{N/2}}\boldsymbol{y}^{(o)} \qquad (2.13)$$

$$(\tilde{x}_{N/2}, \tilde{x}_{N/2+1}, \ldots, \tilde{x}_{N-1}) = \overline{DFT_{N/2}}\boldsymbol{y}^{(e)} - \overline{D_{N/2}DFT_{N/2}}\boldsymbol{y}^{(o)} \qquad (2.14)$$

where $\boldsymbol{y}^{(e)}, \boldsymbol{y}^{(o)} \in \mathbb{R}^{N/2}$ *are the vectors*

$$\boldsymbol{y}^{(e)} = (y_0, y_2, \ldots, y_{N-2}) \qquad\qquad \boldsymbol{y}^{(o)} = (y_1, y_3, \ldots, y_{N-1}).$$

Moreover, $\boldsymbol{x} = IDFT_N\boldsymbol{y}$ *can be computed from* $\boldsymbol{x} = \tilde{\boldsymbol{x}}/N = \overline{DFT_N}\boldsymbol{y}/N$

These theorems can be interpreted as matrix factorizations. Using the block matrix notation from Appendix A on Eqs. (2.11)–(2.12), the DFT matrix can be factorized as

$$(y_0, y_1, \ldots, y_{N/2-1}) = \left(\mathrm{DFT}_{N/2} \ \ D_{N/2}\mathrm{DFT}_{N/2}\right) \begin{pmatrix} \boldsymbol{x}^{(e)} \\ \boldsymbol{x}^{(o)} \end{pmatrix}$$

$$(y_{N/2}, y_{N/2+1}, \ldots, y_{N-1}) = \left(\mathrm{DFT}_{N/2} \ -D_{N/2}\mathrm{DFT}_{N/2}\right) \begin{pmatrix} \boldsymbol{x}^{(e)} \\ \boldsymbol{x}^{(o)} \end{pmatrix}.$$

Combining these and noting that

$$\begin{pmatrix} \mathrm{DFT}_{N/2} & D_{N/2}\mathrm{DFT}_{N/2} \\ \mathrm{DFT}_{N/2} & -D_{N/2}\mathrm{DFT}_{N/2} \end{pmatrix} = \begin{pmatrix} I & D_{N/2} \\ I & -D_{N/2} \end{pmatrix} \begin{pmatrix} \mathrm{DFT}_{N/2} & 0 \\ 0 & \mathrm{DFT}_{N/2} \end{pmatrix},$$

we obtain the following factorizations:

Theorem 16. DFT and IDFT Matrix Factorizations.
We have that

$$DFT_N\boldsymbol{x} = \begin{pmatrix} I & D_{N/2} \\ I & -D_{N/2} \end{pmatrix} \begin{pmatrix} DFT_{N/2} & 0 \\ 0 & DFT_{N/2} \end{pmatrix} \begin{pmatrix} \boldsymbol{x}^{(e)} \\ \boldsymbol{x}^{(o)} \end{pmatrix}$$

$$IDFT_N\boldsymbol{y} = \frac{1}{N} \overline{\begin{pmatrix} I & D_{N/2} \\ I & -D_{N/2} \end{pmatrix}} \; \overline{\begin{pmatrix} DFT_{N/2} & 0 \\ 0 & DFT_{N/2} \end{pmatrix}} \begin{pmatrix} \boldsymbol{y}^{(e)} \\ \boldsymbol{y}^{(o)} \end{pmatrix} \qquad (2.15)$$

Note that we can apply the FFT factorization again to $F_{N/2}$ to obtain

$$\mathrm{DFT}_N\boldsymbol{x} = \begin{pmatrix} I & D_{N/2} \\ I & -D_{N/2} \end{pmatrix} \begin{pmatrix} I & D_{N/4} & 0 & 0 \\ I & -D_{N/4} & 0 & 0 \\ 0 & 0 & I & D_{N/4} \\ 0 & 0 & I & -D_{N/4} \end{pmatrix} \times$$

$$\begin{pmatrix} \mathrm{DFT}_{N/4} & 0 & 0 & 0 \\ 0 & \mathrm{DFT}_{N/4} & 0 & 0 \\ 0 & 0 & \mathrm{DFT}_{N/4} & 0 \\ 0 & 0 & 0 & \mathrm{DFT}_{N/4} \end{pmatrix} \begin{pmatrix} \boldsymbol{x}^{(ee)} \\ \boldsymbol{x}^{(eo)} \\ \boldsymbol{x}^{(oe)} \\ \boldsymbol{x}^{(oo)} \end{pmatrix}$$

where the vectors $\boldsymbol{x}^{(e)}$ and $\boldsymbol{x}^{(o)}$ have been further split into even- and odd-indexed entries. Clearly, if this factorization is repeated, we obtain a factorization

$$\mathrm{DFT}_N = \prod_{k=1}^{\log_2 N} \begin{pmatrix} I & D_{N/2^k} & 0 & 0 & \cdots & 0 & 0 \\ I & -D_{N/2^k} & 0 & 0 & \cdots & 0 & 0 \\ 0 & 0 & I & D_{N/2^k} & \cdots & 0 & 0 \\ 0 & 0 & I & -D_{N/2^k} & \cdots & 0 & 0 \\ \vdots & \vdots & \vdots & \vdots & \vdots & 0 & 0 \\ 0 & 0 & 0 & 0 & \cdots & I & D_{N/2^k} \\ 0 & 0 & 0 & 0 & \cdots & I & -D_{N/2^k} \end{pmatrix} P. \qquad (2.16)$$

The factorization has been repeated until we have a final diagonal matrix with DFT_1 on the diagonal, but clearly $\mathrm{DFT}_1 = 1$, so there is no need to include the final factor. Note that all matrices in this factorization are sparse. A factorization into a product of sparse matrices is the key to many efficient algorithms in linear algebra, such as the computation of eigenvalues and eigenvectors. When we later compute the number of arithmetic operations in this factorization, we will see that this is the case also here.

In Eq. (2.16), P is a permutation matrix which secures that the even-indexed entries come first. Since the even-indexed entries have 0 as the last bit, this is the same as letting the last bit become the first bit. Since we here recursively place even-indexed entries first, P will permute the elements of $\boldsymbol{x}$ by performing a *bit-reversal* of the indices, i.e.

$$P(\boldsymbol{e}_i) = \boldsymbol{e}_j \qquad i = d_1 d_2 \dots d_n \qquad j = d_n d_{n-1} \dots d_1,$$

where we have used the bit representations of i and j. Since $P^2 = I$, a bit-reversal can be computed very efficiently, and *in-place*, i.e. that the output is written into the same memory as the input ($\boldsymbol{x}$), without allocating additional memory. With the vast amounts of memory available in modern computers, it is a good question why in-place computation is desirable. The answer has to do with *memory access*, which is the main bottleneck in many programs: There exist different types of memory, and the fastest type of memory (in terms of memory access) is very small (as is evident in hardware implementations). Algorithms such as the FFT need to be very fast, and therefore require the fastest memory. In-place implementations meets this requirement by using a common memory for the input and output. We will use the function `bit_reversal` in the library to perform in-place bit-reversal. In Exercise 2.31 we will go through its implementation.

Matrix multiplication is usually not computed in-place, i.e. when we compute $\boldsymbol{y} = A\boldsymbol{x}$, different memory is used for $\boldsymbol{x}$ and $\boldsymbol{y}$. For certain simple matrices, however, in-place matrix multiplication is possible. It turns out that all block matrices in the factorization (2.16) also are of this kind, so that the entire FFT can be computed in-place (see Exercise 2.29).

Since matrix multiplication is computed from right to left, bit-reversal (i.e. the P matrix) is computed first in an FFT. Note also that the factorization (2.16) consists of block diagonal matrices, so that the different blocks in each matrix can be applied in parallel. We can thus exploit the parallel processing capabilities of the computer. It turns out that bit-reversal is useful for other DFT algorithms as well. We will look at some of these, and we will therefore split the computation of the DFT into one bit-reversal step, and an algorithm-dependent step represented by a "kernel FFT function", which assumes that the input has been bit-reversed. A simple implementation of the general function can be as follows.

```
function y = fft_impl(x, f)
    x = bit_reversal(x);
    y = f(x);
```

A simple implementation of the kernel FFT function for the factorization (2.16) can be as follows.

```
function y = fft_kernel_standard(x)
    N = size(x, 1);
    if N == 1
        y = x;
    else
        xe = fft_kernel_standard(x(1:(N/2)));
        xo = fft_kernel_standard(x((N/2+1):N));
        D = exp(-2*pi*1j*(0:(N/2-1))'/N);
        xo = xo.*D;
        y = [ xe + xo; xe - xo];
    end
```

Note that, although computations can be performed in-place, this MATLAB implementation does not, since return values and parameters to functions are copied in MATLAB. In Exercise 2.24 we will extend these functions to the general implementations we will use later. We can now run the FFT by combining the general function and the kernel as follows:

```
y = fft_impl(x, @fft_kernel_standard);
```

Note that `fft_kernel_standard` is recursive, i.e. it calls itself. If this is your first encounter with a recursive program, it is worth running through the code manually for a given value of N, such as $N = 4$.

From factorization (2.16) we immediately see two possible implementations of a kernel. First, as we did, we can apply the FFT recursively. Secondly, we can use a loop where we in each iteration compute the product with one matrix from factorization (2.16), from right to left. Inside this loop there must be another loop, where the different blocks in this matrix are applied. We will establish this non-recursive implementation in Exercise 2.29, and see that it leads to a more efficient algorithm.

MATLAB has built-in functions for computing the DFT and the IDFT using the FFT algorithm. The functions are called `fft` and `ifft`, and they make no assumption about the length of the vector, i.e. it may not be of even length. The implementation may however check if the length of the vector is 2^r, and in those cases use variants of the algorithm discussed here.

2.3.1 Reduction in the Number of Arithmetic Operations

Now we will explain why the FFT and IFFT factorizations reduce the number of arithmetic operations when compared to direct DFT and IDFT implementations. We will assume that $x \in \mathbb{R}^N$ with N a power of 2, so that the FFT algorithm can be used recursively, all the way down to vectors of length 1. In many settings this power of 2 assumption can be done. As an example, in compression of sound, one restricts processing to a certain block of the sound data, since the entire sound is too big to be processed in one piece. One then has the freedom to choose the size of the blocks. For optimal speed one often chooses blocks of length 2^r with r some integer in the range 5–10. At the end of this section we will explain how the general FFT can be computed when N is not a power of 2.

We first need some terminology for how we count the number of operations of a given type in an algorithm. In particular we are interested in the limiting behavior when N becomes large, which is the motivation for the following definition.

Definition 17. *Order of an Algorithm.*

Let R_N be the number of operations of a given type (such as multiplication or addition) in an algorithm, where N describes the dimension of the data (such as the size of the matrix or length of the vector), and let f be a positive function. The algorithm is said to be of order $f(N)$, also written $O(f(N))$, if the number of operations grows as $f(N)$ for large N, or more precisely, if

$$\lim_{N\to\infty} \frac{R_N}{f(N)} = 1.$$

Many FFT algorithms count the number of operations exactly. It is easier to obtain the order, however, since this is only a limit. We will therefore restrict ourselves to this. Let us see how we can compute the order of the FFT algorithm. Let M_N and A_N denote the number of real multiplications and real additions, respectively, required by the FFT algorithm. Once the FFT's of order $N/2$ have been computed ($M_{N/2}$ real multiplications and $A_{N/2}$ real additions are needed for each), it is clear from Eqs. (2.11)–(2.12) that an additional N complex additions, and an additional $N/2$ complex multiplications, are required. Since one complex multiplication requires 4 real multiplications and 2 real additions, and one complex addition requires two real additions, we see that we require an additional $2N$ real multiplications, and $2N + N = 3N$ real additions. This means that we have the difference equations

$$M_N = 2M_{N/2} + 2N \qquad\qquad A_N = 2A_{N/2} + 3N. \qquad (2.17)$$

Note that $e^{-2\pi i/N}$ may be computed once and for all and outside the algorithm. This is the reason why we have not counted these operations.

The following example shows how the difference equations (2.17) can be solved. It is not too difficult to argue that $M_N = O(2N\log_2 N)$ and $A_N = O(3N\log_2)$, by noting that there are $\log_2 N$ levels in the FFT, with $2N$ real multiplications and $3N$ real additions at each level. But for $N = 2$ and $N = 4$ we may actually avoid some multiplications, so we should solve these equations by stating initial conditions carefully, in order to obtain exact operation counts. In practice one often has more involved equations than (2.17), for which the solution can not be seen directly, so that one needs to apply systematic mathematical methods.

Operation Count

To use standard solution methods for difference equations to Eqs. (2.17), we first need to write them in a standard form. Assuming that A_N and M_N are powers of 2, we set $N = 2^r$ and $x_r = M_{2^r}$, or $x_r = A_{2^r}$. The difference equations can then be rewritten as $x_r = 2x_{r-1} + 2 \cdot 2^r$ for multiplications, and $x_r = 2x_{r-1} + 3 \cdot 2^r$ for additions, and again be rewritten in the standard forms

$$x_{r+1} - 2x_r = 4 \cdot 2^r \qquad\qquad x_{r+1} - 2x_r = 6 \cdot 2^r.$$

The homogeneous equation $x_{r+1} - 2x_r = 0$ has the general solution $x_r^h = C2^r$. Since the base in the power on the right hand side equals the root in the homogeneous equation, we should in each case guess for a particular solution on the form $(x_p)_r = Ar2^r$. If we do this we find that the first equation has particular solution $(x_p)_r = 2r2^r$, while the second has particular solution $(x_p)_r = 3r2^r$. The general solutions are thus on the form $x_r = 2r2^r + C2^r$, for multiplications, and $x_r = 3r2^r + C2^r$ for additions.

Now let us state initial conditions. For the number of multiplications, Example 2.3 showed that floating point multiplication can be avoided completely for $N = 4$. We can therefore use $M_4 = x_2 = 0$ as an initial value. This gives, $x_r = 2r2^r - 4 \cdot 2^r$, so that $M_N = 2N \log_2 N - 4N$.

For additions we can use $A_2 = x_1 = 4$ as initial value (since $\mathrm{DFT}_2(x_1, x_2) = (x_1 + x_2, x_1 - x_2)$), which gives $x_r = 3r2^r$, so that $A_N = 3N \log_2 N - N$. Our FFT algorithm thus requires slightly more additions than multiplications.

FFT algorithms are often characterized by their *operation count*, i.e. the total number of real additions- and multiplications, i.e. $R_N = M_N + A_N$. We see that $R_N = 5N \log_2 N - 5N$. The operation count for our algorithm is thus of order $O(5N \log_2 N)$, since $\lim_{N \to \infty} \frac{5N \log_2 N - 4N}{5N \log_2 N} = 1$.

In practice one can reduce the number of multiplications further, since $e^{-2\pi i n/N}$ take the simple values $1, -1, -i, i$ for some n. One can also use that $e^{-2\pi i n/N}$ can take the simple values $\pm 1/\sqrt{2} \pm 1/\sqrt{2}i = 1/\sqrt{2}(\pm 1 \pm i)$, which also saves some floating point multiplication, since we can factor out $1/\sqrt{2}$. These observations do not give big reductions in the arithmetic complexity, however, and one can show that the operation count is still $O(5N \log_2 N)$ after using these observations.

It is straightforward to show that the IFFT implementation requires the same operation count as the FFT algorithm.

In contrast, the direct implementation of the DFT requires N^2 complex multiplications and $N(N - 1)$ complex additions. This results in $4N^2$ real multiplications and $2N^2 + 2N(N-1) = 4N^2 - 2N$ real additions. The total operation count is thus $8N^2 - 2N$. The FFT and IFFT thus significantly reduces the number of arithmetic operations. Let us summarize our findings as follows.

Theorem 18. Operation Count for the FFT and IFFT Algorithms.

The N-point FFT and IFFT algorithms we have gone through both require $O(2N \log_2 N)$ real multiplications and $O(3N \log_2 N)$ real additions. In comparison, the number of real multiplications and real additions required by direct implementations of the N-point DFT and IDFT are $O(8N^2)$.

In Exercise 2.30 we present another algorithm, called the Split-radix algorithm, which reduces the operation count further with 20%.

Often we apply the DFT to real data, so we would like to have FFT-algorithms tailored to this, with reduced complexity (since real data has half the dimension of general complex data). By some it has been argued that one can find improved FFT algorithms when one assumes that the data is real. In Exercise 2.28 we address this issue, and conclude that there is little to gain from assuming real input: The general algorithm for complex input can be tailored for real input so that it uses half the number of operations, which harmonizes with the fact that real data has half the dimension of complex data.

As mentioned, the FFT is well suited for parallel computing. Besides this and a reduced operation count, FFT implementations can also apply several programming tricks to speed up computation, see for instance http://cnx.org/content/m12021/latest/ for an overview.

2.3.2 The FFT When N Is Non-prime

It turns out that the idea behind the FFT algorithm for $N = 2^n$ points, easily carries over to the case when N is any composite number, i.e. when $N = N_1N_2$. This make the FFT useful also in settings where we have a dictated number of elements in $\boldsymbol{x}$. The approach we will present in this section will help us as long as N is not a prime number. The case when N is a prime number needs other techniques, which we will consider later.

So, assume that $N = N_1N_2$. Any index k can be written uniquely on the form $N_1k + p$, with $0 \leq k < N_2$, and $0 \leq p < N_1$. We will make the following definition.

Definition 19. *Polyphase Components of a Vector.*
Let $\boldsymbol{x} \in \mathbb{R}^{N_1N_2}$. We denote by $\boldsymbol{x}^{(p)}$ the vector in $\mathbb{R}^{N_2}$ with entries $(\boldsymbol{x}^{(p)})_k = x_{N_1k+p}$. $\boldsymbol{x}^{(p)}$ is also called the p'th *polyphase component* of $\boldsymbol{x}$.

The previous vectors $\boldsymbol{x}^{(e)}$ and $\boldsymbol{x}^{(o)}$ can be seen as special cases of polyphase components. Using the polyphase notation, we can write

$$\text{DFT}_N\boldsymbol{x} = \sum_{k=0}^{N-1} x_k e^{-2\pi ink/N} = \sum_{p=0}^{N_1-1}\sum_{k=0}^{N_2-1} (\boldsymbol{x}^{(p)})_k e^{-2\pi in(N_1k+p)/N}$$

$$= \sum_{p=0}^{N_1-1} e^{-2\pi inp/N} \sum_{k=0}^{N_2-1} (\boldsymbol{x}^{(p)})_k e^{-2\pi ink/N_2}$$

Similarly, any frequency index n can be written uniquely on the form $N_2q + n$, with $0 \leq q < N_1$, and $0 \leq n < N_2$, so that the DFT can also be written as

$$\sum_{p=0}^{N_1-1} e^{-2\pi i(N_2q+n)p/N} \sum_{k=0}^{N_2-1} (\boldsymbol{x}^{(p)})_k e^{-2\pi i(N_2q+n)k/N_2}$$

$$= \sum_{p=0}^{N_1-1} e^{-2\pi iqp/N_1} e^{-2\pi inp/N} \sum_{k=0}^{N_2-1} (\boldsymbol{x}^{(p)})_k e^{-2\pi ink/N_2}.$$

Now, if X is the $N_2 \times N_1$-matrix X where the p'th column is $\boldsymbol{x}^{(p)}$, we recognize the inner sum $\sum_{k=0}^{N_2-1}(\boldsymbol{x}^{(p)})_k e^{-2\pi ink/N_2}$ as matrix multiplication with DFT_{N_2} and X, so that this can be written as $(\text{DFT}_{N_2}X)_{n,p}$. The entire sum can thus be written as

$$\sum_{p=0}^{N_1-1} e^{-2\pi iqp/N_1} e^{-2\pi inp/N} (\text{DFT}_{N_2}X)_{n,p}.$$

Now, define Y as the matrix where X is multiplied component-wise with the matrix with (n,p)-component $e^{-2\pi inp/N}$. The entire sum can then be written as

$$\sum_{p=0}^{N_1-1} e^{-2\pi iqp/N_1} Y_{n,p} = (YF_{N_1})_{n,q}$$

This means that the sum can be written as component (n,q) in the matrix $Y\text{DFT}_{N_1}$. Clearly $Y\text{DFT}_{N_1}$ is the matrix where the DFT is applied to all rows of Y. We have thus shown that component N_2q+n of $F_N\boldsymbol{x}$ equals $(Y\text{DFT}_{N_1})_{n,q}$. This means that $\text{DFT}_N\boldsymbol{x}$

can be obtained by stacking the columns of YDFT_{N_1} on top of one another. We can thus summarize our procedure as follows, which gives a recipe for splitting an FFT into smaller FFT's when N is not a prime number.

Theorem 20. FFT Algorithm When N Is Non-prime.
  When $N = N_1 N_2$, the FFT of a vector $\boldsymbol{x}$ can be computed as follows

- *Form the $N_2 \times N_1$-matrix X, where the p'th column is $\boldsymbol{x}^{(p)}$.*
- *Perform the DFT on all the columns in X, i.e. compute $\text{DFT}_{N_2} X$.*
- *Multiply element (n, p) in the resulting matrix with $e^{-2\pi i n p / N}$, to obtain the matrix Y.*
- *Perform the DFT on all the rows in the resulting matrix, i.e. compute $Y\text{DFT}_{N_1}$.*
- *Form the vector where the columns of the resulting matrix are stacked on top of one another.*

From this algorithm one easily deduces how the IDFT can be computed also: All steps are non-singular, and can be performed by IFFT or multiplication. We thus only need to perform the inverse steps in reverse order.

For the case when N is a prime number, one can use Rader's algorithm [60], see Exercise 3.11. Winograd's FFT algorithm [80] extends Rader's algorithm to work for the case when $N = p^r$. This algorithm is difficult to program, and is rarely used in practice.

Exercise 2.23: Properties of the DFT When N Is Composite

When $N = N_1 N_2$ is composite, the following explains a duality between the polyphase components $\boldsymbol{x}^{(k)}$ with components x_{rN_1+k}, and the polyphase components $\boldsymbol{y}^{(k)}$ of $\boldsymbol{y} = \text{DFT}_N \boldsymbol{x}$ with components y_{rN_2+k}.

a) Assume that all $\boldsymbol{x}^{(k)}$ are constant vectors (i.e. $\boldsymbol{x}$ is N_1-periodic). Show that $\boldsymbol{y}^{(k)} = 0$ for all $k \neq 0$.

b) Assume that $N = N_1 N_2$, and that $\boldsymbol{x}^{(p)} = \boldsymbol{0}$ for $p \neq 0$. Show that the $\boldsymbol{y}^{(p)}$ are constant vectors for all p.

Exercise 2.24: FFT Implementation Which Works for Both Forward and Reverse

Recall that, in Exercise 2.14, we extended the direct DFT implementation so that it accepted a second parameter telling us if the forward or reverse transform should be applied. Extend `fft_impl` and `fft_kernel_standard` in the same way. Again, the forward transform should be used if the `forward` parameter is not present. Assume also that the kernel accepts only one-dimensional data, and that the general function applies the kernel to each column in the input if the input is two-dimensional (so that the FFT can be applied to all channels in a sound with only one call).

It should be straightforward to make the modifications for the reverse transform by consulting the second part of Theorem 16. For simplicity, let `fft_impl` take care of the additional division with N in case of the IDFT. In the following we will assume these signatures for the FFT implementation and the corresponding kernels.

Exercise 2.25: Execution Times for the FFT

Let us compare execution times for the different methods for computing the DFT.

a) Write code which compares the execution times for an N-point DFT for the following three cases: Direct implementation of the DFT (as in Example 2.4), the FFT implementation used in this chapter, and the built-in `fft`-function. Your code should use the sample audio file `castanets.wav`, apply the different DFT implementations to the first $N = 2^r$ samples of the file for $r = 3$ to $r = 15$, store the execution times in a vector, and plot these. You can use the functions `tic` and `toc` to measure the execution time.

b) A problem for large N is that there is such a big difference in the execution times between the two implementations. We can address this by using a loglog-plot. Plot N against execution times using the function `loglog`. How should the fact that the number of arithmetic operations are $8N^2$ and $5N \log_2 N$ be reflected in the plot?

c) It seems that the built-in FFT is much faster than our own FFT implementation, even though they may use similar algorithms. Try to explain what can be the cause of this.

Exercise 2.26: Combining Two FFT's

Let $x_1 = (1, 3, 5, 7)$ and $x_2 = (2, 4, 6, 8)$. Compute $\text{DFT}_4 x_1$ and $\text{DFT}_4 x_2$. Explain how you can compute $\text{DFT}_8(1, 2, 3, 4, 5, 6, 7, 8)$ based on these computations (you don't need to perform the actual computation). What are the benefits of this approach?

Exercise 2.27: FFT Operation Count

When we wrote down the difference equation for the number of multiplications in the FFT algorithm, you could argue that some multiplications were not counted. Which multiplications in the FFT algorithm were not counted when writing down this difference equation? Do you have a suggestion to why these multiplications were not counted?

Exercise 2.28: FFT Algorithm Adapted to Real Data

It is possible to adapt an FFT algorithm to real input. Since $y_{N-n} = \overline{y_n}$ for real input, there is no additional complexity in computing the second half of the DFT coefficients, once the first half has been computed. We will now rewrite Eq. (2.11) for indices n and $N/2 - n$ as

$$
\begin{aligned}
y_n &= (\text{DFT}_{N/2} x^{(e)})_n + e^{-2\pi i n/N} (\text{DFT}_{N/2} x^{(o)})_n \\
y_{N/2-n} &= (\text{DFT}_{N/2} x^{(e)})_{N/2-n} + e^{-2\pi i (N/2-n)/N} (\text{DFT}_{N/2} x^{(o)})_{N/2-n} \\
&= \overline{(\text{DFT}_{N/2} x^{(e)})_n} - e^{2\pi i n/N} \overline{(\text{DFT}_{N/2} x^{(o)})_n} \\
&= \overline{(\text{DFT}_{N/2} x^{(e)})_n} - \overline{e^{-2\pi i n/N} (\text{DFT}_{N/2} x^{(o)})_n}.
\end{aligned}
$$

We see here that, if we already have computed $\mathrm{DFT}_{N/2}\boldsymbol{x}^{(e)}$ and $\mathrm{DFT}_{N/2}\boldsymbol{x}^{(o)}$, we need one additional complex multiplication for each y_n with $0 \leq n < N/4$ (since $e^{-2\pi in/N}$ and $(\mathrm{DFT}_{N/2}\boldsymbol{x}^{(o)})_n$ are complex). No further multiplications are needed in order to compute $y_{N/2-n}$, since we simply conjugate terms before adding them. Again $y_{N/2}$ must be handled explicitly with this approach. For this we can use the formula

$$y_{N/2} = (\mathrm{DFT}_{N/2}\boldsymbol{x}^{(e)})_0 - (D_{N/2}\mathrm{DFT}_{N/2}\boldsymbol{x}^{(o)})_0$$

instead.

a) Conclude from this that an FFT algorithm adapted to real data at each step requires $N/4$ complex additions and $N/2$ complex additions. Conclude from this as before that an algorithm based on real data requires $M_N = O(N \log_2 N)$ real multiplications and $A_N = O\left(\frac{3}{2}N \log_2 N\right)$ real additions (i.e. half the operation count for complex input).

b) Find an IFFT algorithm adapted to vectors with conjugate symmetry, which has the same operation count as this FFT algorithm adapted to real data.

Hint

Consider the vectors $\boldsymbol{z}$, $\boldsymbol{w}$ with entries $z_n = y_n + \overline{y_{N/2-n}} \in \mathbb{C}^{N/2}$ and $w_n = e^{2\pi in/N}(y_n - \overline{y_{N/2-n}}) \in \mathbb{C}^{N/2}$. From the equations above, how can these be used in an IFFT?

Exercise 2.29: Non-recursive FFT Algorithm

Use the factorization in (2.16) to write a kernel function for a non-recursive FFT implementation. In your code, perform the matrix multiplications in Eq. (2.16) from right to left in an (outer) for-loop. For each matrix loop through the different blocks on the diagonal in an (inner) for-loop. Make sure you have the right number of blocks on the diagonal, each block being on the form

$$\begin{pmatrix} I & D_{N/2^k} \\ I & -D_{N/2^k} \end{pmatrix}.$$

It may be a good idea to start by implementing multiplication with such a simple matrix first as these are the building blocks in the algorithm. Do this so that everything is computed in-place. Also compare the execution times with our original FFT algorithm, as we did in Exercise 2.25, and try to explain what you see in this comparison.

Exercise 2.30: The Split-Radix FFT Algorithm

The *split-radix FFT algorithm* is a variant of the FFT algorithm. Until recently it held the record for the lowest operation count for any FFT algorithm. It was first derived in 1968 [81], but largely forgotten until a paper in 1984 [16].

We start by splitting the rightmost $\text{DFT}_{N/2}$ in Eq. (2.15) by using this equation again, to obtain

$$\text{DFT}_N \boldsymbol{x} = \begin{pmatrix} \text{DFT}_{N/2} & D_{N/2} \begin{pmatrix} \text{DFT}_{N/4} & D_{N/4}\text{DFT}_{N/4} \\ \text{DFT}_{N/4} & -D_{N/4}\text{DFT}_{N/4} \end{pmatrix} \\ \text{DFT}_{N/2} & -D_{N/2} \begin{pmatrix} \text{DFT}_{N/4} & D_{N/4}\text{DFT}_{N/4} \\ \text{DFT}_{N/4} & -D_{N/4}\text{DFT}_{N/4} \end{pmatrix} \end{pmatrix} \begin{pmatrix} \boldsymbol{x}^{(e)} \\ \boldsymbol{x}^{(oe)} \\ \boldsymbol{x}^{(oo)} \end{pmatrix}.$$

The term radix describes how an FFT is split into FFT's of smaller sizes, i.e. how the sum in an FFT is split into smaller sums. The FFT algorithm we started this section with is called a radix 2 algorithm, since it splits an FFT of length N into FFT's of length $N/2$. If an algorithm instead splits into FFT's of length $N/4$, it is called a radix 4 FFT algorithm. The algorithm we go through here is called the split radix algorithm, since it uses FFT's of both length $N/2$ and $N/4$.

a) Let $G_{N/4}$ be the $(N/4) \times (N/4)$ diagonal matrix with $e^{-2\pi in/N}$ on the diagonal. Show that $D_{N/2} = \begin{pmatrix} G_{N/4} & \boldsymbol{0} \\ \boldsymbol{0} & -iG_{N/4} \end{pmatrix}$.

b) Let $H_{N/4}$ be the $(N/4) \times (N/4)$ diagonal matrix $G_{D/4}D_{N/4}$. Verify the following rewriting of the equation above:

$$\text{DFT}_N \boldsymbol{x} = \begin{pmatrix} \text{DFT}_{N/2} & \begin{pmatrix} G_{N/4}\text{DFT}_{N/4} & H_{N/4}\text{DFT}_{N/4} \\ -iG_{N/4}\text{DFT}_{N/4} & iH_{N/4}\text{DFT}_{N/4} \end{pmatrix} \\ \text{DFT}_{N/2} & \begin{pmatrix} -G_{N/4}\text{DFT}_{N/4} & -H_{N/4}\text{DFT}_{N/4} \\ iG_{N/4}\text{DFT}_{N/4} & -iH_{N/4}\text{DFT}_{N/4} \end{pmatrix} \end{pmatrix} \begin{pmatrix} \boldsymbol{x}^{(e)} \\ \boldsymbol{x}^{(oe)} \\ \boldsymbol{x}^{(oo)} \end{pmatrix}$$

$$= \begin{pmatrix} I & 0 & G_{N/4} & H_{N/4} \\ 0 & I & -iG_{N/4} & iH_{N/4} \\ I & 0 & -G_{N/4} & -H_{N/4} \\ 0 & I & iG_{N/4} & -iH_{N/4} \end{pmatrix} \begin{pmatrix} \text{DFT}_{N/2} & 0 & 0 \\ 0 & \text{DFT}_{N/4} & 0 \\ 0 & 0 & \text{DFT}_{N/4} \end{pmatrix} \begin{pmatrix} \boldsymbol{x}^{(e)} \\ \boldsymbol{x}^{(oe)} \\ \boldsymbol{x}^{(oo)} \end{pmatrix}$$

$$= \begin{pmatrix} I & \begin{pmatrix} G_{N/4} & H_{N/4} \\ -iG_{N/4} & iH_{N/4} \end{pmatrix} \\ I & -\begin{pmatrix} G_{N/4} & H_{N/4} \\ -iG_{N/4} & iH_{N/4} \end{pmatrix} \end{pmatrix} \begin{pmatrix} \text{DFT}_{N/2}\boldsymbol{x}^{(e)} \\ \text{DFT}_{N/4}\boldsymbol{x}^{(oe)} \\ \text{DFT}_{N/4}\boldsymbol{x}^{(oo)} \end{pmatrix}$$

$$= \begin{pmatrix} \text{DFT}_{N/2}\boldsymbol{x}^{(e)} + \begin{pmatrix} G_{N/4}\text{DFT}_{N/4}\boldsymbol{x}^{(oe)} + H_{N/4}\text{DFT}_{N/4}\boldsymbol{x}^{(oo)} \\ -i\left(G_{N/4}\text{DFT}_{N/4}\boldsymbol{x}^{(oe)} - H_{N/4}\text{DFT}_{N/4}\boldsymbol{x}^{(oo)}\right) \end{pmatrix} \\ \text{DFT}_{N/2}\boldsymbol{x}^{(e)} - \begin{pmatrix} G_{N/4}\text{DFT}_{N/4}\boldsymbol{x}^{(oe)} + H_{N/4}\text{DFT}_{N/4}\boldsymbol{x}^{(oo)} \\ -i\left(G_{N/4}\text{DFT}_{N/4}\boldsymbol{x}^{(oe)} - H_{N/4}\text{DFT}_{N/4}\boldsymbol{x}^{(oo)}\right) \end{pmatrix} \end{pmatrix}$$

c) Explain from the above expression why, once the three FFT's above have been computed, the rest can be computed with $N/2$ complex multiplications, and $2 \times N/4 + N = 3N/2$ complex additions. This is equivalent to $2N$ real multiplications and $N + 3N = 4N$ real additions.

Hint

It is important that $G_{N/4}\text{DFT}_{N/4}\boldsymbol{x}^{(oe)}$ and $H_{N/4}\text{DFT}_{N/4}\boldsymbol{x}^{(oo)}$ are computed first, and the sum and difference of these two afterwards.

d) Due to what we just showed, our new algorithm leads to real multiplication and addition counts which satisfy

$$M_N = M_{N/2} + 2M_{N/4} + 2N \qquad\qquad A_N = A_{N/2} + 2A_{N/4} + 4N$$

Find the general solutions to these difference equations and conclude from these that $M_N = O\left(\frac{4}{3}N \log_2 N\right)$, and $A_N = O\left(\frac{8}{3}N \log_2 N\right)$. The operation count is thus $O\left(4N \log_2 N\right)$, which is a reduction of $N \log_2 N$ from the FFT algorithm.

e) Write a kernel function for the split-radix FFT algorithm (again this should handle both forward and reverse transforms). Are there more or less recursive function calls in this function than in the original FFT algorithm? Also compare the execution times with our original FFT algorithm, as we did in Exercise 2.25. Try to explain what you see in this comparison.

By carefully examining the algorithm we have developed, one can reduce the operation count to $4N \log_2 N - 6N + 8$. This does not reduce the order of the algorithm, but for small N (which often is the case in applications) this reduces the number of operations considerably, since $6N$ is large compared to $4N \log_2 N$ for small N. In addition to having a lower number of operations than the FFT algorithm of Theorem 14, a bigger percentage of the operations are additions for our new algorithm: there are now twice as many additions than multiplications. Since multiplications may be more time-consuming than additions (depending on how the CPU computes floating-point arithmetic), this can be a big advantage.

Exercise 2.31: Bit-Reversal

In this exercise we will make some considerations which will help us explain the code for bit-reversal. This is not a mathematically challenging exercise, but nevertheless a good exercise in how to think when developing an efficient algorithm. We will use the notation i for an index, and j for its bit-reverse. If we bit-reverse k bits, we will write $N = 2^k$ for the number of possible indices.

a) Consider the following code

```
j = 0;
for i = 0:(N-1)
    j
    m = N/2;
    while (m >= 1 && j >= m)
        j = j - m;
        m = m/2;
    end
    j = j + m;
end
```

Explain that the code prints all numbers in $[0, N-1]$ in bit-reversed order (i.e. j). Verify this by running the program, and writing down the bits for all numbers for, say $N = 16$. In particular explain the decrements and increments made to the variable j. The code above thus produces pairs of numbers (i, j), where j is the bit-reverse of i. The function `bit_reversal` in the library applies similar code, and then swaps the values x_i and x_j in $\boldsymbol{x}$, as it should.

Since bit-reverse is its own inverse (i.e. $P^2 = I$), it can be performed by swapping elements i and j. One way to secure that bit-reverse is done only once, is to perform it only when $j > i$. `bit_reversal` makes this check.

b) Explain that $N - j - 1$ is the bit-reverse of $N - i - 1$. Due to this, when $i, j < N/2$, we have that $N - i - 1, N - j - l \geq N/2$, and that `bit_reversal` can swap them. Moreover, all swaps where $i, j \geq N/2$ can be performed immediately when pairs where $i, j < N/2$ are encountered. Explain also that $j < N/2$ if and only if i is even. In the code you can see that the swaps (i, j) and $(N - i - 1, N - j - 1)$ are performed together when i is even, due to this.

c) Assume that $i < N/2$ is odd. Explain that $j \geq N/2$, so that $j > i$. This says that when $i < N/2$ is odd, we can always swap i and j (this is the last swap performed in the code). All swaps where $0 \leq j < N/2$ and $N/2 \leq j < N$ can be performed in this way.

In `bit_reversal`, you can see that the bit-reversal of $2r$ and $2r + 1$ are handled together (i.e. i is increased with 2 in the `for`-loop). The effect of this is that the number of `if`-tests can be reduced, due to the observations from b) and c).

2.4 The Discrete Cosine Transform (DCT)

We passed from Fourier series to the Discrete Fourier Transform by considering the samples of a function, rather than the function itself. The samples were taken uniformly over one period (Fig. 2.1). We can attempt to do something similar with the symmetric extension $\breve{f}$ of f. The idea is as follows:

Idea 21. *Since the Fourier series of $\breve{f}$ in general is a better approximation on $[0, T]$ than the Fourier series of f, is there a discrete transform similar to the DFT (applied to the samples of $\breve{f}$ instead) which captures this advantage?*

We will now see that such a transform indeed exists, and construct it. Recall that, if f has period T, $\breve{f}$ has period $2T$. We will consider two ways of taking uniform samples of $\breve{f}$ on $[0, 2T]$:

1. by including the boundaries (0 and T),
2. by not including the boundaries.

We will first take an even number $M = 2N$ of samples following the second strategy (we will consider the first strategy in Sect. 5.3):

$$\breve{x} = \left\{ f \left(\frac{k + 1/2}{2N} 2T \right) \right\}_{k=0}^{2N-1}$$

The middle points (i.e. $k = N - 1, k = N$) correspond to $t = T \pm T/(2N)$, so that indeed the boundary T is not sampled. Since these middle points are symmetric around T, their values are equal due to the symmetry of $\breve{f}$, so that $\breve{x}$ is symmetric around $N - 1/2$ for these indices. It is easy to see that the other points in time also are symmetric around T, so that the entire $\breve{x}$ is symmetric around $N - 1/2$. $\breve{x}$ is therefore the symmetric extension of its first half, denoted x, in the following sense:

Definition 22. *Symmetric Extension of a Vector.*

By the *symmetric extension* of $\boldsymbol{x} \in \mathbb{R}^N$, we mean the symmetric vector $\breve{\boldsymbol{x}} \in \mathbb{R}^{2N}$ defined by

$$\breve{\boldsymbol{x}}_k = \begin{cases} x_k & 0 \le k < N \\ x_{2N-1-k} & N \le k < 2N-1 \end{cases} \qquad (2.18)$$

In Fig. 2.5 such a symmetric extension is shown. As seen there, but not included in Definition 22, $\breve{\boldsymbol{x}}$ is also repeated periodically in order to obtain a periodic vector. Creating a symmetric extension is thus a two-step process:

- First, "mirror" the vector to obtain a vector in $\mathbb{R}^{2N}$,
- repeat this periodically to obtain a periodic vector.

It may seem unnatural to sample so that the boundaries 0 and T are not included. One good reason to do this is that we obtain a symmetric extension with twice the length of $\boldsymbol{x}$, and this enables us to stay within vectors with lengths being powers of two, for which efficient FFT algorithms are known.

In order to adapt the Discrete Fourier transform to symmetric extensions, note first that, since $\breve{f}$ is symmetric, its Fourier series is a cosine series

$$\sum_{n=0}^{N} a_n \cos(2\pi nt/(2T))$$

Sampling $\cos(2\pi nt/(2T))$ at $t = (k+1/2)2T/(2N)$ as described we get the cosine vector

$$\cos(2\pi n(k+1/2)/(2N))$$

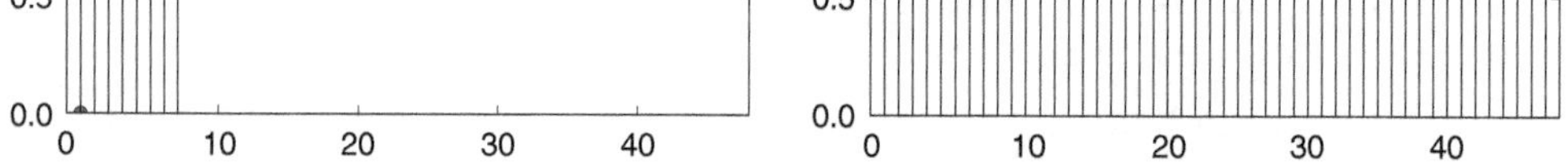

Fig. 2.5 A vector and its symmetric extension

for $n = 0, \ldots, N-1$. The $\cos(2\pi nt/(2T))$ are clearly orthogonal as functions on $[0, 2T]$, since each cosine is a sum of different Fourier basis functions, which are orthogonal. To see if the vectors $\cos(2\pi n(k+1/2)/(2N))$ also are orthogonal, write

$$\cos(2\pi n(k+1/2)/(2N))$$
$$= \frac{1}{2} \left(e^{2\pi in(k+1/2)/(2N)} + e^{-2\pi in(k+1/2)/(2N)} \right)$$
$$= \frac{1}{2} \left(e^{\pi in/(2N)} e^{2\pi ink/(2N)} + e^{-\pi in/(2N)} e^{2\pi i(2N-n)k/(2N)} \right).$$

Orthogonality of the cosine vectors in $\mathbb{R}^{2N}$ for $n = 0, \ldots, N-1$ now follows from orthogonality of $\{e^{2\pi i n k/(2N)}\}_{n=0}^{2N-1}$. Since the Fourier matrix is unitary, and since the norm of $e^{2\pi i n k/(2N)}$ is $\sqrt{2N}$, the norm of the cosine vector is $\sqrt{\frac{1}{4}(2N + 2N)} = \sqrt{N}$ for $n \neq 0$. For $n = 0$ the norm is clearly $\sqrt{2N}$. Since the space of vectors in $\mathbb{R}^{2N}$ symmetric about $N - 1/2$ clearly is N-dimensional, and since there are N cosine vectors, we have proved the following result.

Theorem 23. Orthonormal Basis for Symmetric Extensions in $\mathbb{R}^{2N}$.
  We have that

$$\left\{ \frac{1}{\sqrt{2N}} \cos\left(\frac{2\pi \cdot 0(k + 1/2)}{2N} \right), \left\{ \frac{1}{\sqrt{N}} \cos\left(\frac{2\pi n(k + 1/2)}{2N} \right) \right\}_{n=1}^{N-1} \right\} \tag{2.19}$$

is an orthonormal basis for the set of symmetric extensions in $\mathbb{R}^{2N}$.

Since symmetric extensions are equal on the first and second half, the cosine vectors give rise to an orthonormal basis for $\mathbb{R}^N$ as well:

Theorem 24. The Basis $\mathcal{D}_N$ in $\mathbb{R}^N$.
  We have that

$$\left\{ \frac{1}{\sqrt{N}} \cos\left(\frac{2\pi \cdot 0(k + 1/2)}{2N} \right), \left\{ \sqrt{\frac{2}{N}} \cos\left(\frac{2\pi n(k + 1/2)}{2N} \right) \right\}_{n=1}^{N-1} \right\} \tag{2.20}$$

is an orthonormal basis for $\mathbb{R}^N$, denoted $\mathcal{D}_N$. We denote the vectors by $\boldsymbol{d}_n$.

We remark that, also in this case we can appeal to the next chapter to give a more insightful proof of orthogonality: Also $\boldsymbol{d}_n$ turn out to be eigenvectors for certain symmetric matrices with distinct eigenvalues.

Using the basis $\mathcal{D}_N$ instead of the Fourier basis we can make the following definitions, which parallels that of the DFT:

Definition 25. *Discrete Cosine Transform.*
  The change of coordinates from the standard basis of $\mathbb{R}^N$ to the basis $\mathcal{D}_N$ is called the *discrete cosine transform* (or DCT). The corresponding $N \times N$ change of coordinate matrix, DCT_N, is also called the (N-point) DCT matrix. If $\boldsymbol{x}$ is a vector in R^N, its coordinates $\boldsymbol{y} = (y_0, y_1, \ldots, y_{N-1})$ relative to the basis $\mathcal{D}_N$ are called the DCT coefficients of $\boldsymbol{x}$ (in other words, $\boldsymbol{y} = \mathrm{DCT}_N \boldsymbol{x}$). We will call $\boldsymbol{x} = (\mathrm{DCT}_N)^T \boldsymbol{y}$ the inverse DCT or (IDCT) of $\boldsymbol{x}$.

It is smart to apply the DCT instead of the DFT when possible, since the DCT captures the Fourier series of $\breve{f}$, rather than that of f (which the DFT captures). The point again is that the first Fourier series often provides better approximations to f. The Fourier series are captured only by its samples, so we also have to choose the sample rate carefully, as always.

By construction the DCT is an orthogonal transform, As for the Fourier basis, the vectors in $\mathcal{D}_N$ are called synthesis vectors, since we can write

$$\boldsymbol{x} = y_0 \boldsymbol{d}_0 + y_1 \boldsymbol{d}_1 + \cdots + y_{N-1} \boldsymbol{d}_{N-1}. \tag{2.21}$$

Following the same reasoning as for the DFT, DCT_N^{-1} is the matrix where the $\boldsymbol{d}_n$ are columns. Orthogonality thus implies that the $\boldsymbol{d}_n$ are the rows in the DCT matrix.

The functions in MATLAB for computing the DCT and IDCT are called dct and idct, respectively. Replacing the calls to fft with calls to these DCT counterparts, we can repeat the sound experiments in Examples 2.5–2.6. You may not hear much improvements in these simple experiments, but in theory the DCT should be able to approximate sound better.

Similarly to the DFT, one can think of the DCT in terms of interpolating with sinusoids, rather than complex exponentials.

Theorem 26. Interpolation with the Basis $\mathcal{D}_N$.
Let f be a function defined on the interval $[0, T]$, and let $\boldsymbol{x}$ be the sampled vector

$$x_k = f((k + 1/2)2T/(2N)) \quad \text{for } k = 0, 1, \ldots, N - 1.$$

There is exactly one linear combination $g(t)$ on the form

$$\sum_{n=0}^{N-1} y_n \cos(2\pi n t/(2T))$$

which satisfies the conditions

$$g((k + 1/2)2T/(2N)) = f((k + 1/2)2T/(2N)), \quad k = 0, 1, \ldots, N - 1,$$

and the y_n can be found by computing a DCT.

Proof. This follows by inserting $t = (k + 1/2)2T/(2N)$ in the equation

$$g(t) = \sum_{n=0}^{N-1} y_n \cos(2\pi n t/(2T))$$

to arrive at the equations

$$f(kT/N) = \sum_{n=0}^{N-1} y_n \cos(2\pi n(k + 1/2)/(2N)) \qquad 0 \le k \le N - 1.$$

Scaling with the factors $1/\sqrt{N}$ and $\sqrt{2/N}$, this gives us an equation system for finding the y_n with the non-singular DCT matrix as coefficient matrix. The result follows. $\quad\square$

In Fig. 2.6 we have plotted the sinusoids used for interpolation for $T = 1$, as well as the corresponding sample points. The sample points in the upper left plot correspond to the first row in the DCT matrix, the sample points in the upper right plot to the second row, and so on. As n increases, the functions oscillate more and more. y_5 says how much the sinusoid of highest frequency contributes.

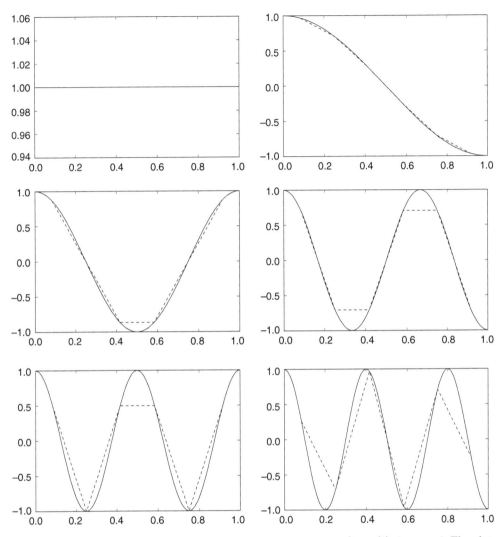

Fig. 2.6 The six different sinusoids used in DCT for $N = 6$, i.e. $\cos(2\pi nt/2)$, $0 \le n < 6$. The plots also show piecewise linear functions (dashed lines) between the sample points $\frac{2k+1}{2N}$ $0 \le k < 6$

2.4.1 Cosine Matrices

Note that we can write

$$\text{DCT}_N = \sqrt{\frac{2}{N}} \begin{pmatrix} 1/\sqrt{2} & 0 & \cdots & 0 \\ 0 & 1 & \cdots & 0 \\ \vdots & \vdots & \vdots & \vdots \\ 0 & 0 & \cdots & 1 \end{pmatrix} \left(\cos(2\pi n(k + 1/2)/(2N))\right). \tag{2.22}$$

Since DCT_N is orthogonal, it is immediate that

$$\left(\cos\left(\tfrac{2\pi n(k+1/2)}{2N}\right)\right)^{-1} = \frac{2}{N}\left(\cos\left(\tfrac{2\pi k(n+1/2)}{2N}\right)\right)\begin{pmatrix} 1/2 & 0 & \cdots & 0 \\ 0 & 1 & \cdots & 0 \\ \vdots & \vdots & \vdots & \vdots \\ 0 & 0 & \cdots & 1 \end{pmatrix} \quad (2.23)$$

$$\left(\cos\left(\tfrac{2\pi k(n+1/2)}{2N}\right)\right)^{-1} = \frac{2}{N}\begin{pmatrix} 1/2 & 0 & \cdots & 0 \\ 0 & 1 & \cdots & 0 \\ \vdots & \vdots & \vdots & \vdots \\ 0 & 0 & \cdots & 1 \end{pmatrix}\left(\cos\left(\tfrac{2\pi n(k+1/2)}{2N}\right)\right). \quad (2.24)$$

In other words, not only can DCT_N be directly expressed in terms of a cosine-matrix, but our developments helped us to express the inverse of a cosine matrix in terms of other cosine-matrices. In the literature different types of cosine matrices have been useful:

Type-I cosine matrices: with entries $\cos(2\pi nk/(2(N-1)))$.
Type-II cosine matrices: with entries $\cos(2\pi n(k+1/2)/(2N))$.
Type-III cosine matrices: with entries $\cos(2\pi(n+1/2)k/(2N))$.
Type-IV cosine matrices: with entries $\cos(2\pi(n+1/2)(k+1/2)/(2N))$.

These thus differ only in how cosine functions are sampled in time, and which frequencies we use. The analysis above concerns type-II cosine-matrices. It will turn out that not all of these cosine-matrices are orthogonal, but that we in all cases, as we did above, can express their inverses in terms of cosine matrices of other types, and that any cosine matrix is easily expressed in terms of an orthogonal matrix. These orthogonal matrices will be called $\text{DCT}_N^{(I)}$, $\text{DCT}_N^{(II)}$, $\text{DCT}_N^{(III)}$, and $\text{DCT}_N^{(IV)}$, respectively. The DCT_N we constructed above is thus $\text{DCT}_N^{(II)}$. This is perhaps the most used, and the type is therefore often dropped when referring to them. We will consider some of the other cosine matrices later in the book: In Chap. 5 we will encounter type-I cosine matrices, in connection with a wavelet extension strategy. Type-IV cosine matrices will be encountered in Sect. 7.3.

The different cosine matrices can all be associated with some extension strategy for the vector. Oppenheim and Schafer [54] contains a review of these. The exercises will address implementation aspects on some of the DCT variants. Wang [79] goes through implementation aspects of all of them.

Example 2.32: Computing Lower Order DCT's

As with Example 2.3, exact expressions for the DCT can be written down in a few specific cases. It turns out that the case $N = 4$, as considered in Example 2.3, does not give the same type of nice, exact values, so let us instead consider the case $N = 2$, which gives

$$\text{DCT}_4 = \begin{pmatrix} \frac{1}{\sqrt{2}}\cos(0) & \frac{1}{\sqrt{2}}\cos(0) \\ \cos\left(\frac{\pi}{2}\left(0+\frac{1}{2}\right)\right) & \cos\left(\frac{\pi}{2}\left(1+\frac{1}{2}\right)\right) \end{pmatrix} = \begin{pmatrix} \frac{1}{\sqrt{2}} & \frac{1}{\sqrt{2}} \\ \frac{1}{\sqrt{2}} & -\frac{1}{\sqrt{2}} \end{pmatrix}.$$

Exercise 2.33: Writing Down Lower Order DCT's

As in Example 2.32, state the exact Cartesian form of the DCT matrix for the case $N = 3$.

Exercise 2.34: Type-IV DCT

Show that the vectors $\{\cos(2\pi(n+1/2)(k+1/2)/(2N))\}_{n=0}^{N-1}$ in $\mathbb{R}^N$ are orthogonal, with lengths $\sqrt{N/2}$. This means that the matrix with entries $\sqrt{\frac{2}{N}}\cos(2\pi(n+1/2)(k+1/2)/(2N))$ is orthogonal. Since this matrix also is symmetric, it is its own inverse. This is the matrix $\mathrm{DCT}_N^{(IV)}$, which can be used for type IV cosine matrices.

Hint

Compare with the orthogonal vectors d_n, used in the DCT.

Exercise 2.35: Modified Discrete Cosine Transform

The *Modified Discrete Cosine Transform* (MDCT) is defined as the $N \times (2N)$-matrix M with elements $M_{n,k} = \cos(2\pi(n+1/2)(k+1/2+N/2)/(2N))$. This exercise will take you through the details of multiplication with this matrix.

a) Show that

$$M = \sqrt{\frac{N}{2}}\mathrm{DCT}_N^{(IV)}\begin{pmatrix} \mathbf{0} & A \\ B & \mathbf{0} \end{pmatrix}$$

where A and B are the $(N/2) \times N$-matrices

$$A = \begin{pmatrix} \cdots\cdots & 0 & -1 & -1 & 0 & \cdots\cdots \\ \vdots & \vdots & \vdots & \vdots & \vdots & \vdots & \vdots & \vdots \\ 0 & -1 & \cdots\cdots & \cdots\cdots & -1 & 0 \\ -1 & 0 & \cdots\cdots & \cdots\cdots & 0 & -1 \end{pmatrix} = \left(-(I_{N/2})^{\leftrightarrow} \ -I_{N/2}\right)$$

$$B = \begin{pmatrix} 1 & 0 & \cdots\cdots & \cdots\cdots & 0 & -1 \\ 0 & 1 & \cdots\cdots & \cdots\cdots & -1 & 0 \\ \vdots & \vdots & \vdots & \vdots & \vdots & \vdots & \vdots & \vdots \\ \cdots\cdots & 0 & 1 & -1 & 0 & \cdots\cdots \end{pmatrix} = \left(I_{N/2} \ -(I_{N/2})^{\leftrightarrow}\right),$$

where where $A^{\leftrightarrow}$ is the matrix A with the columns reversed. Due to this, any Type-IV DCT algorithm can be used to compute the MDCT.

b) The MDCT is singular, since it is not a square matrix. We will show here that it still can be used in connection with non-singular transformations. We first define the IMDCT as the matrix M^T/N. Transposing what we obtained in a) gives

$$\frac{1}{\sqrt{2N}} \begin{pmatrix} \mathbf{0} & B^T \\ A^T & \mathbf{0} \end{pmatrix} \text{DCT}_N^{(IV)}$$

for the IMDCT, which thus also has an efficient implementation. Show that if

$$\boldsymbol{x}_0 = (x_0,\ldots,x_{N-1}) \qquad \boldsymbol{x}_1 = (x_N,\ldots,x_{2N-1}) \qquad \boldsymbol{x}_2 = (x_{2N},\ldots,x_{3N-1})$$

and

$$\boldsymbol{y}_{0,1} = M\begin{pmatrix}\boldsymbol{x}_0\\\boldsymbol{x}_1\end{pmatrix} \qquad\qquad \boldsymbol{y}_{1,2} = M\begin{pmatrix}\boldsymbol{x}_1\\\boldsymbol{x}_2\end{pmatrix}$$

(i.e. we compute two MDCT's where half of the data overlap), then

$$\boldsymbol{x}_1 = \{\text{IMDCT}(\boldsymbol{y}_{0,1})\}_{k=N}^{2N-1} + \{\text{IMDCT}(\boldsymbol{y}_{1,2})\}_{k=0}^{N-1}.$$

Thus, even though the MDCT itself is singular, the input can still be recovered from two overlapping IMDCT's.

2.5 Efficient Implementations of the DCT

The DCT is by definition very related to the DFT. This leads one to believe that it too has an efficient algorithm. The following theorem addresses, and is much used in practical implementations of DCT. It can also be used for practical implementation of the DFT, as we will see in Exercise 2.37. The result, and the following results in this section, are stated in terms of the cosine matrix C_N (where the entries are $(C_N)_{n,k} = \cos\left(2\pi\frac{n}{2N}\left(k+\frac{1}{2}\right)\right)$, rather than the DCT matrix (i.e. we do not scale the rows). The reason is that C_N appears to be most practical for stating algorithms. When computing the DCT, we simply need to scale at the end, after using the statements below.

Theorem 27. Computing the DCT from a DFT.
Let $\boldsymbol{y} = C_N\boldsymbol{x}$. Then we have that

$$y_n = \left(\cos\left(\pi\frac{n}{2N}\right)\Re((DFT_N\boldsymbol{x}^{(1)})_n) + \sin\left(\pi\frac{n}{2N}\right)\Im((DFT_N\boldsymbol{x}^{(1)})_n)\right), \qquad (2.25)$$

where $\boldsymbol{x}^{(1)} \in \mathbb{R}^N$ is defined by

$$(\boldsymbol{x}^{(1)})_k = x_{2k} \text{ for } 0 \le k \le N/2 - 1$$
$$(\boldsymbol{x}^{(1)})_{N-k-1} = x_{2k+1} \text{ for } 0 \le k \le N/2 - 1,$$

Proof. Using the definition of C_N, and splitting the computation of $\boldsymbol{y} = C_N\boldsymbol{x}$ into two sums, corresponding to the even and odd indices as follows:

$$y_n = \sum_{k=0}^{N-1} x_k \cos\left(\frac{2\pi n(k+1/2)}{2N}\right)$$

$$= \sum_{k=0}^{N/2-1} x_{2k} \cos\left(\frac{2\pi n(2k+1/2)}{2N}\right) + \sum_{k=0}^{N/2-1} x_{2k+1} \cos\left(\frac{2\pi n(2k+1+1/2)}{2N}\right).$$

If we reverse the indices in the second sum, this sum becomes

$$\sum_{k=0}^{N/2-1} x_{N-2k-1} \cos\left(\frac{2\pi n(N-2k-1+1/2)}{2N}\right).$$

If we then also shift the indices with $N/2$ in this sum, we get

$$\sum_{k=N/2}^{N-1} x_{2N-2k-1} \cos\left(\frac{2\pi n(2N-2k-1+1/2)}{2N}\right)$$

$$= \sum_{k=N/2}^{N-1} x_{2N-2k-1} \cos\left(\frac{2\pi n(2k+1/2)}{2N}\right),$$

where we used that cos is symmetric and periodic with period 2π. We see that we now have the same cos-terms in the two sums. If we thus define the vector $\boldsymbol{x}^{(1)}$ as in the text of the theorem, we see that we can write

$$y_n = \sum_{k=0}^{N-1} (\boldsymbol{x}^{(1)})_k \cos\left(\frac{2\pi n(2k+1/2)}{2N}\right)$$

$$= \Re\left(\sum_{k=0}^{N-1} (\boldsymbol{x}^{(1)})_k e^{-2\pi in\left(2k+\frac{1}{2}\right)/(2N)}\right)$$

$$= \Re\left(e^{-\pi in/(2N)} \sum_{k=0}^{N-1} (\boldsymbol{x}^{(1)})_k e^{-2\pi ink/N}\right)$$

$$= \Re\left(e^{-\pi in/(2N)} (\text{DFT}_N \boldsymbol{x}^{(1)})_n\right)$$

$$= \left(\cos\left(\pi\frac{n}{2N}\right) \Re((\text{DFT}_N \boldsymbol{x}^{(1)})_n) + \sin\left(\pi\frac{n}{2N}\right) \Im((\text{DFT}_N \boldsymbol{x}^{(1)})_n)\right),$$

where we have recognized the N-point DFT. This completes the proof. $\square$

If we in the proof above define the $N \times N$-diagonal matrix Q_N by $Q_{n,n} = e^{-\pi in/(2N)}$, the result can also be written on the more compact form

$$\boldsymbol{y} = C_N\boldsymbol{x} = \Re\left(Q_N\text{DFT}_N\boldsymbol{x}^{(1)}\right).$$

We will, however, not use this form, since there is complex arithmetic involved, contrary to Eq. (2.25). Code which uses Eq. (2.25) to compute the DCT, using the function `fft_impl` from Sect. 2.3, can look as follows:

```
function y = dct_impl(x)
    N = length(x);
    if N == 1
        y = x;
    else
        x1 = [x(1:2:N, :); x(N:(-2):1, :)];
        y = fft_impl(x1, @fft_kernel_standard);
        cosvec = cos(pi*((0:(N-1))')/(2*N));
        sinvec = sin(pi*((0:(N-1))')/(2*N));
        for s2 = 1:size(x, 2)
            y(:, s2) = cosvec.*real(y(:, s2)) + sinvec.*imag(y(:, s2));
        end
        y(1, :) = sqrt(1/N)*y(1, :);
        y(2:N, :) = sqrt(2/N)*y(2:N, :);
    end
end
```

In the code, the vector $\boldsymbol{x}^{(1)}$ is created first by rearranging the components, and it is sent as input to `fft_impl`. After this we take real parts and imaginary parts, and multiply with the cos- and sin-terms in Eq. (2.25).

2.5.1 Efficient Implementations of the IDCT

As with the FFT, it is straightforward to modify the DCT implementation in order to compute the IDCT. To see how we can do this, write from Theorem 27, for $n \geq 1$

$$y_n = \left(\cos\left(\pi \frac{n}{2N} \right) \Re((\mathrm{DFT}_N \boldsymbol{x}^{(1)})_n) + \sin\left(\pi \frac{n}{2N} \right) \Im((\mathrm{DFT}_N \boldsymbol{x}^{(1)})_n) \right)$$

$$y_{N-n} = \left(\cos\left(\pi \frac{N-n}{2N} \right) \Re((\mathrm{DFT}_N \boldsymbol{x}^{(1)})_{N-n}) + \sin\left(\pi \frac{N-n}{2N} \right) \Im((\mathrm{DFT}_N \boldsymbol{x}^{(1)})_{N-n}) \right)$$

$$= \left(\sin\left(\pi \frac{n}{2N} \right) \Re((\mathrm{DFT}_N \boldsymbol{x}^{(1)})_n) - \cos\left(\pi \frac{n}{2N} \right) \Im((\mathrm{DFT}_N \boldsymbol{x}^{(1)})_n) \right), \qquad (2.26)$$

where we have used the symmetry of DFT_N for real vectors. These two equations enable us to determine $\Re((\mathrm{DFT}_N \boldsymbol{x}^{(1)})_n)$ and $\Im((\mathrm{DFT}_N \boldsymbol{x}^{(1)})_n)$ from y_n and y_{N-n}. We get

$$\cos\left(\pi \frac{n}{2N} \right) y_n + \sin\left(\pi \frac{n}{2N} \right) y_{N-n} = \Re((\mathrm{DFT}_N \boldsymbol{x}^{(1)})_n)$$

$$\sin\left(\pi \frac{n}{2N} \right) y_n - \cos\left(\pi \frac{n}{2N} \right) y_{N-n} = \Im((\mathrm{DFT}_N \boldsymbol{x}^{(1)})_n).$$

Adding we get

$$(\mathrm{DFT}_N \boldsymbol{x}^{(1)})_n = \cos\left(\pi \frac{n}{2N} \right) y_n + \sin\left(\pi \frac{n}{2N} \right) y_{N-n}$$

$$+ i(\sin\left(\pi \frac{n}{2N} \right) y_n - \cos\left(\pi \frac{n}{2N} \right) y_{N-n})$$

$$= (\cos\left(\pi \frac{n}{2N} \right) + i \sin\left(\pi \frac{n}{2N} \right))(y_n - i y_{N-n})$$

$$= e^{\pi i n/(2N)}(y_n - i y_{N-n}).$$

This means that $(\mathrm{DFT}_N \boldsymbol{x}^{(1)})_n = e^{\pi i n/(2N)}(y_n - i y_{N-n}) = (y_n - i y_{N-n})/Q_{n,n}$ for $n \geq 1$. Since $\Im((\mathrm{DFT}_N \boldsymbol{x}^{(1)})_0) = 0$ we have that $(\mathrm{DFT}_N \boldsymbol{x}^{(1)})_0 = \frac{1}{d_{0,N}} y_0 = y_0/Q_{0,0}$. This means that $\boldsymbol{x}^{(1)}$ can be recovered by taking the IDFT of the vector with component 0 being $y_0/Q_{0,0}$, and the remaining components being $(y_n - iy_{N-n})/Q_{n,n}$:

Theorem 28. IDCT Algorithm.
Let $\boldsymbol{x} = (C_N)^{-1}\boldsymbol{y}$. and let $\boldsymbol{z}$ be the vector with component 0 being $y_0/Q_{0,0}$, and the remaining components being $(y_n - iy_{N-n})/Q_{n,n}$. Then we have that

$$\boldsymbol{x}^{(1)} = IDFT_N\boldsymbol{z},$$

where $\boldsymbol{x}^{(1)}$ is defined as in Theorem 27.

The implementation of IDCT can thus go as follows:

```
function x = idct_impl(y)
    N = size(y, 1);
    if N == 1
        x = y;
    else
        y(1, :) = y(1, :)/sqrt(1/N);
        y(2:N, :) = y(2:N, :)/sqrt(2/N);
        Q = exp(-pi*1i*((0:(N-1))')/(2*N));
        y1 = zeros(size(y)); y1(1, :) = y(1, :)/Q(1);
        for s2 = 1:size(y, 2)
            y1(2:N, s2) = (y(2:N, s2)-1i*y(N:(-1):2, s2))./Q(2:N);
        end
        y1 = fft_impl(y1, @fft_kernel_standard, 0);
        x = zeros(size(y));
        x(1:2:N, :) = real(y1(1:(N/2), :));
        x(2:2:N, :) = real(y1(N:(-1):(N/2+1), :));
    end
end
```

2.5.2 Reduction in the Number of Arithmetic Operations

It turns out that the DCT and IDCT implementations give the same type of reductions in the number multiplications as the FFT and IFFT:

Theorem 29. Number of Multiplications Required by the DCT and IDCT Algorithms.
The DCT and the IDCT can be implemented using any FFT and IFFT algorithms, and with operation count of the same order as these. In particular, when the standard FFT algorithms of Sect. 2.3 are used, we obtain an operation counts of $O(5N\log_2/2)$. In comparison, the operation count for a direct implementation of the N-point DCT/IDCT is $2N^2$.

Note that we divide the previous operation counts by 2 since the DCT applies an FFT to real input only, and the operation count for the FFT can be halved when we adapt to real data, see Exercise 2.28.

Proof. By Theorem 18, the number of multiplications required by the standard FFT algorithm from Sect. 2.3 adapted to real data is $O(N\log_2 N)$, while the number of additions is $O(3N\log_2 N/2)$. By Theorem 27, two additional multiplications and one addition are required for each index (so that we have $2N$ extra real multiplications and N extra real additions in total), but this does not affect the operation count, since $O(N\log_2 N + 2N) = O(N\log_2 N)$. Since the operation counts for the IFFT is the same as for the FFT, we only need to count the additional multiplications needed in forming the vector $z = (y_n - iy_{N-n})/Q_{n,n}$. Clearly, this also does not affect the order of the algorithm. $\square$

Since the DCT and IDCT can be implemented using the FFT and IFFT, it has the same advantages as the FFT. Much literature is devoted to reducing the number of multiplications in the DFT and the DCT even further (see [28] for one of the most recent developments).

Another remark on the operation count is in order: we have not counted the operations of taking sin and cos in the DCT. The reason is that these values can be pre-computed, since we take the sine and cosine in a specific set of values. This is contrary to multiplication and addition, since these involve input values not known before runtime. Although our code does not use pre-computed such vectors, any optimized algorithm should do so.

Exercise 2.36: Trick for Reducing the Number of Multiplications with the DCT

In this exercise we will take a look at a small trick which reduces the number of multiplications we need for DCT algorithm from Theorem 27. This exercise does not reduce the order of the DCT algorithms, but we will see in Exercise 2.37 how the result can be used to achieve this.

a) Assume that x is a real signal. Equation (2.26) said that

$$y_n = \cos\left(\pi \frac{n}{2N}\right) \Re((\mathrm{DFT}_N x^{(1)})_n) + \sin\left(\pi \frac{n}{2N}\right) \Im((\mathrm{DFT}_N x^{(1)})_n)$$
$$y_{N-n} = \sin\left(\pi \frac{n}{2N}\right) \Re((\mathrm{DFT}_N x^{(1)})_n) - \cos\left(\pi \frac{n}{2N}\right) \Im((\mathrm{DFT}_N x^{(1)})_n)$$

for the n'th and $N - n$'th coefficient of the DCT. This can also be rewritten as

$$y_n = \left(\Re((\mathrm{DFT}_N x^{(1)})_n) + \Im((\mathrm{DFT}_N x^{(1)})_n)\right) \cos\left(\pi \frac{n}{2N}\right)$$
$$- \Im((\mathrm{DFT}_N x^{(1)})_n)(\cos\left(\pi \frac{n}{2N}\right) - \sin\left(\pi \frac{n}{2N}\right))$$
$$y_{N-n} = -\left(\Re((\mathrm{DFT}_N x^{(1)})_n) + \Im((\mathrm{DFT}_N x^{(1)})_n)\right) \cos\left(\pi \frac{n}{2N}\right)$$
$$+ \Re((\mathrm{DFT}_N x^{(1)})_n)(\sin\left(\pi \frac{n}{2N}\right) + \cos\left(\pi \frac{n}{2N}\right)).$$

Explain that the first two equations require 4 multiplications to compute y_n and y_{N-n}, and that the last two equations require 3 multiplications to compute y_n and y_{N-n}.

b) Explain why the trick in a) reduces the number of multiplications in a DCT, from $2N$ to $3N/2$.

c) Explain why the trick in a) can be used to reduce the number of multiplications in an IDCT with the same number.

Hint

match the expression $e^{\pi i n/(2N)}(y_n - i y_{N-n})$ you encountered in the IDCT with the rewriting you did in b).

d) Show that the penalty of the trick we here have used to reduce the number of multiplications, is an increase in the number of additions from N to $3N/2$. Why can this trick still be useful?

Exercise 2.37: An Efficient Joint Implementation of the DCT and the FFT

In this exercise we will explain another joint implementation of the DFT and the DCT, which has the benefit of a lower multiplication count, at the expense of a higher addition count. It also has the benefit that it is specialized to real vectors, with a very structured implementation (this may not be the case for the quickest FFT implementations, as one often sacrifices clarity of code in pursuit of higher computational speed). a) of this exercise can be skipped, as it is difficult and quite technical. For further details of the algorithm the reader is referred to [77].

a) Let $y = \text{DFT}_N x$ be the N-point DFT of the real vector x. Show that

$$
\Re(y_n) = \begin{cases} \Re((\text{DFT}_{N/2}x^{(e)})_n) + (C_{N/4}z)_n & 0 \le n \le N/4 - 1 \\ \Re((\text{DFT}_{N/2}x^{(e)})_n) & n = N/4 \\ \Re((\text{DFT}_{N/2}x^{(e)})_n) - (C_{N/4}z)_{N/2-n} & N/4 + 1 \le n \le N/2 - 1 \end{cases}
$$

$$
\Im(y_n) = \begin{cases} \Im((\text{DFT}_{N/2}x^{(e)})_n) & n = 0 \\ \Im((\text{DFT}_{N/2}x^{(e)})_n) + (C_{N/4}w)_{N/4-n} & 1 \le n \le N/4 - 1 \\ \Im((\text{DFT}_{N/2}x^{(e)})_n) + (C_{N/4}w)_{n-N/4} & N/4 \le n \le N/2 - 1 \end{cases}
$$

where $x^{(e)}$ is as defined in Theorem 14, where $z, w \in \mathbb{R}^{N/4}$ defined by

$$
\begin{aligned}
z_k &= x_{2k+1} + x_{N-2k-1} & 0 \le k \le N/4 - 1, \\
w_k &= (-1)^k (x_{N-2k-1} - x_{2k+1}) & 0 \le k \le N/4 - 1,
\end{aligned}
$$

Explain from this how you can make an algorithm which reduces an FFT of length N to an FFT of length $N/2$ (on $x^{(e)}$), and two DCT's of length $N/4$ (on z and w). We will call this algorithm the revised FFT algorithm.

a) says nothing about the coefficients y_n for $n > \frac{N}{2}$. These are obtained in the same way as before through symmetry. a) also says nothing about $y_{N/2}$. This can be obtained with the same formula as in Theorem 14.

Let us now compute the number of arithmetic operations our revised algorithm needs. Denote by the number of real multiplications needed by the revised N-point FFT algorithm

b) Explain from the algorithm in a) that

$$
M_N = 2(M_{N/4} + 3N/8) + M_{N/2} \qquad A_N = 2(A_{N/4} + 3N/8) + A_{N/2} + 3N/2
$$

Hint

$3N/8$ should come from the extra additions/multiplications (see Exercise 2.36) you need to compute when you run the algorithm from Theorem 27 for $C_{N/4}$. Note also that the equations in a) require no extra multiplications, but that there are six equations involved, each needing $N/4$ additions, so that we need $6N/4 = 3N/2$ extra additions.

c) Explain why $x_r = M_{2^r}$ is the solution to the difference equation

$$x_{r+2} - x_{r+1} - 2x_r = 3 \times 2^r,$$

and that $x_r = A_{2^r}$ is the solution to

$$x_{r+2} - x_{r+1} - 2x_r = 9 \times 2^r.$$

and show that the general solution to these are $x_r = \frac{1}{2}r2^r + C2^r + D(-1)^r$ for multiplications, and $x_r = \frac{3}{2}r2^r + C2^r + D(-1)^r$ for additions.

d) Explain why, regardless of initial conditions to the difference equations, $M_N = O\left(\frac{1}{2}N\log_2 N\right)$ and $A_N = O\left(\frac{3}{2}N\log_2 N\right)$ both for the revised FFT and the revised DCT. The total number of operations is thus $O(2N\log_2 N)$, i.e. half the operation count of the split-radix algorithm. The orders of these algorithms are thus the same, since we here have adapted to read data.

e) Explain that, if you had not employed the trick from Exercise 2.36, we would instead have obtained $M_N = O\left(\frac{2}{3}\log_2 N\right)$, and $A_N = O\left(\frac{4}{3}\log_2 N\right)$, which equal the orders for the number of multiplications/additions for the split-radix algorithm. In particular, the order of the operation count remains the same, but the trick from Exercise 2.36 turned a bigger percentage of the arithmetic operations into additions.

The algorithm we here have developed thus is constructed from the beginning to apply for real data only. Another advantage of the new algorithm is that it can be used to compute both the DCT and the DFT.

Exercise 2.38: Implementation of the IFFT/IDCT

In Exercise 2.37 we did not write down corresponding algorithms for the revised IFFT and IDCT algorithms. Use the equations for $\Re(y_n)$ and $\Im(y_n)$ in that exercise to show that

a)

$$\Re(y_n) - \Re(y_{N/2-n}) = 2(C_{N/4}\boldsymbol{z})_n$$
$$\Im(y_n) + \Im(y_{N/2-n}) = 2(C_{N/4}\boldsymbol{w})_{N/4-n}$$

for $1 \le n \le N/4-1$. Explain how one can compute $\boldsymbol{z}$ and $\boldsymbol{w}$ from this using two IDCT's of length $N/4$.

b)

$$\Re(y_n) + \Re(y_{N/2-n}) = 2\Re((\mathrm{DFT}_{N/2}\boldsymbol{x}^{(e)})_n)$$
$$\Im(y_n) - \Im(y_{N/2-n}) = 2\Im((\mathrm{DFT}_{N/2}\boldsymbol{x}^{(e)})_n),$$

and explain how one can compute $\boldsymbol{x}^{(e)}$ from this using an IFFT of length $N/2$.

2.6 Summary

We defined the analogs of Fourier analysis and Fourier series for vectors, i.e. Discrete Fourier analysis and the Discrete Fourier Transform. We looked at the properties of this transform and its relation to Fourier series. We also saw that the sampling theorem guaranteed that there is no loss in considering the samples of a function, as long as the sampling rate is high enough compared to the highest frequency.

We obtained an implementation of the DFT, called the FFT, with a much lower operation count (also called the *flop count*) than direct matrix multiplication. The FFT has been cited as one of the ten most important algorithms of the twentieth century [7]. The original paper [10] by Cooley and Tukey dates back to 1965, and handles the case when N is composite. In the literature, one has been interested in FFT algorithms with as low operation count as possible. The split-radix algorithm from Exercise 2.30 was by many thought to be optimal in this respect (with operation count $O(4N\log_2 N)$). Recently, however, Frigo and Johnson [28] showed that the operation count could be reduced to $O(34N\log_2(N)/9)$.

It may seem strange that the operation count includes both additions and multiplications: Aren't multiplications more time-consuming than additions? This is certainly the case when you consider how this is done mechanically, as floating point multiplication can be considered as a combination of many floating point additions. Many have in fact attempted to rewrite expressions so that the multiplication count is reduced, at the cost of more additions. Winograd's algorithm [80] is an example of this. However, most modern CPU's have more complex hardware dedicated to multiplication, so that floating point multiplication often can be performed in one cycle, just as one for addition. In other words, if we run a test program on a computer, it may be difficult to detect any differences in performance between addition and multiplication. Another important aspect of the FFT is memory use. We saw that it was possible to implement the FFT so that the output is computed into the same memory as the input, so that the FFT algorithm does not require extra memory.

We defined the DCT as a discrete version of the Fourier series for symmetric functions. We also showed how to obtain an efficient implementation of the DCT by reusing an FFT implementation. A majority of modern image and video coding standards, including the MP3 standard, are based on the DCT or the MDCT (Exercise 2.35). The MDCT was introduced in [58], and is used to obtain a higher spectral resolution of sound in the more advanced version of the MP3 standard (layer III), as well in more recent standards such as AC-3. The MP3 standard document [26] does not go into the theory of the MDCT, it only presents what is needed in order for an implementation.

Libraries exist which go into lengths to provide efficient implementation of the FFT and the DCT. FFTW, short for *Fastest Fourier Transform in the West* [22], is perhaps the best known of these.

The sampling theorem is also one of the most important results of the last century. It was discovered by Nyquist and Shannon [66], but also by others independently. Usually it is not presented in a finite-dimensional setting (but see [20, p. 15]). Much more theory on sampling exists, in particular for the case when samples are not taken uniformly, or when there is some kind on non-linearity built into the sampling process. For a recent review of such results, see [18].

At the end of Chap. 1 we mentioned that the Continuous time Fourier transform was the parallel to Fourier series for non-periodic functions. Similarly, the *Discrete time Fourier transform* (DTFT) is the parallel to the DFT for non-periodic vectors. We will encounter this in the next chapter. The general version of the sampling theorem in proved in the literature using an argument which combines the DTFT and the CTFT. We will go through this in the summary of Chap. 6.

What You Should Have Learned in This Chapter

- The definition of the Fourier basis and its orthonormality.
- The definition of the Discrete Fourier Transform as a change of coordinates to the Fourier basis, its inverse, and its unitarity.
- Properties of the DFT, such as conjugate symmetry when the vector is real, how it treats delayed vectors, or vectors multiplied with a complex exponential.
- Translation between DFT index and frequency. In particular DFT indices for high and low frequencies.
- How one can use the DFT to adjust frequencies in sound.
- How the FFT algorithm works by splitting into two FFT's of half the length. Simple FFT implementation.
- Reduction in the operation count with the FFT.

Chapter 3
Discrete Time Filters

In Sect. 1.5 we defined filters as operations in continuous time which preserved frequencies. Such operations are important since they can change the frequency content in many ways. They are difficult to use computationally, however, since they are defined for all instances in time. This will now be addressed by changing focus to *discrete-time*. Filters will now be required to operate on a possibly infinite vector $\boldsymbol{x} = (x_n)_{n=-\infty}^{\infty}$ of values, corresponding to concrete instances in time. Such filters will be called *discrete time filters*, and they too will be required to be frequency preserving. We will see that discrete time filters make analog filters computable, similarly to how the DFT made Fourier series computable in Chap. 2. The DFT will be a central ingredient also now.

For the remainder of the book capital letter S will usually denote discrete time filters, contrary to lowercase letter s used for analog filters previously. Also, $\boldsymbol{x}$ will denote the *input vector* to a discrete time filter, and $\boldsymbol{z}$ the corresponding *output vector*, so that $\boldsymbol{z} = S(\boldsymbol{x})$. A discrete-time filter can also be viewed as a (possibly infinite) matrix, and we will therefore mostly write $\boldsymbol{z} = S\boldsymbol{x}$ instead of $\boldsymbol{z} = S(\boldsymbol{x})$.

In the next sections we will go through the basics of discrete time filters. They are defined in the frequency domain, and we will use mappings to jump between time and frequency in order to find expressions for filters in the time domain. The analysis will be split in two. First we specialize to periodic input, as we did for analog filters in Sect. 1.5, for which the DFT will turn out to be a mapping between time and frequency. Moving to the general case of non-periodic input, we will need a generalization of the notion of a frequency decomposition. Using this we will establish the Discrete time Fourier Transform, or DTFT, which also will serve as a mapping between time and frequency. The last sections of the chapter deal with important special cases of discrete time filters.

3.1 Discrete Time Filters on Periodic Vectors

If $\boldsymbol{x}$ is N-periodic it can be represented as a vector $(x_0, x_1, \ldots, x_{N-1})$ in $\mathbb{R}^N$, and we know that the set of N-periodic vectors is spanned by the Fourier basis. The frequency

© Springer Nature Switzerland AG 2019
Ø. Ryan, *Linear Algebra, Signal Processing, and Wavelets - A Unified Approach*,
Springer Undergraduate Texts in Mathematics and Technology,
https://doi.org/10.1007/978-3-030-01812-2_3

preserving requirement for a filter on periodic vectors is then that, for each $0 \leq n < N$, there exists a value $\lambda_{S,n}$ so that

$$S(e^{2\pi i k n/N}) = \lambda_{S,n} e^{2\pi i k n/N}. \tag{3.1}$$

The frequency response thus takes the form of a vector

$$(\lambda_{S,0}, \lambda_{S,1}, \ldots, \lambda_{S,N-1}),$$

which we will denote by $\boldsymbol{\lambda}_S$. From Eq. (3.1) it immediately follows that the output from a filter is N-periodic when the input is, so that the restriction to N-periodic vectors can be represented by the linear transformation

$$(x_0, x_1, \ldots, x_{N-1}) \to (z_0, z_1, \ldots, z_{N-1}).$$

Thus, S corresponds to an $N \times N$-matrix. By definition the Fourier basis is an orthonormal basis of eigenvectors for S, and since F_N^H is the unitary matrix where the columns are ϕ_n, it follows that S is a filter if and only if it can be diagonalized as

$$S = F_N^H D F_N, \tag{3.2}$$

with D diagonal and having the eigenvalues/frequency response $\boldsymbol{\lambda}_S$ on the diagonal.[1] From this it is also straightforward to see how the matrix S must look. We have that

$$S_{k,l} = \frac{1}{N} \sum_n e^{2\pi i k n/N} \lambda_{S,n} e^{-2\pi i l n/N} = \frac{1}{N} \sum_n \lambda_{S,n} e^{2\pi i (k-l)n/N}.$$

This expression clearly depends only on $(k-l) \bmod N$, with $\bmod$ denoting the remainder modulo N, so that S must be constant along each diagonal. Such matrices are called Toeplitz matrices. Matrices where $S_{k,l}$ only depends on $(k-l) \bmod N$ are also called circulant Toeplitz matrices, circulant meaning that each row and column of the matrix 'wraps over' at the edges:

Definition 1. *Toeplitz Matrices.*

An $N \times N$-matrix S is called a *Toeplitz matrix* if its entries are constant along each diagonal. A Toeplitz matrix is said to be circulant if in addition

$$S_{(k+s) \bmod N,(l+s) \bmod N} = S_{k,l}$$

for all k, l in $[0, N-1]$, and all s.

Toeplitz matrices are very popular in the literature and have many applications. A Toeplitz matrix is uniquely identified by the values on its diagonals, and a circulant Toeplitz matrix is uniquely identified by the values on the main diagonal, and on the diagonals under it. These correspond to the values in the first column. Denoting this column by $\boldsymbol{s}$, we have that

$$s_k = S_{k,0} = \frac{1}{N} \sum_n \lambda_{S,n} e^{2\pi i k n/N}.$$

This is the expression for an IDFT, so that we have proved the following.

---

[1] Recall that the orthogonal diagonalization of S takes the form $S = PDP^T$, where P contains as columns an orthonormal set of eigenvectors, and D is diagonal with the eigenvectors listed on the diagonal (see Section 7.1 in [32]).

Theorem 2. Connection Between Frequency Response and the Matrix.

If s is the first column of a circulant Toeplitz matrix S, then the frequency response is

$$\lambda_S = DFT_N s. \tag{3.3}$$

Conversely,

$$s = IDFT_N \lambda_S. \tag{3.4}$$

Since λ_S is the vector of eigenvalues of S, this result states that there is no need to find the roots of the characteristic polynomial for circulant Toeplitz matrices: Only a simple DFT is required to find the eigenvalues.

In signal processing, the frequency content of a vector (i.e. its DFT) is also referred to as its *spectrum*. This may be somewhat confusing from a linear algebra perspective, where the spectrum denotes the eigenvalues of a matrix. But due to Theorem 2 this is not so confusing after all if we interpret the spectrum of a vector as the spectrum (i.e. eigenvalues) of the corresponding filter.

We can also write

$$(S\boldsymbol{x})_n = \sum_{k=0}^{N-1} S_{n,k} x_k = \sum_{k=0}^{N-1} S_{k,0} \breve{x}_{n-k} = \sum_{k=0}^{N-1} s_k \breve{x}_{n-k},$$

where $\breve{x}$ is the periodic extension of $\boldsymbol{x} \in \mathbb{R}^N$. When $\boldsymbol{s} \in \mathbb{R}^N$ and $\boldsymbol{x} \in \mathbb{R}^N$, the vector with components $\sum_{k=0}^{N-1} s_k \breve{x}_{n-k}$ is also called the *circular convolution* of $\boldsymbol{s}$ and $\boldsymbol{x}$, and is denoted $\boldsymbol{s} \circledast \boldsymbol{x}$. Thus, any filter restricted to periodic vectors can be written as a circular convolution. If $\boldsymbol{s}$ and $\boldsymbol{x}$ are uniform samples of functions g and f on $[0, T]$, $\boldsymbol{s} \circledast \boldsymbol{x}$ can be thought of as a Riemann sum for the integral $\int_0^T g(u) f(t - u) du$, which in Sect. 1.5 was defined as the circular convolution of f and g. In the exercises we will explore some further aspects of circular convolution, in particular its usefulness in computing the DFT.

Including the general time-invariance property, we can summarize the properties for filters on periodic vectors as follows.

Theorem 3. Characterizations of Filters on Periodic Vectors.

The following are equivalent for an $N \times N$-matrix S:

1. *S is a filter,*
2. *$S = (F_N)^H D F_N$ for a diagonal matrix D,*
3. *S is a circulant Toeplitz matrix,*
4. *S is linear and time-invariant,*
5. *$S\boldsymbol{x} = \boldsymbol{s} \circledast \boldsymbol{x}$ for all $\boldsymbol{x}$, and where $\boldsymbol{s}$ is the first column of S.*

Due to this result and Proposition 26, it follows that also the set of circulant Toeplitz matrices is a commutative algebra. A circulant Toeplitz matrix is non-singular if and only if 0 is not an eigenvalue, i.e. if and only if all $\lambda_{S,n} \neq 0$. If the inverse exists, it is clear that it too is circulant Toeplitz, since then $S^{-1} = (F_N)^H D^{-1} F_N$.

Proof. In the above discussion we have covered that 1. implies all the other four. Also, 2. clearly implies 1. Noting that the set of filters have dimension N, as well as the set of circulant Toeplitz matrices (since they are characterized by the values in the first column), 3. implies 1. Also 5. clearly implies 3. It only remains to establish that time invariance implies any of the other statements. If $Se_0 = s$, time-invariance implies that Se_d equals s delayed (cyclically) with d entries. But since Se_d is column d in S, this implies that S is a circulant Toeplitz matrix. □

Now, let us denote by $v_1 \circ v_2$ the pointwise product of two vectors, i.e.

$$(v_1 \circ v_2)_n = (v_1)_n (v_2)_n.$$

For periodic vectors, "convolution in time corresponds to multiplication in frequency", as established in Sect. 1.5, means that $\lambda_{S_1 S_2} = \lambda_{S_1} \circ \lambda_{S_2}$ whenever S_1 and S_2 are circulant Toeplitz matrices. Combining this with Theorem 2, and noting that the first column of $S_1 S_2$ is $s_1 \circledast s_2$, the following is clear.

Proposition 4. *We have that*

$$DFT_N(s_1 \circledast s_2) = (DFT_N s_1) \circ (DFT_N s_2)$$

When S is sparse, it can be used efficiently for computing: If S has k nonzero diagonals, the matrix has Nk nonzero entries, so that kN multiplications and $(k-1)N$ additions are required for matrix multiplication. This is much less than the N^2 multiplications and $(N-1)N$ additions required for general matrix multiplication. For this reason we will have the primary focus on such filters.

Example 3.1: Comparison with Direct Computation of Eigenvalues

Consider the simple matrix

$$S = \begin{pmatrix} 4 & 1 \\ 1 & 4 \end{pmatrix}.$$

It is straightforward to compute the eigenvalues and eigenvectors of this matrix the way you learned in your first course in linear algebra. However, this is also a circulant Toeplitz matrix, so that we can apply a DFT to the first column to obtain the eigenvalues. The Fourier basis vectors for $N = 2$ are $(1,1)/\sqrt{2}$ and $(1,-1)/\sqrt{2}$, which also are the eigenvectors of S. The eigenvalues are

$$DFT_N s = \begin{pmatrix} 1 & 1 \\ 1 & -1 \end{pmatrix} \begin{pmatrix} 4 \\ 1 \end{pmatrix} = \begin{pmatrix} 5 \\ 3 \end{pmatrix}$$

The eigenvalues are thus 5 and 3. A computer may not return the eigenvectors exactly as the Fourier basis vectors, since the eigenvectors are unique only up to a scalar: The computer may for instance switch the signs. We have no control over what the computer chooses to do, since some underlying numerical algorithm is used, which we can't influence.

Exercise 3.2: Some Linear Operations Which Are Not Filters

a) In Example 1.2 we looked at time reversal as an operation on digital sound. In $\mathbb{R}^N$ this can be defined as the linear mapping which sends the vector e_k to e_{N-1-k} for all $0 \leq k \leq N - 1$. Write down the matrix for time reversal, and explain from this why time reversal is not a filter.

b) Consider the linear mapping S which keeps every second component in $\mathbb{R}^N$, i.e. $S(e_{2k}) = e_{2k}$, and $S(e_{2k-1}) = \mathbf{0}$. Write down the matrix for this, and explain why this operation is not a filter.

Exercise 3.3: Eigenvectors and Eigenvalues

Consider the matrix

$$S = \begin{pmatrix} 4 & 1 & 3 & 1 \\ 1 & 4 & 1 & 3 \\ 3 & 1 & 4 & 1 \\ 1 & 3 & 1 & 4 \end{pmatrix}.$$

a) Compute the eigenvalues and eigenvectors of S using the results of this section. You should only need to perform one DFT in order to achieve this.

b) Verify the result from a) by computing the eigenvectors and eigenvalues the way you taught in your first course in linear algebra. This should be a much more tedious task.

c) Use a computer to find the eigenvectors and eigenvalues of S also. For some reason some of the eigenvectors seem to be different from the Fourier basis vectors, which you would expect from the theory in this section. Try to find an explanation for this.

Exercise 3.4: Connection Between Continuous Time- and Discrete Time Filters

Let s be a continuous time filter with frequency response $\lambda_s(\nu)$, and assume that $f \in V_{M,T}$ (so that also $g = s(f) \in V_{M,T}$). Let

$$x = (f(0 \cdot T/N), f(1 \cdot T/N), \ldots, f((N-1)T/N))$$
$$z = (g(0 \cdot T/N), g(1 \cdot T/N), \ldots, g((N-1)T/N))$$

be vectors of $N = 2M + 1$ uniform samples from f and g. Show that the operation $S : x \to z$ (i.e. the operation which sends the samples of the input to the samples of the output) is well-defined on $\mathbb{R}^N$, and corresponds to a filter with frequency response $\lambda_{S,n} = \lambda_s(n/T)$.

This explains how we can implement the restriction of a continuous time filter to the Fourier spaces, by means of sampling, a discrete time filter, and interpolation. This also states that any discrete time filter can be expressed in terms of a continuous time filter.

Exercise 3.5: Replacing a Circular Convolution with a Longer One

Let $s, x \in \mathbb{R}^N$. For $M \geq 2N - 1$ define the vectors

$$\tilde{s} = (s_0, s_1, \ldots, s_{N-1}, 0, \ldots, 0, s_1, \ldots, s_{N-1})^T$$
$$\tilde{x} = (x_0, x_1, \ldots, x_{N-1}, 0, \ldots, 0)^T$$

in $\mathbb{R}^M$ (in $\tilde{s}$, $M - (2N - 1)$ zeros have been inserted in the middle). Show that the first N elements of $\tilde{s} \circledast \tilde{x}$ equal $s \circledast x$, so that a circular convolution can be computed as part of a longer circular convolution. A good strategy may now be to choose M as a power of two, since circular convolution can be computed with the help of the DFT, and since there are efficient DFT algorithms when the number of points is a power of two.

Exercise 3.6: Replacing a Sparse Circular Convolution with Shorter Ones

Let $s, x \in \mathbb{R}^N$, and assume that s has at most k nonzero entries, gathered in one segment.

a) Show that there exists a $\tilde{s} \in \mathbb{R}^{M+k-1}$ and a number a so that, for any r,

$$(s \circledast x)_{r,r+1\ldots,r+M-1} = (\tilde{s} \circledast \tilde{x})_{0,\ldots,M-1},$$

where

$$\tilde{x} = (x_{a+r}, x_{a+r+1}, \ldots, x_{a+r+M+k-2}).$$

In other words, any set of M consecutive entries in $s \circledast x$ can be obtained from a circular convolution of size $M + k - 1$. Thus, a circular convolution of size N can be computed as $\frac{N}{M}$ circular convolutions of size $M + k - 1$.

b) It is natural to implement the shorter circular convolutions in terms of the FFT. Use this strategy when $M + k - 1 = 2^r$ to find the number of operations needed to compute the $\frac{N}{M}$ circular convolutions from a), and write a program which finds the r which gives the lowest number of operations. Create a plot where k is plotted against the number of operations, and where the number of operations corresponding to the optimal r and also against the number of operations required by a direct implementation of circular convolution.

Exercise 3.7: Dualities Between Circular Convolution and Multiplication

Show that

$$IDFT_N(\boldsymbol{\lambda}_1 \circ \boldsymbol{\lambda}_2) = (IDFT_N\boldsymbol{\lambda}_1) \circledast (IDFT_N\boldsymbol{\lambda}_2)$$

$$DFT_N(\boldsymbol{s}_1 \circ \boldsymbol{s}_2) = \frac{1}{N}(DFT_N\boldsymbol{s}_1) \circledast (DFT_N\boldsymbol{s}_2)$$

$$IDFT_N(\boldsymbol{\lambda}_1 \circledast \boldsymbol{\lambda}_2) = N(IDFT_N\boldsymbol{\lambda}_1) \circ (IDFT_N\boldsymbol{\lambda}_2)$$

Note that there is a factor in front of the last two equations, so that one must be a bit careful here.

Exercise 3.8: Finding Sparse Solutions to Under-Determined Linear Systems

Let $\boldsymbol{x} \in \mathbb{C}^N$ be a vector where we know that at most s components are nonzero. When s is small such vectors are called *sparse*. Denote the indices of these nonzero components by S, and let $\boldsymbol{y} = DFT_N\boldsymbol{x}$. In this exercise we will find a procedure for finding $\boldsymbol{x}$ from the values $y_0, \ldots, y_{2s-1}$, i.e. we can recover all components of a sparse vector from an under-determined set of measurements. This procedure is also called a *Reed-Solomon code*. In some sense this parallels the sampling theorem (which in a similar way fills in the values between the samples), with an assumption on the highest frequency replaced by an assumption on sparsity. Many other results also exist on how to recover sparse vectors, and these results give rise to the field of *compressed sensing*. See [20] for a review on this field.

a) Let $\boldsymbol{z}$ be the vector with components $z_k = \frac{1}{N}\prod_{n \in S}\left(1 - e^{-2\pi in/N}e^{2\pi ik/N}\right)$. Show that $z_k = 0$ for $k \in S$, and also that $\boldsymbol{x} \circ \boldsymbol{z} = \boldsymbol{0}$.

b) Let $\boldsymbol{w} = DFT_N\boldsymbol{z}$. Show that $w_0 = 1$ and that $w_n = 0$ for $n > s$. Conclude from this, that $\boldsymbol{x} \circ \boldsymbol{z} = \boldsymbol{0}$, and due the previous exercise that $w_1, w_2, \ldots, w_s$ must fulfill the equations

$$
\begin{aligned}
y_s &+ w_1 y_{s-1} + \cdots + w_s y_0 &= 0 \\
y_{s+1} &+ w_1 y_s &+ \cdots + w_s y_1 &= 0 \\
\vdots \quad & \qquad \vdots \qquad \ddots \qquad \vdots \quad &= 0 \\
y_{2s-1} &+ w_1 y_{2s-2} + \cdots + w_s y_{s-1} &= 0
\end{aligned}
$$

We now have s equations with s unknowns. By construction this has a solution, but there may not be a unique solution. In the last part of the exercise we will show how to find $\boldsymbol{x}$, regardless of which solution we choose.

c) Assume that $v_1, v_2, \ldots, v_s$ is any solution to the system in b), and extend this to a vector $\boldsymbol{v} \in \mathbb{R}^N$ by setting $v_0 = 1$, $v_{s+1} = \cdots = v_{N-1} = 0$. Show that $(\boldsymbol{v} \circledast \boldsymbol{y})_n = 0$ for $s \leq n \leq 2s-1$, and conclude from the previous exercise that $(DFT_N((IDFT_N\boldsymbol{v}) \circ \boldsymbol{x}))_n = 0$ for $s \leq n \leq 2s - 1$.

The vector $(IDFT_N v) \circ x$ has at most s nonzero components (since x has). If we take the columns from DFT_N with indices in S, and the rows with indices between s and $2s - 1$, the resulting $s \times s$-matrix was shown in Exercise 2.17 to be non-singular.

d) Explain that $((IDFT_N v) \circ x)_n = 0$ for $n \in S$, and conclude that $(IDFT_N v) \circ x = \mathbf{0}$.

e) In particular we must have that $(IDFT_N v)_k = 0$ for $k \in S$, so that

$$\frac{1}{N} \sum_{n=0}^{N-1} v_n e^{2\pi i k n / N} = \frac{1}{N} \sum_{n=0}^{s} v_n e^{2\pi i k n / N} = 0$$

for $k \in S$. This is a polynomial in $e^{2\pi i k / N}$ of degree at most s. Explain how one can find the set S from this polynomial.

f) Explain how you can recover x from $y_0, \dots, y_{2s-1}$ when the set S is known.

g) Write a function

```
recover_x(y, N)
```

which uses the procedure obtained by combining b), e), and f) to compute and return $x \in \mathbb{R}^N$ under the assumption that x has at most s nonzero components, and that y contains the first $2s$ components in $y = DFT_N x$. Test that the code works correctly on a vector x of your own choosing.

Exercise 3.9: Windows

We mentioned in Sect. 2.2 that we obtain some undesirable effects in the frequency representation when we restrict to a block of the signal. Assume that $x \in \mathbb{R}^M$, and that

$$w = \{w_0, \dots, w_{N-1}\} \text{ with } N < M.$$

We call $(w_0 x_0, \dots, w_{N-1} x_{N-1}, 0, \dots, 0) \in \mathbb{R}^M$ a *windowed signal* and $w \in \mathbb{R}^N$ a *window* of length N.

a) Use Exercise 3.7 to show that the DFT of the windowed signal is

$$\frac{1}{M} (DFT_M (w_0, \dots, w_{N-1}, 0, \dots, 0)) \circledast (DFT_M x)$$

If $DFT_M (w_0, w_1, \dots, w_{N-1}, 0, \dots, 0)$ is close to $(M, 0, \dots, 0)$, this will be close to $y = DFT_M x$. In other words, a good window should satisfy

$$DFT_M (w_0, w_1, \dots, w_{N-1}, 0, \dots, 0) \approx (M, 0, \dots, 0).$$

There is a loss when we pass from the signal to the windowed signal, since we can only construct a DFT close to, not equal to, $(M, 0, \dots, 0)$. We will not go into techniques for how to find values w_i which are close to satisfying this, only evaluate three such in the rest of the exercise.

b) The *rectangular window* is defined by $\boldsymbol{w} = \{\underline{1}, 1, \ldots, 1\}$. Show that

$$(DFT_M(1, \ldots, 1, 0, \ldots, 0))_n = \sum_{k=0}^{N-1} e^{-2\pi i k n/M} = \frac{1 - e^{-2\pi i n N/M}}{1 - e^{-2\pi i n/M}},$$

and use this to check whether $DFT_M(w_0, w_1, \ldots, w_{N-1}, 0, \ldots, 0) \approx (M, 0, \ldots, 0)$

c) The *Hamming window* is defined by

$$w_n = 2(0.54 - 0.46 \cos(2\pi n/(N-1))).$$

Make a plot of $DFT_M(w_0, w_1, \ldots, w_{N-1}, 0, \ldots, 0)$, and compare with $(M, 0, \ldots, 0)$ to see if the Hamming window is a good window.

d) The *Hanning window* is defined by

$$w_n = 1 - \cos(2\pi n/(N-1)).$$

Repeat the analysis you did above for the rectangular and Hamming windows, for a Hanning window for $N = 32$.

The Hanning window is used in the MP3 standard, where it is applied to blocks which overlap, contrary to the non-overlapping block pattern we have used. After the Hanning window has been applied, the MP3 standard applies an FFT to the windowed signal in order to make a frequency analysis of that part of the sound.

A good survey on the windows mentioned here, as well as other ones, can be found in [25]. One lesson to be learned from this exercise is that the windowed signal may not give a good frequency representation. One must therefore be careful when splitting a sound into blocks, as this alters the frequency representation. You can use the function `forw_comp_rev_DFT` from Example 2.5 to experiment with this. This function accepts a named parameter N, which can be used to split the DFT of a sound into blocks of length N, and eliminate low frequencies before taking an IDFT to reconstruct and listen to the new sound (see also Example 3.31).

Exercise 3.10: *The Multiplicative Group of Integers Modulo N*

In number theory it is known that, when N is a prime number, the numbers $1, \ldots, N-1$ is a *group* under multiplication modulo N. In particular this means that for each $1 \leq a \leq N-1$ there exists a unique $1 \leq b \leq N-1$ so that $ab = 1 \mod N$ (the number 1 acts as a *unit*). b us then called the inverse of a, and we write $b = a^{-1}$ (this is not the same as $1/a$!). Also, it is known that there exists an integer g so that the set $\{g^q \mod N\}_{q=0}^{N-2}$ constitutes all numbers $1, \ldots, N-1$ (the order these numbers appear in may have changed, though). g is called a *generator* for the group.

a) Find generators for the multiplicative group of integers modulo 11, 23 and 41. Are the generators for each group unique?

b) Write a function

```
reorder(x, g)
```

which, for a vector $\boldsymbol{x}$ of length N, and a generator g for the numbers $1, \ldots, N-1$, returns the vector $(x_{g^0}, x_{g^1}, \ldots, x_{g^{N-2}})$.

In the next exercise you will see how the multiplicative group of integers modulo N relates to circular convolution.

Exercise 3.11: The FFT Algorithm When N Is Prime: Rader's Algorithm

It is possible to show that, if g is a generator, then g^{-1} is also a generator. This means that $\{g^{-p} \mod N\}_{p=0}^{N-2}$ also constitute the numbers $1, \ldots, N-1$ (where g^{-p} is defined as $(g^{-1})^p$).

Replacing $n = g^{-p}$ and $k = g^q$ in the FFT algorithm with N prime, we can rephrase it as (we pull $k = 0$ out of the sum, since the number 0 is not an element in our group)

$$y_0 = \sum_{k=0}^{N-1} x_k$$

$$y_{g^{-p}} = x_0 + \sum_{q=0}^{N-2} x_{g^q} e^{-2\pi i g^{-(p-q)}/N}, \ 0 \le p \le N - 2.$$

Define

$$\boldsymbol{a} = (x_{g^0}, x_{g^1}, \ldots, x_{g^{N-2}}) \in \mathbb{R}^{N-1}$$
$$\boldsymbol{b} = (e^{-2\pi i g^0/N}, e^{-2\pi i g^{-1}/N}, \ldots, e^{-2\pi i g^{-(N-2)}/N}) \in \mathbb{R}^{N-1}$$
$$\boldsymbol{c} = (y_{g^0}, y_{g^{-1}}, \ldots, y_{g^{-(N-2)}}) \in \mathbb{R}^{N-1}$$

Explain that $\boldsymbol{c} = x_0 + \boldsymbol{a} \circledast \boldsymbol{b}$, where x_0 is added to every component in the vector.

This explains how to compute an N-point DFT (with N prime) using $(N-1)$-point circular convolution. Since a circular convolution can be computed using a DFT/IDFT of the same size, so this method effectively reduces an N-point DFT to an $(N-1)$-point DFT. Since $N - 1$ is not prime, we can use the algorithm for the FFT for composite N, to reduce the problem to DFT's of lengths being smaller prime numbers. Another possibility is to use Exercise 3.5 to replace the circular convolution with a longer one, of length a power of two, effectively reducing the problem to a DFT of length a power of two.

3.2 General Discrete Time Filters

Assume now that the input to a filter is not periodic. An example of such a vector is $e^{i\omega n}$ with ω being irrational. For a general ω, $e^{i\omega n}$ is called a *discrete-time frequency*, and ω is called *angular frequency*. Since $e^{i\omega n}$ is 2π-periodic in ω, angular frequencies are usually restricted to $[0, 2\pi)$ or $[-\pi, \pi)$. A general discrete time filter is required to preserve all angular frequencies, i.e.

$$S e^{i\omega n} = \lambda_S(\omega) e^{i\omega n}, \tag{3.5}$$

i.e. the frequency response is now denoted $\lambda_S(\omega)$. The advantage of restricting ω to $[-\pi, \pi)$ is that low frequencies (i.e. $\omega \approx 0$) are located together (both the positive and negative ω). High frequencies on the other hand (i.e. $|\omega| \approx \pi$) account for two segments of $[-\pi, \pi)$ (see also Observation 10).

Note that the Fourier basis can be considered as a subset of the $e^{i\omega n}$, with $\omega = \{2\pi n/N\}_{n=0}^{N-1}$, and that a filter with frequency response $\lambda_S(\omega)$ can be considered as a filter on periodic vectors with frequency response $\lambda_{S,n} = \lambda_S(2\pi n/N)$.

Let us argue, without addressing to what extent it holds, that the $e^{i\omega n}$ can be used as building blocks for many infinite vectors $(x_k)_{k=-\infty}^{\infty}$. Any vector with a finite number of nonzero components can in particular be visualized as the Fourier coefficients of a unique function f in some Fourier space $V_{M,T}$ (choose M so high that $[-M, M]$ contains all nonzero components of $\boldsymbol{x}$). If we set $T = 2\pi$ the vector can thus be obtained from the formula for the Fourier coefficients, i.e.

$$x_n = \frac{1}{2\pi} \int_0^{2\pi} f(t)e^{-int}dt.$$

Setting $\omega = -t$, the integral can be seen as a limit of Riemann sums of the $e^{in\omega}$, with contribution of angular frequency ω being $f(-\omega) = \sum_{k=-M}^{M} x_k e^{-ik\omega}$ (since the x_n are the Fourier coefficients of f). Thus,

$$\hat{\boldsymbol{x}}(\omega) = \sum_{k=-M}^{M} x_k e^{-ik\omega} \tag{3.6}$$

can be interpreted as the frequency decomposition of $\boldsymbol{x}$ into angular frequencies. Equation (3.6) is called the *Discrete-time Fourier Transform*, or DTFT, of $\boldsymbol{x}$. Using our results on Fourier series, its inverse is clearly

$$x_n = \frac{1}{2\pi} \int_0^{2\pi} \hat{\boldsymbol{x}}(\omega)e^{in\omega}d\omega, \tag{3.7}$$

and is called the *inverse DTFT*, or IDTFT. The DTFT and IDTFT serve as transformations between the time- and frequency domains for general vectors, just like the DFT and IDFT for periodic vectors, and we see that their forms are similar. Since the DTFT and IDFT essentially are deduced by turning the Fourier series relationship between periodic functions and vectors around, properties for the DTFT can be established from the properties of Fourier series in Theorem 17. One example is the relationship between delay in one domain and multiplication with a complex exponential in the other domain.

Using the DTFT and the IDTFT as bridges between time and frequency, we can find a general expression for a discrete time filter in the time domain, similarly to how we did this for filters on periodic vectors in Sect. 3.1. We simply combine Eqs. (3.6), (3.5), and (3.7) as follows.

$$Sx = S\left(\left(\frac{1}{2\pi} \int_0^{2\pi} \hat{\boldsymbol{x}}(\omega)e^{in\omega}d\omega\right)_{n=-\infty}^{\infty}\right)$$

$$= \left(\frac{1}{2\pi} \int_0^{2\pi} \lambda_S(\omega)\hat{\boldsymbol{x}}(\omega)e^{in\omega}d\omega\right)_{n=-\infty}^{\infty}$$

$$= \left(\frac{1}{2\pi} \int_0^{2\pi} \lambda_S(\omega)\left(\sum_k x_k e^{-ik\omega}\right)e^{in\omega}d\omega\right)_{n=-\infty}^{\infty}$$

$$= \left(\sum_k \left(\frac{1}{2\pi} \int_0^{2\pi} \lambda_S(\omega) e^{i(n-k)\omega} d\omega \right) x_k \right)_{n=-\infty}^{\infty}$$

$$= \left(\sum_k t_{n-k} x_k \right)_{n=-\infty}^{\infty},$$

where we defined

$$t_k = \frac{1}{2\pi} \int_0^{2\pi} \lambda_S(\omega) e^{ik\omega} d\omega. \tag{3.8}$$

We have also moved S inside the integral between the first and second lines, replacing S with the frequency response. Again this is justified by viewing the integral as the limit of a Riemann sum of angular frequencies. This motivates the following definition.

Definition 5. By the *convolution* of t and x, denoted $t * x$, we mean the vector z with components

$$z_n = \sum_k t_k x_{n-k}. \tag{3.9}$$

The t_k are called the *filter coefficients* of S.

Any filter can thus be written as a convolution. Clearly a convolution can be viewed as an infinite Toeplitz matrix without circulation in the entries. Transposing an infinite matrix can be defined as for finite matrices, and this leads to another filter, denoted S^T, obtained by mirroring the filter coefficients about zero. We will write these matrices as

$$S = \begin{pmatrix} \ddots & \cdots & \cdots & \cdots & \ddots \\ \vdots & t_0 & t_{-1} & t_{-2} & \vdots \\ \vdots & t_1 & \underline{t_0} & t_{-1} & \vdots \\ \vdots & t_2 & t_1 & t_0 & \vdots \\ \ddots & \cdots & \cdots & \cdots & \ddots \end{pmatrix} \qquad S^T = \begin{pmatrix} \ddots & \cdots & \cdots & \cdots & \ddots \\ \vdots & t_0 & t_1 & t_2 & \vdots \\ \vdots & t_{-1} & \underline{t_0} & t_1 & \vdots \\ \vdots & t_{-2} & t_{-1} & t_0 & \vdots \\ \ddots & \cdots & \cdots & \cdots & \ddots \end{pmatrix},$$

where we have underlined the entry with index $(0,0)$, to avoid misunderstandings on the placements of the t_k. We will follow this convention for infinite vectors and matrices in the following.

Due to Eq. (3.8) the vector t is the inverse DTFT of the frequency response, so that also the frequency response is the DTFT of t, i.e.

$$\lambda_S(\omega) = \sum_k t_k e^{-ik\omega}, \tag{3.10}$$

In particular $\lambda_S(2\pi n/N) = \sum_k t_k e^{-2\pi ink/N}$, which parallels

$$\lambda_S(2\pi n/N) = \sum_{k=0}^{N-1} s_k e^{-2\pi ink/N}$$

from the last section.

Since the frequency response can be written as a DTFT, the following hold.

Theorem 6. Properties of the Frequency Response.
Assume that S is a filter with real coefficients. We have that

- $\lambda_S(-\omega) = \overline{\lambda_S(\omega)}$.
- *If the frequency response of S is $\lambda_S(\omega)$, then the frequency response of S^T is $\overline{\lambda_S(\omega)}$.*
- *If S is symmetric, $\lambda_S(\omega)$ is real. Also, if S is antisymmetric (i.e. $t_{-k} = -t_k$ for all k), $\lambda_S(\omega)$ is purely imaginary.*

Proof. Properties 1. and 3. follow in the same way as in the proof of Theorem 7. Since transposing the matrix corresponds to mirroring the filter coefficients about 0, Property 2. follows easily from a DTFT. □

For the filter (3.9) we will use the compact notation

$$S = \{\ldots, t_{-2}, t_{-1}, \underline{t_0}, t_1, t_2, \ldots\},$$

where the coefficient with index 0 again has been underlined. This will simplify notation in many cases. There may be an infinite number of nonzero filter coefficients t_k. The sum (3.9) is then infinite, and may diverge. We will, however, mostly assume that there is a finite number of nonzero t_k. Such filters are called *Finite Impulse Response filters*, or simply *FIR filters*. In signal processing the *impulse response* is commonly referred to as the output resulting from the input e_0, also called an impulse. This is clearly the vector $\boldsymbol{t} = (t_n)_{n=-\infty}^{\infty}$. If S is a FIR filter and $k_{\min}$, $k_{\max}$ are the smallest and biggest k so that $t_k \neq 0$, i.e. so that

$$z_n = \sum_{k=k_{min}}^{k_{max}} t_k x_{n-k}, \qquad (3.11)$$

we will also write

$$S = \{t_{k_{min}}, \ldots, t_{-1}, \underline{t_0}, t_1, \ldots, t_{k_{max}}\},$$

i.e. only list the nonzero t_k's. The range of k so that $t_k \neq 0$ can be anything, but is typically an interval around 0 for FIR filters. This means that z_n is calculated by combining x_k's with k close to n. Filters where $t_k \neq 0$ only for $k \leq 0$ are called *causal*. z_n only depends on x_n, x_{n-1}, $x_{n-2}, \ldots$, for causal filters, i.e. previous input. By the *length of a FIR filter S* we will mean the number $l(S) = k_{max} - k_{min}$. A filter which is not FIR is said to have *Infinite Impulse Response*, and called an IIR filter. Most of our filters will be assumed to be FIR, but we will also address IIR filters at the end of the chapter.

Theorem 3 concerns $N \times N$-matrices, but most of it holds for FIR filters and infinite vectors as well:

- The (infinite) matrix of a filter is always an (infinite) Toeplitz matrix,
- filters are time-invariant,
- filters can be written as convolutions.

It is also useful to state how the filter coefficients $\boldsymbol{t}$ and the vector $\boldsymbol{s} \in \mathbb{R}^N$ form the last section are related. This is answered by the following proposition.

Proposition 7. Filters as Matrices.

Assume that a discrete timer filter has filter coefficients t, and consider its restriction to periodic vectors. The first column s of the corresponding circulant Toeplitz matrix is given by

$$s_k = \begin{cases} t_k, & \text{if } 0 \le k < N/2; \\ t_{k-N} & \text{if } N/2 \le k \le N-1. \end{cases} \tag{3.12}$$

In other words, the first column can be obtained by placing the coefficients t_k with positive indices at the beginning, and the coefficients with negative indices at the end.

Proof. Let us first consider the special case

$$z_n = \frac{1}{4}(x_{n-1} + 2x_n + x_{n+1}). \tag{3.13}$$

This has compact notation $\{1/4, 1/2, 1/4\}$. To find the corresponding $N \times N$-matrix, for $1 \le n \le N-2$ it is clear from $\overline{\text{Eq.}}$ (3.13) that row n has the value $1/4$ in column $n-1$, the value $1/2$ in column n, and the value $1/4$ in column $n+1$. For the first and last rows we must be a bit more careful, since the indices -1 and N are outside the legal range of the indices. This is where the periodicity helps us out, since

$$z_0 = \frac{1}{4}(x_{-1} + 2x_0 + x_1) = \frac{1}{4}(2x_0 + x_1 + x_{N-1})$$

$$z_{N-1} = \frac{1}{4}(x_{N-2} + 2x_{N-1} + x_N) = \frac{1}{4}(x_0 + x_{N-2} + 2x_{N-1}).$$

From this we see that row 0 has the entry $1/4$ in columns 1 and $N-1$, and the entry $1/2$ in column 0, while row $N-1$ has the entry $1/4$ in columns 0 and $N-2$, and the entry $1/2$ in column $N-1$. In summary, the $N \times N$ circulant Toeplitz matrix corresponding to the filter is

$$\frac{1}{4}\begin{pmatrix} 2 & 1 & 0 & 0 & \cdots & 0 & 0 & 0 & 1 \\ 1 & 2 & 1 & 0 & \cdots & 0 & 0 & 0 & 0 \\ 0 & 1 & 2 & 1 & \cdots & 0 & 0 & 0 & 0 \\ \vdots & \vdots & \vdots & \vdots & \vdots & \vdots & \vdots & \vdots & \vdots \\ 0 & 0 & 0 & 0 & \cdots & 0 & 1 & 2 & 1 \\ 1 & 0 & 0 & 0 & \cdots & 0 & 0 & 1 & 2 \end{pmatrix}. \tag{3.14}$$

Thus, the first column s is related to t as stated in the text of the proposition. It is straightforward to generalize this proof to all FIR filters, but we omit the details. $\square$

Multiplication with the matrix S from Eq. (3.14) can be implemented in vectorized form as follows.

```
z(1)       = x(2)/4 + x(1)/2 + x(N)/4;
z(2:(N-1)) = x(3:N)/4 + x(2:(N-1))/2 + x(1:(N-2))/4;
z(N)       = x(1))/4 + x(N)/2 + x(N-1)/4;
```

This code is much more efficient than constructing the entire matrix and performing matrix multiplication. This is the case in particular for large N, for which allocating the matrix can lead to out of memory situations. Proposition 7 is useful since it explains how convolution and circular convolution are related If we above assumed that the filter is FIR, and that N is so large that all t_k with $|k| \ge N/2$ are zero, we would obtain no "collision" when placing the filter coefficients in the first column, so that all entries in the matrix equal a filter coefficient.

A discrete time filter with frequency response $\lambda_S(\omega)$ has different realizations in terms of matrices for different N. For some N the corresponding matrix may be non-singular, while for other N it may not: Since the eigenvalues of the matrix $\lambda_S(2\pi n/N)$, it all depends on where the zeros of $\lambda_S(\omega)$ are located.

Example 3.12: Time Delay

The simplest filter we can think of is the identity. An equally simple filter is *time delay* with d elements. This is denoted by E_d, and defined by $E_d x = z$, where z is the vector with components $z_k = x_{k-d}$. In this case there is only one non-zero filter coefficient, namely $t_d = 1$. The corresponding circulant Toeplitz matrix has only one (circulating) nonzero diagonal with ones, and it is easily verified that

$$\lambda_S(\omega) = e^{-id\omega},$$

so that $|\lambda_S(\omega)| = 1$ for any ω, i.e. time delays do not change the magnitude of frequencies. That S is time-invariant is clearly the same as $SE_d = E_d S$ for any d. Note also that $E_{d_1} E_{d_2} = E_{d_2} E_{d_1} = E_{d_1+d_2}$ for any d_1 and d_2, so that time delays form a group under composition.

Example 3.13: Finding the Compact Filter Notation of a Matrix

For the circulant Toeplitz matrix

$$\begin{pmatrix} 2 & 1 & 0 & 3 \\ 3 & 2 & 1 & 0 \\ 0 & 3 & 2 & 1 \\ 1 & 0 & 3 & 2 \end{pmatrix}$$

we see that $N = 4$, and that $s_0 = 2$, $s_1 = 3$, $s_2 = 0$, and $s_3 = 1$. This gives that $t_0 = s_0 = 2$, $t_1 = s_1 = 3$, $t_{-2} = s_2 = 0$, and $t_{-1} = s_3 = 1$. By including only the t_k which are nonzero, the operation can be written as

$$z_n = t_{-1} x_{n-(-1)} + t_0 x_n + t_1 x_{n-1} + t_2 x_{n-2} = x_{n+1} + 2x_0 + 3x_{n-1},$$

and the compact filter notation is $S = \{1, \underline{2}, 3\}$.

Example 3.14: Plotting a Frequency Response

The frequency response of the filter (3.13) is

$$\lambda_S(\omega) = \frac{1}{4} e^{i\omega} + \frac{1}{2} e^0 + \frac{1}{4} e^{-i\omega} = \frac{1}{2} + \frac{1}{4}(e^{i\omega} + e^{-i\omega}) = \frac{1}{2} + \frac{1}{2}\cos(\omega).$$

Figure 3.1 shows plots of the frequency response over the intervals $[0, 2\pi)$ and $[-\pi, \pi)$.

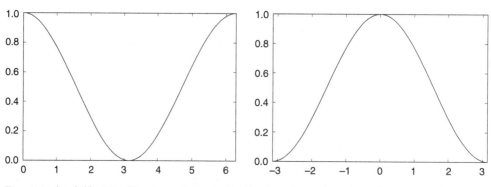

Fig. 3.1 $|\lambda_S(\omega)|$ of the filter from Formula (3.13), plotted over $[0, 2\pi]$ and $[-\pi, \pi]$

The plots clearly show that the high frequencies are made smaller by the filter, and it is therefore called a *low-pass filter*. These are investigated further in the next section.

A more general way to plot the frequency response of a FIR filter is to combine Proposition 7 with Theorem 2: If t is the set of filter coefficients, Eq. (3.3) gives that

```
omega = 2*pi*(0:(N-1))/N;
s = [t; zeros(N - length(t), 1)];
plot(omega, abs(fft(s)));
```

Some comments are in order. First of all we have restricted to $[0, 2\pi)$. Secondly, t was expanded to a longer vector s, to increase the number of plot points. Finally, the expanded vector s is correct only up to a delay, due to Proposition 7, and as we have seen a delay of the filter does not affect the magnitude of the frequency response. Here only the magnitude is plotted, as we often do.

Example 3.15: Computing the Output of a Filter

Certain vectors are easy to express in terms of the Fourier basis. This enables us to compute the output of such vectors from filters easily. Let us consider the filter

$$z_n = \frac{1}{6}(x_{n+2} + 4x_{n+1} + 6x_n + 4x_{n-1} + x_{n-2}),$$

and see how we can compute Sx when

$$x = (\cos(2\pi 5 \cdot 0/N), \cos(2\pi 5 \cdot 1/N), \ldots, \cos(2\pi 5 \cdot (N-1)/N)).$$

We note first that

$$\sqrt{N}\phi_5 = \left(e^{2\pi i 5 \cdot 0/N}, e^{2\pi i 5 \cdot 1/N}, \ldots, e^{2\pi i 5 \cdot (N-1)/N}\right)$$

$$\sqrt{N}\phi_{N-5} = \left(e^{-2\pi i 5 \cdot 0/N}, e^{-2\pi i 5 \cdot 1/N}, \ldots, e^{-2\pi i 5 \cdot (N-1)/N}\right),$$

Since $e^{2\pi i 5k/N} + e^{-2\pi i 5k/N} = 2\cos(2\pi 5k/N)$, we get by adding the two vectors that $x = \frac{1}{2}\sqrt{N}(\phi_5 + \phi_{N-5})$. Since the ϕ_n are eigenvectors, we have expressed x as a sum of eigenvectors. The corresponding eigenvalues are given by the vector frequency response, so let us compute this. If $N = 8$, computing $S\boldsymbol{x}$ means to multiply with the 8×8 circulant Toeplitz matrix

$$\frac{1}{6}\begin{pmatrix} 6 & 4 & 1 & 0 & 0 & 0 & 1 & 4 \\ 4 & 6 & 4 & 1 & 0 & 0 & 0 & 1 \\ 1 & 4 & 6 & 4 & 1 & 0 & 0 & 0 \\ 0 & 1 & 4 & 6 & 4 & 1 & 0 & 0 \\ 0 & 0 & 1 & 4 & 6 & 4 & 1 & 0 \\ 0 & 0 & 0 & 1 & 4 & 6 & 4 & 1 \\ 1 & 0 & 0 & 0 & 1 & 4 & 6 & 4 \\ 4 & 1 & 0 & 0 & 0 & 1 & 4 & 6 \end{pmatrix}$$

We now see that

$$\lambda_S(2\pi n/N) = \frac{1}{6}(6 + 4e^{-2\pi in/N} + e^{-2\pi i2n/N} + e^{-2\pi i(N-2)n/N} + 4e^{-2\pi i(N-1)n/N})$$

$$= \frac{1}{6}(6 + 4e^{2\pi in/N} + 4e^{-2\pi in/N} + e^{2\pi i2n/N} + e^{-2\pi i2n/N})$$

$$= 1 + \frac{4}{3}\cos(2\pi n/N) + \frac{1}{3}\cos(4\pi n/N).$$

The two values of this we need are

$$\lambda_S(2\pi5/N) = 1 + \frac{4}{3}\cos(2\pi5/N) + \frac{1}{3}\cos(4\pi5/N)$$

$$\lambda_S(2\pi(N-5)/N) = 1 + \frac{4}{3}\cos(2\pi(N-5)/N) + \frac{1}{3}\cos(4\pi(N-5)/N)$$

$$= 1 + \frac{4}{3}\cos(2\pi5/N) + \frac{1}{3}\cos(4\pi5/N).$$

Since these are equal, $\boldsymbol{x}$ is a sum of eigenvectors with equal eigenvalues. This means that $\boldsymbol{x}$ itself also is an eigenvector, with the same eigenvalue, so that

$$S\boldsymbol{x} = \left(1 + \frac{4}{3}\cos(2\pi5/N) + \frac{1}{3}\cos(4\pi5/N)\right)\boldsymbol{x}.$$

Example 3.16: Adding Echo

An echo is a delayed and softer copy of the original sound. If $\boldsymbol{x}$ is this original sound, the sound $\boldsymbol{z}$ with samples given by

```
[N, nchannels] = size(x);
z = zeros(N, nchannels);
z(1:d,:) = x(1:d,:);                    % No echo at the start
z((d+1):N,:) = x((d+1):N,:)+c*x(1:(N-d),:); % Add echo
z = z/max(max(abs(z)));                 % Scale to within [-1,1]
```

includes an echo of $\boldsymbol{x}$. d is an integer which represents the delay in samples. If you need a delay in t seconds, set $d = tf_s$ (and round to the nearest integer), with f_s the sample rate. c is called the *damping factor*. An echo is usually weaker than the original sound, so that $c < 1$. The sample sound with echo added with $d = 10{,}000$ and $c = 0.5$ can be found in the file castanetsecho.wav.

The compact notation for a filter which adds echo is

$$S = \{\underline{1}, 0, \ldots, 0, c\},$$

where the damping factor c appears at index d. The frequency response of this is $\lambda_S(\omega) = 1 + ce^{-id\omega}$, which is not real. We have plotted this in Fig. 3.2.

We see that the response varies between $1 - c$ and $1 + c$. The oscillation is controlled by the delay d.

Example 3.17: Computing a Composite Filter

Assume that the filters S_1 and S_2 have the frequency responses $\lambda_{S_1}(\omega) = \cos(2\omega)$, $\lambda_{S_2}(\omega) = 1 + 3\cos\omega$, and let us characterize the filter $S = S_1 S_2$. We first notice that, since both frequency responses are real, all S_1, S_2, and $S = S_1 S_2$ are symmetric. We rewrite the frequency responses as

$$\lambda_{S_1}(\omega) = \frac{1}{2}(e^{2i\omega} + e^{-2i\omega}) = \frac{1}{2}e^{2i\omega} + \frac{1}{2}e^{-2i\omega}$$
$$\lambda_{S_2}(\omega) = 1 + \frac{3}{2}(e^{i\omega} + e^{-i\omega}) = \frac{3}{2}e^{i\omega} + 1 + \frac{3}{2}e^{-i\omega}.$$

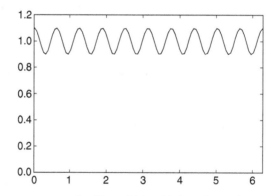

Fig. 3.2 The frequency response of a filter which adds an echo with damping factor $c = 0.1$ and delay $d = 10$

Using that convolution in time corresponds to multiplication in frequency we obtain

$$\lambda_{S_1 S_2}(\omega) = \lambda_{S_1}(\omega)\lambda_{S_2}(\omega) = \left(\frac{1}{2}e^{2i\omega} + \frac{1}{2}e^{-2i\omega}\right)\left(\frac{3}{2}e^{i\omega} + 1 + \frac{3}{2}e^{-i\omega}\right)$$
$$= \frac{3}{4}e^{3i\omega} + \frac{1}{2}e^{2i\omega} + \frac{3}{4}e^{i\omega} + \frac{3}{4}e^{-i\omega} + \frac{1}{2}e^{-2i\omega} + \frac{3}{4}e^{-3i\omega}$$

From this expression we see that the filter coefficients of S are $t_{\pm 1} = 3/4$, $t_{\pm 2} = 1/2$, $t_{\pm 3} = 3/4$. All other filter coefficients are 0. Using Theorem 7, a corresponding circulant Toeplitz matrix can be obtained from $s_1 = 3/4$, $s_2 = 1/2$, and $s_3 = 3/4$, while $s_{N-1} = 3/4$, $s_{N-2} = 1/2$, and $s_{N-3} = 3/4$ (all other s_k are 0).

3.2.1 A Second Approach to Finite Input

In order to make the convolution (3.9) computable, we have to somehow restrict to finite input. This was handled in the previous section by assuming that x is periodic. A second approach is to assume that the input is padded with zeros in both directions. Assume that $t_0, \ldots, t_{M-1}$ and $x_0, \ldots, x_{N-1}$ are the only nonzero elements, i.e. we can view them as vectors in $\mathbb{R}^M$ and $\mathbb{R}^N$, respectively. From the expression $z_n = \sum_k t_k x_{n-k}$ it is clear that we only need sum over k so that $0 \leq k < M$, $0 \leq n - k < N$. As a consequence only $z_0, \ldots, z_{M+N-2}$ can be nonzero, so that

$$(\ldots, 0, t_0, \ldots, t_{M-1}, 0, \ldots) * (\ldots, 0, x_0, \ldots, x_{N-1}, 0, \ldots) = (\ldots, 0, z_0, \ldots, z_{M+N-2}, 0, \ldots).$$
$$(3.15)$$

Convolution of zero-padded input can thus be viewed as an operation from $\mathbb{R}^M \times \mathbb{R}^N$ to $\mathbb{R}^{M+N-1}$ defined by

$$(t_0, \ldots, t_{M-1}) \times (x_0, \ldots, x_{N-1}) \to (z_0, \ldots, z_{M+N-2}).$$

Also this finite dimensional counterpart will be called convolution, as opposed to circular convolution. It is not too difficult so see that the $(M+N-1) \times N$-matrix corresponding to such convolution has the same repeating pattern on diagonals as in a Toeplitz matrix, but with no circulation in the entries.

The following loop will compute the entries $z_{M-1}, \ldots, z_{N-1}$ in a such a convolution:

```
for n = (M-1):(N-1)
    for k = 0:(M-1)
        z(n + 1) = z(n + 1) + t(k + 1)*x(n - k + 1);
    end
end
```

As for Eq. (3.14), specialized formulas are needed for the first and last entries, but we have omitted them here. As always we favor vectorized code, and this can be achieved either by observing that the inner loop can be written in terms of a scalar product, or similarly to how we vectorized equation (3.14):

```
for k = 0:(M-1)
    z(M:N) = z(M:N) + t(k + 1)*x((M-k):(N-k));
end
```

Although vectorized versions are preferred, we have to take into account that a filter may be applied in real time, with the input becoming available continuously, and that we would like to compute the output as soon as enough input elements are available. The last piece of code fails in this respect, since it computes nothing before all input is available.

MATLAB has the built-in function `conv` for computing the convolution of two finite-dimensional vectors x and y. This function is much used and is highly optimized, as we shall see in the exercises.

There is a very nice connection between convolution and polynomials, which is straightforward to show:

Proposition 8. Convolution and Polynomials.
Assume that

$$p(x) = a_M x^M + a_{M-1} x_{M-1} + \ldots + a_1 x + a_0$$
$$q(x) = b_N x^N + b_{N-1} x_{N-1} + \ldots + b_1 x + b_0$$
$$p(x)q(x) = c_{M+N} x^{M+N} + c_{M+N-1} x_{M+N-1} + \ldots + c_1 x + c_0.$$

Then we have that

$$(a_0, a_1, \ldots, a_M) * (b_0, b_1 \ldots, b_N) = (c_0, c_1, \ldots, c_{M+N}).$$

As a consequence, any factorization of a polynomial can be interpreted as a factorization into convolutions. In particular, since the roots of a real polynomial come in conjugate pairs, such polynomials can be factored into a product of real second degree polynomials, and consequently any convolution with real coefficients can be factored into successive convolutions on the form $(t_0, t_1, t_2) * x$.

3.2.2 Connection Between Convolution and Circular Convolution

Both convolution and circular convolution are used frequently, so that efficient implementations for both are needed. The next proposition expresses circular convolution in terms of convolution. As a consequence, an efficient implementation of convolution can be translated to one for circular convolution.

Proposition 9. Using Convolution to Compute Circular Convolution.
Assume that
$$S = \{t_{-L}, \ldots, \underline{t_0}, \ldots, t_L\},$$
and that $s \in \mathbb{R}^N$ is the first column in the corresponding circulant Toeplitz matrix. If $x \in \mathbb{R}^N$, then $s \circledast x$ can be computed as follows:

- *Form the vector $\tilde{x} = (x_{N-L}, \cdots, x_{N-1}, x_0, \cdots, x_{N-1}, x_0, \cdots, x_{L-1}) \in \mathbb{R}^{N+2L}$.*
- *Compute the convolution $\tilde{z} = t * \tilde{x} \in \mathbb{R}^{M+N+2L-1}$.*
- *We have that $s \circledast x = (\tilde{z}_{2L}, \ldots, \tilde{z}_{M+N-2})$.*

We will consider an implementation for this result in the exercises. The result applies only when there are equally many filter coefficients with negative and positive indices, such as for symmetric filters. It is a simple exercise to generalize the result to other filters, however. It is another exercise to turn the proposition around, i.e. express convolution in terms of circular convolution. Since circular convolution can be implemented efficiently using the DFT (using for instance Exercise 3.6), these two facts can be used as a basis for efficient implementations of both types of convolutions.

Proof. When $x \in \mathbb{R}^N$, the operation $x \to t*x$ can be represented by an $(M+N-1)\times N$ matrix. It is easy to see that this matrix has element $(i+s, i)$ equal to t_s, for $0 \le i < M$, $0 \le s < N$. In the left part of Fig. 3.3 such a matrix is shown for $M = 5$. The (constant) nonzero diagonals are shown as diagonal lines.

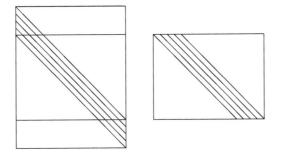

Fig. 3.3 Matrix for the operation $x \to t*x$ (left), as well as this matrix with the first and last $2L$ rows dropped (right)

Now, form the vector $\tilde{x} \in \mathbb{R}^{N+2L}$ as in the text of the theorem. Convolving $(t_{-L}, \ldots, t_L)$ with vectors in $\mathbb{R}^{N+2L}$ can similarly be represented by an $(M + N + 2L - 1) \times (N + 2L)$-matrix. The rows from $2L$ up to and including $M + N - 2$ in this matrix (we have marked these with horizontal lines above) make up a new matrix $\tilde{S}$, shown in the right part of Fig. 3.3 ($\tilde{S}$ is an $N \times (N + 2L)$ matrix).

We need to show that $Sx = \tilde{S}\tilde{x}$. We have that $\tilde{S}\tilde{x}$ equals the matrix shown in the left part of Fig. 3.4 multiplied with

$$(x_{N-L}, \ldots, x_{N-1}, x_0, \ldots, x_{N-1}, x_0, \ldots, x_{L-1}).$$

(we inserted extra vertical lines in the matrix where circulation occurs), which equals the matrix shown in the right part of Fig. 3.4 multiplied with $(x_0, \ldots, x_{N-1})$. We see that this is Sx, and the proof is complete.

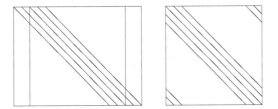

Fig. 3.4 The matrix we multiply with $(x_{N-L}, \ldots, x_{N-1}, x_0, \ldots, x_{N-1}, x_0, \ldots, x_{L-1})$ (left), and the matrix we multiply with $(x_0, \ldots, x_{N-1})$ (right)

□

Exercise 3.18: Passing Between Compact Filter Notation and Matrix Form

a) Assume that a filter is defined by the formula

$$z_n = \frac{1}{4}x_{n+1} + \frac{1}{4}x_n + \frac{1}{4}x_{n-1} + \frac{1}{4}x_{n-2}.$$

Write down the corresponding compact filter notation, and the corresponding circulant Toeplitz matrix when $N = 8$.

b) Given the circulant Toeplitz matrix

$$S = \begin{pmatrix} 1 & 2 & 0 & 0 \\ 0 & 1 & 2 & 0 \\ 0 & 0 & 1 & 2 \\ 2 & 0 & 0 & 1 \end{pmatrix}.$$

Write down the corresponding compact filter notation.

Exercise 3.19: Composing Two Filters

Assume that the filters S_1 and S_2 have the frequency responses $\lambda_{S_1}(\omega) = 2 + 4\cos(\omega)$, $\lambda_{S_2}(\omega) = 3\sin(2\omega)$.

a) Compute and plot the frequency response of the filter $S_1 S_2$.

b) Write down the filter coefficients t_k for the filter $S_1 S_2$.

Exercise 3.20: Factoring a Filter

Factor the frequency response of the filter $S = \{1, 5, 10, 6\}$ as the product of two polynomials in $e^{-i\omega}$, one of order 1 and one of order 2. Use this to factor S as a product of two filters, one with two filter coefficients, and one with three filter coefficients.

Exercise 3.21: Plotting a Simple Frequency Response

Let S be the filter defined by the equation

$$z_n = \frac{1}{4}x_{n+1} + \frac{1}{4}x_n + \frac{1}{4}x_{n-1} + \frac{1}{4}x_{n-2},$$

as in Exercise 3.18. Compute and plot (the magnitude of) $\lambda_S(\omega)$.

Exercise 3.22: Adding Two Echo Filters

Consider the two filters $S_1 = \{\underline{1}, 0, \ldots, 0, c\}$ and $S_2 = \{\underline{1}, 0, \ldots, 0, -c\}$. Both of these can be interpreted as filters which add an echo. Show that $\frac{1}{2}(S_1 + S_2) = I$. What is the interpretation of this relation in terms of echos?

Exercise 3.23: Filters with Coefficients Being Powers

Assume that $S = \{\underline{1}, c, c^2, \ldots, c^k\}$. Compute and plot $\lambda_S(\omega)$ when $k = 4$ and $k = 8$, for $c = 0.5$. How do the choices of k and c influence the frequency response?

Exercise 3.24: Convolution and Polynomials

Compute the convolution of $(1, 2, 1)$ with itself, and interpret the result in terms of a product of polynomials. Compute also the circular convolution of $(1, 2, 1)$ with itself.

Exercise 3.25: Execution Times for Convolution

Implement code which computes $t * x$ in the vectorized- and non-vectorized ways described in Sect. 3.2.1 (i.e. as a single loop over k or n with the other variable vectorized, or a double loop). As your t, take k randomly generated numbers. Compare execution times with the `conv` function, for different values of k. Present the result as a plot where k runs along the x-axis, and execution times run along the y-axis. Your result will depend on how vectorization is performed by the computing environment.

Exercise 3.26: Implementing Convolution In-Place

Filters are often implemented in hardware, and hardware implementations have strong limitations on the number of local variables. Assume that a filter t has n nonzero coefficients, all known to the hardware, and that the vector x is input. Show that $t * x$ can be implemented in-place (i.e. that $t * x$ can be computed and stored directly in x) by using $n - 1$ local variables. You can assume that two local variables can be swapped without using an additional local variable. It follows that, since any real convolution can be split into real convolutions with three filter coefficients, only 2 local variables are needed in order to compute any convolution in-place.

Exercise 3.27: Circular Convolution When There Is a Different Number of Coefficients with Positive and Negative Indices

Assume that $S = \{t_{-E}, \ldots, t_0, \ldots, t_F\}$. Formulate a generalization of Proposition 9 for such filters, i.e. to filters where there may be a different number of filter coefficients with positive and negative indices. You should only need to make some small changes to the proof.

Exercise 3.28: Implementing Circular Convolution Through Convolution

Implement a function which takes two parameters t and x, and uses Proposition 9 and convolution to compute the circular convolution of $t = (t_{-L}, \ldots, t_0, \ldots, t_L)$ and x. Write the function so that, if x has many columns, the circular convolution with each column is computed.

Hint

The function `filter_impl` in the library is slightly more general than the one you are asked to implement here, and can serve as a template for your implementation.

Exercise 3.29: Expressing Convolution in Terms of Circular Convolution

Find a similar statement to that in Proposition 9, which explains how to express convolution in terms of circular convolution.

Hint

Add some zeros at the start and end of the vector. How many zeros do you need to add?

3.3 Low-Pass and High-Pass Filters

Some filters have the desirable property that they favor certain frequencies, while annihilating others. Such filters have their own names.

Definition 10. *Low-Pass and High-Pass Filters.*
A filter S is called

- a *low-pass filter* if $\lambda_S(\omega)$ is large for ω close to 0, and $\lambda_S(\omega) \approx 0$ for ω close to π (i.e. S keeps low frequencies and annihilates high frequencies),
- a *high-pass filter* if $\lambda_S(\omega)$ is large for ω close to π, and $\lambda_S(\omega) \approx 0$ for ω close to 0 (i.e. S keeps high frequencies and annihilates low frequencies),
- a *band-pass filter* if $\lambda_S(\omega)$ is large within some interval $[a, b] \subset [0, 2\pi]$, and $\lambda_S(\omega) \approx 0$ outside this interval.

This definition should be considered vague when it comes to what we mean by "ω close to $0, \pi$", and "$\lambda_S(\omega)$ is large or ≈ 0": in practice the frequency responses of these filters may be quite far from what is commonly referred to as *ideal low-pass- or high-pass filters*, where the frequency response only assumes the values 0 and 1. One common application of low-pass filters is to reduce treble in sound, which is a common option in audio systems. The treble in a sound is generated by the fast oscillations (high frequencies) in the signal. Another option in audio systems is to reduce the bass, which corresponds to reducing the low frequencies, making high-pass filters suitable.

It turns out that there is a simple way to jump between low-pass and high-pass filters:

Proposition 11. Passing Between Low-Pass- and High-Pass Filters.
Assume that S_2 is obtained by adding an alternating sign to the filter coefficients of S_1. If S_1 is a low-pass filter, then S_2 is a high-pass filter, and vice versa.

Proof. Let S_1 have filter coefficients t_k, S_2 filter coefficients $(-1)^k t_k$. The frequency response of S_2 is

$$\lambda_{S_2}(\omega) = \sum_k (-1)^k t_k e^{-i\omega k} = \sum_k (e^{-i\pi})^k t_k e^{-i\omega k}$$

$$= \sum_k e^{-i\pi k} t_k e^{-i\omega k} = \sum_k t_k e^{-i(\omega+\pi)k} = \lambda_{S_1}(\omega + \pi).$$

where we set $e^{-i\pi} = -1$. Adding π to ω means that the points 0 and π are swapped, and this means that a low-pass filter is turned to a high-pass filter and vice versa. □

Example 3.30: Moving Average Filters

A general way of reducing variations in sound is to replace one number by the average of itself and its neighbors. If $\mathbf{z} = (z_i)_{i=0}^{N-1}$ is the sound signal produced by taking the average of three successive samples, we have that

$$z_n = \frac{1}{3}(x_{n+1} + x_n + x_{n-1}),$$

i.e. $S = \{1/3, 1/3, 1/3\}$. This is also called a *moving average filter* (with three elements). If we set $N = 4$, the corresponding circulant Toeplitz matrix for the filter is

$$S = \frac{1}{3}\begin{pmatrix} 1 & 1 & 0 & 1 \\ 1 & 1 & 1 & 0 \\ 0 & 1 & 1 & 1 \\ 1 & 0 & 1 & 1 \end{pmatrix}$$

The frequency response is

$$\lambda_S(\omega) = (e^{i\omega} + 1 + e^{-i\omega})/3 = (1 + 2\cos(\omega))/3.$$

More generally we can construct the moving average filter of $2L + 1$ elements, which is $S = \{1, \cdots, \underline{1}, \cdots, 1\}/(2L+1)$ (there is symmetry around t_0). Let us verify that these filters are low-pass filters.

Clearly the first column of S is $\boldsymbol{s} = (\underbrace{1, \ldots, 1}_{L+1 \text{ times}}, 0, \ldots, 0, \underbrace{1, \ldots, 1}_{L \text{ times}})/(2L+1)$. In Example 2.2 we computed that

$$DFT_N\left(\underbrace{1, \ldots, 1}_{L+1 \text{ times}}, 0, \ldots, 0, \underbrace{1, \ldots, 1}_{L \text{ times}}\right) = \boldsymbol{y}$$

where $\boldsymbol{y}$ had components

$$y_n = \frac{\sin(\pi n(2L+1)/N)}{\sin(\pi n/N)}.$$

Since $\boldsymbol{\lambda}_S = \mathrm{DFT}_N \boldsymbol{s}$, dividing by $2L + 1$ and inserting $\omega = 2\pi n/N$ gives that

$$\lambda_S(\omega) = \frac{1}{2L+1} \frac{\sin((2L+1)\omega/2)}{\sin(\omega/2)}.$$

We clearly have

$$0 \le \frac{1}{2L+1} \frac{\sin((2L+1)\omega/2)}{\sin(\omega/2)} \le 1,$$

and the frequency response approaches 1 as $\omega \to 0$, so that it peaks at 0. This peak gets narrower and narrower as L increases. This filter thus "keeps" only the lowest frequencies. It is also seen that the frequency response is small for $\omega \approx \pi$. In fact it is straightforward to see that $|\lambda_S(\pi)| = 1/(2L+1)$. The frequency responses for the moving average filters corresponding to $L = 1$, $L = 5$, and $L = 20$ are shown in Fig. 3.5.

In conclusion, moving average filters are low-pass filters, but they are far from ideal such, since not all higher frequencies are annihilated, and since small frequencies also are changed.

Example 3.31: Ideal Low-Pass Filters

By definition, the ideal low-pass filter keeps frequencies near 0 unchanged, and completely removes frequencies near π. $S = (F_N)^H D F_N$ is an ideal low-pass filter when D is diagonal with

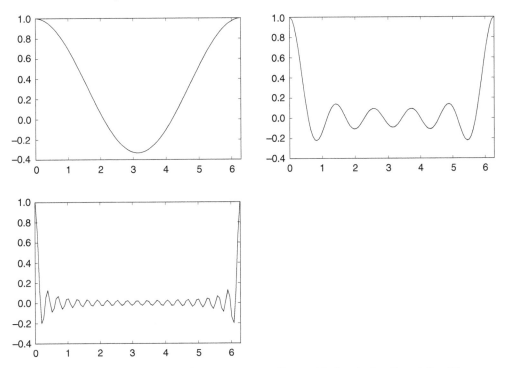

Fig. 3.5 The frequency response of moving average filters with $L = 1$, $L = 5$, and $L = 20$

$$(\underbrace{1,\ldots,1}_{L+1 \text{ times}},0,\ldots,0,\underbrace{1,\ldots,1}_{L \text{ times}})$$

on the diagonal. If the filter should keep the angular frequencies $|\omega| \leq \omega_c$ only, where ω_c is the "cutoff" frequency, we should choose L so that $\omega_c = 2\pi L/N$. An ideal high-pass filters similarly corresponds to a diagonal matrix with

$$(\underbrace{0,\ldots,0}_{N/2-L \text{ times}},\underbrace{1,\ldots,1}_{2L+1 \text{ times}},\underbrace{0,\ldots,0}_{N/2-L-1 \text{ times}})$$

on the diagonal.

Let us compute the filter coefficients for the ideal low-pass filter. Again, in Example 2.2 we computed the DFT of the vector above, and it followed from Theorem 7 that the IDFT of this vector equals its DFT, up to a factor $1/N$. This means that $s = IDFT_N \lambda_S$ is

$$\frac{1}{N}\frac{\sin(\pi k(2L+1)/N)}{\sin(\pi k/N)}.$$

The filter coefficients are thus N points uniformly spaced between 0 and 1 on the curve $\frac{1}{N}\frac{\sin(\pi t(2L+1)/2)}{\sin(\pi t/2)}$. This curve has been encountered many other places in the book. Moving average filters and ideal low-pass filters thus have a duality between vectors which contain only zeros and ones on one side (i.e. windows), and the vector $\frac{1}{N}\frac{\sin(\pi k(2L+1)/N)}{\sin(\pi k/N)}$ on the other side: filters of the one type correspond to frequency responses of the other type, and vice versa.

The extreme cases for L are

- $L = 1$: Only the lowest frequency is kept. All filter coefficients are equal to $\frac{1}{N}$.
- $L = N$: All frequencies are kept. The filter equals the identity matrix.

Between these two extremes, s_0 is the biggest coefficient, while the others decrease towards 0 along the curve we stated. The bigger L and N are, the quicker they decrease to zero. All filter coefficients are usually nonzero for this filter, since this curve is zero only at certain points. The filter is thus not a FIR filter. Many filters which are not FIR still have efficient implementations, but for this filter it turns out to be difficult to find one. The best thing we can do is probably to use a DFT for computing the filter, followed by an inverse DFT.

The function `forw_comp_rev_DFT` from Example 2.5 accepts named parameters L and `lower`, where L is as described above, and where `lower` states whether the lowest or the highest frequencies should be kept. We can use the function to listen to the lower frequencies in the audio sample file.

- For $L = 13{,}000$, the result can be found in the file castanetslowerfreq7.wav.
- For $L = 5000$, the result can be found in the file castanetslowerfreq3.wav.

With $L = 13{,}000$ you can hear the disturbance in the sound, but we have not lost that much even if about 90% of the DFT coefficients are dropped. The quality is much poorer when $L = 5000$ (here we keep less than 5% of the DFT coefficients). However we can still recognize the song, and this suggests that most of the frequency information is contained in the lower frequencies.

Let us then listen to higher frequencies instead.

- For $L = 140{,}000$, the result can be found in the file castanetshigherfreq7.wav.
- For $L = 100{,}000$ result can be found in the file castanetshigherfreq3.wav.

Both sounds are quite unrecognizable. We find that we need very high values of L to hear anything, suggesting again that most information is contained in the lowest frequencies.

Example 3.32: Windowing an Ideal Low-Pass Filter

In order to decrease the operation count of the ideal low-pass filter, one could apply a rectangular window to the filter coefficients (see Exercise 3.9), i.e. consider the filter

$$\left\{ \frac{1}{N} \frac{\sin(\pi k (2L+1)/N)}{\sin(\pi k/N)} \right\}_{k=-N_0}^{N_0}.$$

In light of that exercise, this may not be the best strategy—applying a window different from the rectangular one may better preserve the frequency response of the ideal low-pass filter.

Consider the ideal low-pass filter with $N = 128$, $L = 32$ (i.e. the filter removes frequencies $\omega > \pi/2$). In Fig. 3.6 we show the corresponding frequency responses. N_0 has been chosen so that the given percentage of all coefficients are included.

This shows that we should be careful when we omit filter coefficients: if we drop too many, the frequency response is far away from that of an ideal low-pass filter. In particular, we see that the new frequency response oscillates wildly near the discontinuity of the ideal low-pass filter.

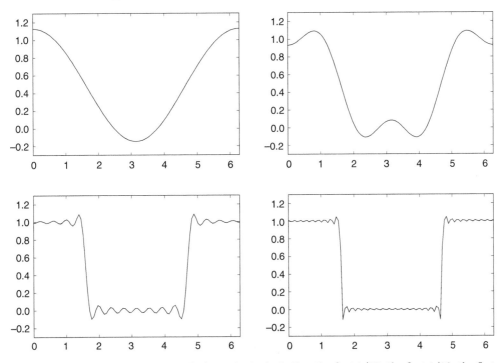

Fig. 3.6 The frequency response which results by including the first $1/32$, the first $1/16$, the first $1/4$, and all of the filter coefficients for the ideal low-pass filter

Example 3.33: Band-Pass Filters in the MP3 Standard

It turns out that the MP3 standard uses 32 different filters in an attempt to split sound into 32 disjoint frequency bands, with each band on the form

$$\pm[k\omega/32, (k+1)\omega/32], \ 0 \le k < 32.$$

The *center frequency* in each band is thus $\omega_c = (2k+1)\omega/64$. All the filters are constructed from a (low-pass) prototype filter with 512 coefficients t_k. The frequency response of the prototype filter is shown left in Fig. 3.7. Note that the value is 2 at the origin. Apart from the *transition band* (where the values are between 0 and 1), the filter is seen to be close to on an ideal low-pass filter extracting the frequency band $[-\pi/64, \pi/64]$. In Exercise 3.39 you will be asked to show that the frequency response of the filter with coefficients $\cos(k\omega_c)t_k$ is

$$\frac{1}{2}(\lambda_S(\omega - \omega_c) + \lambda_S(\omega + \omega_c)).$$

This means that, using a center frequency $\omega_c = (2k+1)\omega/64$, the new filter is close to an ideal filter w.r.t. the two bands

$$\pm(2k+1)\omega/64 + [-\pi/64, \pi/64] = \pm[k\omega/32, (k+1)\omega/32].$$

Since we divide by 2 when summing the frequency responses, the value of the frequency response will be close to 1 in the two bands. 5 of these frequency responses are shown in the right part of Fig. 3.7. If you apply all 32 filters in successive order, with the most low-pass filters first, the result can be found in the file mp3bands.wav. You should interpret the result as bands of low frequencies first, followed by the high frequencies.

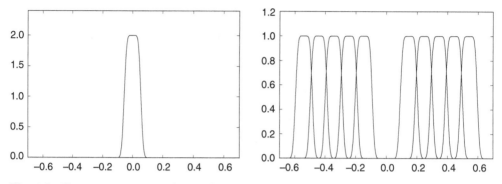

Fig. 3.7 Frequency responses of some filters used in the MP3 standard (right). They are simply shifts of that of the prototype filter (left)

π corresponds to the frequency 22.05 KHz (the highest representable frequency equals half the sampling rate $f_s = 44.1$ KHz).

In Exercise 3.9 we mentioned that the MP3 standard also applies a window to the data, followed by an FFT. This is actually performed in parallel to the application of the filters. Applying filters and an FFT in parallel to the data may seem strange. In the MP3 standard [26, p. 109] this is explained by "the lack of spectral selectivity obtained at low frequencies" by the filters above. In other words, the FFT can give more precise frequency information than the filters can.

We will not go into how the prototype filter of the MP3 standard is constructed, only note that it seems to avoid some undesirable effects in the frequency response, such as the oscillations near the discontinuities we obtained when we applied a rectangular window to the coefficients of the ideal low-pass filter. Avoiding these oscillations can be important for our perception.

Applying 32 filters to sound in parallel seems to be an operation of high complexity. Efficient algorithms exist for this, however, since all the filters are constructed in a similar way by multiplying with a cosine. We will return to this later.

Example 3.34: Low-Pass Filters Deduced from Pascal's Triangle

When computing an average, it is reasonable to let the middle sample count more than the neighbors. So, an alternative to that of moving averages is to compute

$$z_n = (x_{n-1} + 2x_n + x_{n+1})/4$$

The coefficients $1, 2, 1$ are taken from row 2 in Pascal's triangle. It turns out that this is a good choice of coefficients. In fact, other rows from Pascal's triangle are also good choices. To explain why, let $S = \{\underline{1/2}, 1/2\}$ be the moving average filter of two elements.

The frequency response of S is

$$\lambda_S(\omega) = \frac{1}{2}(1 + e^{-i\omega}) = e^{-i\omega/2}\cos(\omega/2).$$

If we apply this filter k times, Theorem 6 states that

$$\lambda_{S^k}(\omega) = \frac{1}{2^k}(1 + e^{-i\omega})^k = e^{-ik\omega/2}\cos^k(\omega/2),$$

which is a polynomial in $e^{-i\omega}$ with the coefficients taken from Pascal's triangle (the values in Pascal's triangle are the coefficients of x in the expression $(1 + x)^k$, i.e. the binomial coefficients $\binom{k}{r}$ for $0 \leq r \leq k$). Thus, the filter coefficients of S^k are rows in Pascal's triangle. The reason why the filters S^k are more desirable than moving average filters is that, since $(1 + e^{-i\omega})^k$ is a factor in its frequency response, it has a zero of multiplicity k at $\omega = \pi$. This implies that the frequency response is very flat for $\omega \approx \pi$ when k increases, i.e. the filter is good at removing the highest frequencies. This can be seen in Fig. 3.8, where we have plotted the magnitude of the frequency response when $k = 5$, and when $k = 30$. Clearly the latter frequency response is much flatter for $\omega \approx \pi$. On the other side, the filters of Example 3.30 satisfied $|\lambda_S(\pi)| = 1/(2L + 1)$, with a frequency response not very flat near π (see Fig. 3.5).

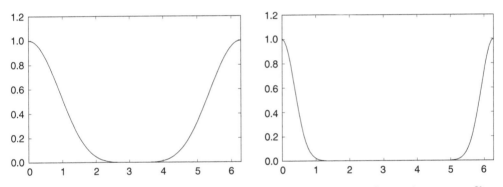

Fig. 3.8 The frequency response of filters corresponding to iterating the moving average filter $\{1/2, 1/2\}$ $k = 5$ and $k = 30$ times (i.e. using row k in Pascal's triangle)

While using S^k gives a desirable behavior for $\omega \approx \pi$, we see that the behavior is not so desirable for small frequencies $\omega \approx 0$: Only frequencies very close to 0 are kept unaltered. It should be possible to produce better low-pass filters than this also, as the frequency responses of the filters in the MP3 standard suggest.

Filtering with S^k can be computed by iterating the filter $\{1/2, 1/2\}$ k times:

```
z = x(:, 1);
for kval=1:k
    z = conv(z,[1/2 1/2]);
end
```

This code disregards the circularity of S, and we introduce a time delay. These issues will, however, not be audible when we listen to the output. In Exercise 3.36 you will be asked to perform these steps our sample audio file. If you do this you should hear that the sound gets softer when you increase k: For $k = 32$ the sound can be found in the file castanetstreble32.wav, for $k = 256$ it can be found in the file castanetstreble256.wav.

Example 3.35: High-Pass Filters Deduced from Pascal's Triangle

If we apply Proposition 11 to the low-pass filter deduced from the fourth row in Pascal's triangle we obtain

$$z_n = \frac{1}{16}(x_{n-2} - 4x_{n-1} + 6x_n - 4x_{n+1} + x_{n+2})$$

Clearly the high-pass filter arising from row k in Pascal's triangle can be written as S^k, with $S = \frac{1}{2}\{\underline{1}, -1\}$. In other words, we can use convolution as in the previous example to compute the output from such filters. In Exercise 3.36 you will be asked to apply these to the audio sample file. The new sound will be difficult to hear for large k, and we will explain why later. For $k = 1$ the sound can be found in the file castanetsbass1.wav, for $k = 2$ it can be found in the file castanetsbass2.wav. Even if the sound is quite low, you can hear that more of the bass has disappeared for $k = 2$.

The frequency response we obtain from using row 5 of Pascal's triangle is shown in Fig. 3.9. It is just the frequency response of the corresponding low-pass filter shifted with π.

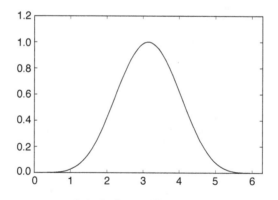

Fig. 3.9 The frequency response of the high-pass filter deduced from row 5 of Pascal's triangle

Exercise 3.36: Applying Low-Pass- and High-Pass Filters Deduced from Pascal's Triangle to the Audio Sample File

a) Write code where you apply the low-pass and high-pass filters in Examples 3.34 and 3.35 to the audio sample file, and verify that the sounds you get are the same as in these examples. How high must k be in order for you to hear difference from the actual sound? How high can you choose k and still recognize the sound at all? If you solved Exercise 3.28, you can also use the function `filter_impl` to perform the filtering, rather than using convolution (which, as mentioned, discards circularity).

b) In your code, it is not necessary to scale the values after applying the low-pass or high-pass filters so that values fit inside $[-1, 1]$. Explain why this is the case.

Exercise 3.37: Constructing a High-Pass Filter

Consider again Example 3.31. Find an expression for a filter so that only frequencies so that $|\omega - \pi| < \omega_c$ are kept, i.e. the filter should only keep angular frequencies close to π (i.e. here we construct a high-pass filter).

Exercise 3.38: Combining Low-Pass and High-Pass Filters

In this exercise we will investigate how we can combine low-pass and high-pass filters to produce other filters

a) Assume that S_1 and S_2 are low-pass filters. Is $S_1 S_2$ also a low-pass filter? What if both S_1 and S_2 are high-pass filters?

b) Assume that one of S_1, S_2 is a high-pass filter, and that the other is a low-pass filter. Is $S_1 S_2$ low-pass or high-pass?

Exercise 3.39: Filters in the MP3 Standard

Assume that t_k are the filter coefficients of a filter S_1, and that S_2 is the filter with filter coefficients $\cos(k\omega_c)t_k$, where $\omega_c \in [0, \pi)$. Show that

$$\lambda_{S_2}(\omega) = \frac{1}{2}(\lambda_{S_1}(\omega - \omega_c) + \lambda_{S_1}(\omega + \omega_c)).$$

In other words, when we multiply (modulate) the filter coefficients with a cosine, the new frequency response can be obtained by shifting the old frequency response with ω_c in both directions, and taking the average of the two. We saw that the filters in the MP3 standard were constructed in this way. The collection of these filters are also called a *cosine modulated filter bank*. It is also possible to construct *DFT-modulated filter banks* where the shift in frequency is obtained by multiplying with a complex exponential instead, but this produces filters with complex coefficients.

Exercise 3.40: Code Example

a) Explain what the code below does, line by line.

```
[x, fs] = audioread('sounds/castanets.wav');
[N, nchannels] = size(x);
z = zeros(N, nchannels);
for n=2:(N-1)
    z(n,:) = 2*x(n+1,:) + 4*x(n,:) + 2*x(n-1,:);
end
z(1,:) = 2*x(2,:) + 4*x(1,:) + 2*x(N,:);
z(N,:) = 2*x(1,:) + 4*x(N,:) + 2*x(N-1,:);
z = z/max(abs(z));
playerobj=audioplayer(z, fs);
playblocking(playerobj)
```

Comment in particular on what happens in the three lines directly after the `for`-loop, and why we do this. What kind of changes in the sound do you expect to hear?

b) Write down the compact filter notation for the filter which is used in the code, and write down a 5×5 circulant Toeplitz matrix which corresponds to this filter. Plot the (continuous) frequency response. Is the filter a low-pass- or a high-pass filter?

c) Another filter is given by the circulant Toeplitz matrix

$$\begin{pmatrix} 4 & -2 & 0 & 0 & -2 \\ -2 & 4 & -2 & 0 & 0 \\ 0 & -2 & 4 & -2 & 0 \\ 0 & 0 & -2 & 4 & -2 \\ -2 & 0 & 0 & -2 & 4 \end{pmatrix}.$$

Express a connection between the frequency responses of this filter and the filter from b). Is the new filter a low-pass- or a high-pass filter?

3.4 IIR Filters

In the beginning of this chapter we showed that filters could be written on the form

$$z_n = \sum_k t_k x_{n-k}. \tag{3.16}$$

The filters up to now have all been FIR, i.e. with a finite number of nonzero t_k. We will now give a short introduction to IIR filters, where the number of nonzero t_k is infinite. It will turn out that many IIR filters (but not all) can be realized in terms of finite order difference equations of the form

$$\sum_{0 \le k \le M} a_k z_{n-k} = \sum_k b_k x_{n-k} \tag{3.17}$$

We will here therefore use the notation S for such difference equations, and change the notation for filters on the form (3.16) to S'. Any difference equation with constant coefficients can be written as a lower triangular system, which here is

$$
\begin{pmatrix}
a_0 & 0 & \cdots & 0 & 0 \\
a_1 & a_0 & \cdots & 0 & 0 \\
\vdots & \vdots & \ddots & \vdots & \vdots \\
0 & 0 & \cdots & a_0 & 0 \\
0 & 0 & \cdots & a_1 & a_0
\end{pmatrix}
\begin{pmatrix}
z_0 \\
\vdots \\
z_{N-1}
\end{pmatrix}
=
\begin{pmatrix}
\cdots & b_1 & \underline{b_0} & b_{-1} & \cdots \\
& \vdots & b_2 & b_1 & b_0 & \vdots \\
& \vdots & \cdots\cdots\cdots & \ddots
\end{pmatrix}
\begin{pmatrix}
\vdots \\
x_{-1} \\
\underline{x_0} \\
x_1 \\
\vdots
\end{pmatrix}.
$$

We will assume that a_0 and a_M are nonzero, so that M initial values $z_0, \ldots, z_{M-1}$ are needed to compute the remaining output $\{z_n\}_{n \geq M}$. Our point will be that iterating (3.17) as the finite sum

$$
z_n = \frac{1}{a_0} \left(\sum_k b_k x_{n-k} - \sum_{1 \leq k \leq M} a_k z_{n-k} \right).
$$

is much simpler than computing the infinite sum (3.16).

Since a filter is characterized by its frequency response, let us use results on difference equations to compute the output of (3.17) when the input is an angular frequency $x_n = e^{i\omega n}$. The right hand side of (3.17) is then

$$
\sum_k b_k e^{i\omega(n-k)} = \left(\sum_k b_k e^{-i\omega k} \right) e^{i\omega n}.
$$

In order to find a particular solution to a difference equation, we have learned that we should try a z_n similar to what is on the right hand side. This is here a power, so we try a power with the same base, i.e. $z_n^{(p)} = Be^{i\omega n}$. Inserting this on the left hand side gives

$$
\sum_{0 \leq k \leq M} a_k z_{n-k}^{(p)} = B \sum_{0 \leq k \leq M} a_k e^{i\omega(n-k)} = Be^{i\omega n} \sum_{0 \leq k \leq M} a_k e^{-i\omega k}.
$$

Comparing the left and right hand sides gives the value of B, and we obtain

$$
z_n^{(p)} = \frac{\sum_k b_k e^{-i\omega k}}{\sum_{0 \leq k \leq M} a_k e^{-i\omega k}} e^{i\omega n}.
$$

We now turn to the general solution $z^{(h)}$ of the homogeneous equation

$$
\sum_{0 \leq k \leq M} a_k z_{n-k} = 0.
$$

The *characteristic equation* of this is $\sum_{k=0}^{M} a_{M-k} r^k = r^M \sum_{k=0}^{M} a_k r^{-k}$. Since a_0 and a_M are nonzero, there are M nonzero roots in the characteristic equation, equal to the roots of $\sum_{k=0}^{M} a_k r^{-k} = 0$. Let us denote these by $r_0, \ldots, r_{M-1}$. In the following we will assume that these are distinct (the analysis is slightly different otherwise), so that the general solution to the homogeneous equation is

$$
z_n^{(h)} = \sum_{i=0}^{M-1} A_i r_i^n
$$

The general solution to (3.17) is now $z_n = z_n^{(p)} + z_n^{(h)}$, and it is also straightforward to find the A_i from initial values $z_0, \ldots, z_{M-1}$. Let us summarize as follows.

Proposition 12. The Output from Difference Equations.

Assume that there are distinct zeros $r_0, \ldots, r_{M-1}$ in the characteristic equation of (3.17), and define

$$\lambda_S(\omega) = \frac{\sum_k b_k e^{-i\omega k}}{\sum_{0 \le k \le M} a_k e^{-i\omega k}}.$$

If the input to (3.17) is $x_n = e^{i\omega n}$, then the output will be

$$z_n = \lambda_S(\omega) e^{i\omega n} + \sum_{i=0}^{M-1} A_i r_i^n$$

for $n \ge M$, where the A_i can be found from the initial values $z_0, \ldots, z_{M-1}$ by solving

$$\begin{pmatrix} r_0^0 & \cdots & r_{M-1}^0 \\ \vdots & \ddots & \vdots \\ r_0^{M-1} & \cdots & r_{M-1}^{M-1} \end{pmatrix} \begin{pmatrix} A_0 \\ \vdots \\ A_{M-1} \end{pmatrix} = \begin{pmatrix} z_0 \\ \vdots \\ z_{M-1} \end{pmatrix} - \lambda_S(\omega) \begin{pmatrix} e^{i\omega 0} \\ \vdots \\ e^{i\omega(M-1)} \end{pmatrix}.$$

Although S now may not be a filter, we still used the notation $\lambda_S(\omega)$ to represent the contribution from angular frequency ω in the output. Assume now more generally that x is a finite sum of angular frequencies $\sum_k c_k e^{i\omega_k n}$, and let S' be the filter with frequency response $\lambda_S(\omega)$. Linearly extending the arguments above we obtain that

$$z_n = (S'x)_n + \sum_{i=0}^{M-1} A_i r_i^n, \tag{3.18}$$

for $n \ge M$, where the A_i and the initial values $z_0, \ldots, z_{M-1}$ satisfy

$$\begin{pmatrix} r_0^0 & \cdots & r_{M-1}^0 \\ \vdots & \ddots & \vdots \\ r_0^{M-1} & \cdots & r_{M-1}^{M-1} \end{pmatrix} \begin{pmatrix} A_0 \\ \vdots \\ A_{M-1} \end{pmatrix} = \begin{pmatrix} z_0 - (S'x)_0 \\ \vdots \\ z_{M-1} - (S'x)_{M-1} \end{pmatrix}. \tag{3.19}$$

So, when does (3.17) give rise to a filter? The terms $A_i r_i^n$ represent aliasing terms, i.e. frequencies that may not be part of the input. We therefore require that all $A_i = 0$. Since the coefficient matrix in (3.19) is clearly non-singular, this is equivalent to that $z_k = (S'x)_k$ for $0 \le k < M$. Let us summarize as follows.

Theorem 13. *Let S' be the filter with frequency response $\lambda_S(\omega)$. If the initial values satisfy $z_k = (S'x)_k$ for $0 \le k < M$, then $S = S'$. In particular, S is a filter as well.*

Thus, the initial values can not be chosen arbitrarily if one requires a filter: they must be considered as a function of the input. As we have seen those initial values can be computed if we have a frequency decomposition of the input, something we usually don't have. In practice the initial values may not be defined as to give a filter.

If all $|r_i| < 1$, however, we see that the aliasing terms $A_i r_i^n$ diminish as $n \to \infty$, so that there is "no aliasing in the limit", and the output will be bounded for any input

$e^{i\omega n}$. Such systems are also called *BIBO-stable*, where BIBO stands for Bounded Input Bounded Output. If on the other hand some $|r_i| \geq 1$, however, then $A_i r_i^n$ will go to infinity, unless the initial values imply that $A_i = 0$. Even when $A_i = 0$ the output will typically go to infinity, due to round-off errors in the initial values or in the coefficients.

To see that many IIR filters have a simpler realization in terms of difference equations, substitute $z = e^{i\omega}$ and compare the frequency responses of S and S':

$$\sum_k t_k z^{-k} = \frac{\sum_k b_k z^{-k}}{\sum_{0 \leq k \leq M} a_k z^{-k}}, \qquad (3.20)$$

The infinite sum on the left should thus equal a quotient of two low degree polynomials. This is easy to obtain by considering geometric series (see Example 3.41).

In complex analysis a sum on the form $\sum_k t_k z^{-k}$ is called a *Laurent series*. Taylor series of functions are examples of such where only negative k contribute. It is known [23] that they converge absolutely inside some annular region (called the region of convergence, or ROC), and diverge outside this region. If the unit circle is a subset of the ROC, the output will be bounded for all angular frequencies. Laurent series can be differentiated and integrated term-wise in the ROC. Since differentiation (and many integrals) of the right hand side of (3.20) gives expressions of the same type, it is straightforward to come up with many more examples of Laurent series. Also, starting with a right hand side in (3.20), one can find the corresponding Laurent series (and thereby an IIR filter) similarly to how one finds a Taylor series.

If an IIR filter has a realization (3.17), it is easy to see that its transpose has a similar realization which can be obtained by reversing the a_k and the b_k.

Even in cases when we start with a FIR filter, a factorization (3.20) can be useful, see Example 3.42. Again, such a factorization amounts to writing a polynomial as a quotient of simple polynomials.

For other IIR filters, however, it may be difficult or impossible to find a realization in terms of difference equations. The power series must then converge to a function which is quotient of polynomials, but not all functions are on this form.

The difference equation form of a filter can be greatly simplified. Assuming again that there are distinct zeros in the characteristic equation, and assuming that only positive k contribute in the sums, we can use polynomial division to write

$$\frac{\sum_k b_k z^{-k}}{\sum_{0 \leq k \leq M} a_k z^{-k}} = P(z^{-1}) + \sum_i \frac{a_i}{1 + b_i z^{-1}} + \sum_j \frac{c_i + d_i z^{-1}}{1 + e_i z^{-1} + f_i z^{-2}}.$$

Here $P(z^{-1})$ corresponds to a FIR filter, while the remaining terms have realizations on the form

$$z_n + b_i z_{n-1} = a_i x_n$$
$$z_n + e_i z_{n-1} + f_i z_{n-2} = c_i x_n + d_i x_{n-1}.$$

Thus, while FIR filters could be factored into convolutions where each component has at most three filter coefficients, a realization in terms of difference equations can be written as a sum of a FIR filter (which again can be factored further), and simpler such realizations where the left and right hand sides have at most three coefficients.

Example 3.41: A Simple IIR Filter

Consider the IIR filter

$$z_n = \sum_{k=0}^{\infty} a^k x_{n-k},$$

where $|a| < 1$. We recognize $\sum a^k z^{-k}$ as a geometric series with sum $\frac{1}{1-az^{-1}}$ for $|z| > |a|$. A realization of this is clearly

$$z_n - az_{n-1} = x_n,$$

where we need one initial value. This form is much simpler than the form we started with.

For an IIR filter it may not be that the output of an angular frequency is finite, but note that scaling the filter coefficients with a^k like above will scale the ROC accordingly. In other words, it is straightforward to find a corresponding filter which produces finite output for the angular frequencies.

Example 3.42: Moving Average Filter

Consider again the moving average filter S from Example 3.30:

$$z_n = \frac{1}{2L+1}(x_{n+L} + \cdots + x_n + \cdots + x_{n-L}).$$

We recognize $\frac{1}{2L+1}\sum_{k=-L}^{L} z^{-k}$ as a finite geometric series with sum

$$\frac{1}{2L+1}z^L\frac{1-z^{-2L-1}}{1-z^{-1}} = \frac{1}{2L+1}\frac{z^L - z^{-L-1}}{1-z^{-1}}.$$

A realization of this is clearly

$$z_n - z_{n-1} = \frac{1}{2L+1}(x_{n+L} - x_{n-(L+1)}).$$

Again one initial value is needed. It is easy to obtain this formula directly as well, since the terms $x_{n-L}, \ldots, x_{n+L-1}$ cancel when we compute $z_n - z_{n-1}$.

Example 3.43: IIR Filters Applied to Periodic Vectors

Filters preserve periodicity in vectors, regardless if they are FIR or IIR. Let x be a vector with period N, and suppose we have an IIR filter with realization (3.17). For periodic input it is straightforward to compute its frequency decomposition and compute the initial conditions from that.

If there are very few nonzero coefficients a_k and b_k in the realization, however, another strategy is better. Let A and B be the circulant Toeplitz matrices constructed from those coefficients, respectively, and suppose that N is so large that there are no "collisions" in these matrices. It is straightforward to see that solving

$$Az = Bx$$

for z (x and z are now N-dimensional counterparts) computes the output z correctly. This strategy involves no specific computation of the initial values, and is computationally simple in that we only need to multiply with a sparse matrix, and solve a system with a sparse coefficient matrix.

Exercise 3.44: All-Pass Filters

All-pass filters are filters where $|\lambda_S(\omega)| = 1$ for all ω. Such filters are uniquely determined by the phase of the frequency response. Any all-pass FIR filter must clearly be on the form αE_d with $|\alpha| = 1$. It is, however, possible to find other all-pass filters by considering realizations in terms of difference equations.

a) If α is any real number, consider the filter S given by the realization

$$z_n - \alpha z_{n-1} = \alpha x_n - x_{n-1}.$$

Write down the frequency response of this realization, and show that it is an all-pass filter.

b) A filter is said to be *linear phase* if the phase (argument) of the frequency response is linear in ω, i.e. $\lambda_S(\omega) = Ce^{i(r\omega+s)}$ for some real r, s, and C. Any delay filter has linear phase since $\lambda_{E_d}(\omega) = e^{-id\omega}$. Show that S and the filter $\{-\alpha^2, \underline{2\alpha}, -1\}$ have frequency responses with the same phase, and explain why it follows from this that S has nonlinear phase as long as $|\alpha| \neq 1$. Since delay filters have linear phase, these filters can not be delays, and thus represent a new kind of all-pass filters.

Hint

Consult Exercise 3.49 on linear phase filters.

c) Explain that S has an inverse, and that a realization of the inverse system is

$$-\alpha z_n + z_{n-1} = -x_n + \alpha x_{n-1}.$$

d) Explain that combining two all-pass filters as above with different values for α will also give an all-pass filter, different from the ones we combined. This gives some flexibility in how one can construct all-pass filters as well.

Exercise 3.45: The IIR Filter for the Karplus-Strong Algorithm

Let us rewrite the difference equation in Exercise 1.12 as

$$z_n - \frac{1}{2}(z_{n-p} + z_{n-(p+1)}) = 0$$

a) Explain that the frequency response of this realization is 0. Therefore, this realization is only interesting if the initial conditions are chosen so that we don't obtain a filter.

b) Plot the zeros of the characteristic equation for $p = 20$. Is the filter BIBO stable?

c) What happens with the zeros of the characteristic equation when you increase p? Attempt to explain why the resulting sound changes more slowly to a static sound when you increase p.

d) A related IIR filter is

$$z_n - \frac{1}{2}(z_{n-p} + z_{n-(p+1)}) = x_n.$$

Plot the frequency response of this filter for $p = 20$. Include only values in the response between 0 and 10.

3.5 Symmetric Filters and the DCT

We have argued that the samples of a symmetric extension can be viewed as a better approximation to the function. It is therefore desirable to adapt filters to such extensions. Due to the following result, this is easiest to do if the filter itself is symmetric.

Theorem 14. Symmetric Filters Preserve Symmetric Extensions.
  Is S is a symmetric $(2N) \times (2N)$-filter then S preserves symmetric extensions (see Definition 22). This means that its restriction to symmetric extensions can be viewed as a linear mapping $S_r : \mathbb{R}^N \to \mathbb{R}^N$ defined by $S_r \boldsymbol{x} = \boldsymbol{z}$ whenever $S\breve{\boldsymbol{x}} = \breve{z}$. S_r is also called the symmetric restriction *of S*

Proof. Since the frequency response of S is the DFT of the filter coefficients, and since these coefficients are real and symmetric, it follows from 1. and 2. of Theorem 7 that the frequency response satisfies $\lambda_{S,n} = \lambda_{S,2N-n}$. It follows that

$$S\left(\cos\left(\frac{2\pi n(k+1/2)}{2N}\right)\right)$$

$$= S\left(\frac{1}{2}\left(e^{2\pi i(n/(2N))(k+1/2)} + e^{-2\pi i(n/(2N))(k+1/2)}\right)\right)$$

$$= \frac{1}{2}\left(e^{\pi i n/(2N)}S\left(e^{2\pi i n k/(2N)}\right) + e^{-\pi i n/(2N)}S\left(e^{-2\pi i n k/(2N)}\right)\right)$$

$$= \frac{1}{2}\left(e^{\pi i n/(2N)}\lambda_{S,n}e^{2\pi i n k/(2N)} + e^{-\pi i n/(2N)}\lambda_{S,2N-n}e^{-2\pi i n k/(2N)}\right)$$

$$= \frac{1}{2} \left(\lambda_{S,n} e^{2\pi i (n/(2N))(k+1/2)} + \lambda_{S,2N-n} e^{-2\pi i (n/(2N))(k+1/2)} \right)$$

$$= \lambda_{S,n} \frac{1}{2} \left(e^{2\pi i (n/(2N))(k+1/2)} + e^{-2\pi i (n/(2N))(k+1/2)} \right)$$

$$= \lambda_{S,n} \cos \left(\frac{2\pi n (k+1/2)}{2N} \right).$$

Here we have used that $e^{2\pi i n k/(2N)}$ is an eigenvector of S with eigenvalue $\lambda_{S,n}$, and $e^{-2\pi i n k/(2N)} = e^{2\pi i (2N-n)k/(2N)}$ is an eigenvector of S with eigenvalue $\lambda_{S,2N-n}$. Since the cosine vectors form a basis for the set of all symmetric extensions, it follows that S preserves all symmetric extensions. That S_r is linear is obvious. $\square$

It is straightforward to find an expression for S_r from S:

Theorem 15. Expression for S_r.
Assume that $S : \mathbb{R}^{2N} \to \mathbb{R}^{2N}$ is a symmetric filter, and that

$$S = \begin{pmatrix} S_1 & S_2 \\ S_3 & S_4 \end{pmatrix},$$

where all S_i are $N \times N$-matrices. Then S_r is symmetric, and $S_r = S_1 + (S_2)^{\leftrightarrow}$, where $(S_2)^{\leftrightarrow}$ is the matrix S_2 with the columns reversed.

Proof. With S as in the text of the theorem, we compute

$$S_r \boldsymbol{x} = \begin{pmatrix} S_1 & S_2 \end{pmatrix} \begin{pmatrix} x_0 \\ \vdots \\ x_{N-1} \\ \hline x_{N-1} \\ \vdots \\ x_0 \end{pmatrix} = S_1 \begin{pmatrix} x_0 \\ \vdots \\ x_{N-1} \end{pmatrix} + S_2 \begin{pmatrix} x_{N-1} \\ \vdots \\ x_0 \end{pmatrix}$$

$$= S_1 \begin{pmatrix} x_0 \\ \vdots \\ x_{N-1} \end{pmatrix} + (S_2)^{\leftrightarrow} \begin{pmatrix} x_0 \\ \vdots \\ x_{N-1} \end{pmatrix} = (S_1 + (S_2)^{\leftrightarrow}) \boldsymbol{x},$$

so that $S_r = S_1 + (S_2)^{\leftrightarrow}$. Since S is symmetric, S_1 is also symmetric. $(S_2)^{\leftrightarrow}$ is also symmetric, since it is constant on cross-diagonals. It follows that S is also symmetric, and this completes the proof. $\square$

In the block matrix factorization of S, S_2 contains the "circulating" part of the matrix, and forming $(S_2)^{\leftrightarrow}$ means that the circulating parts switch corners. Note that S_r is not a digital filter, since its matrix is not circulant. In particular, its eigenvectors are not pure tones. From the proof of Theorem 14 it can be seen that the cosine vectors $\boldsymbol{d}_n$ are eigenvectors of S_r. We can thus use the DCT to diagonalize S_r, similarly to how we used the Fourier matrix to diagonalize S.

Corollary 16. Basis of Eigenvectors for S_r.

If S is a symmetric $(2N) \times (2N)$-filter then $S_r = (DCT_N)^T D DCT_N$, where D is the diagonal matrix with $\lambda_{S,0}, \lambda_{S,1}, \ldots, \lambda_{S,N-1}$ on the diagonal.

The reason we transposed the first DCT in $(DCT_N)^T D DCT_N$ is that the d_n are rows in DCT_N, so that we must transpose in order to obtain eigenvectors in the columns.

3.5.1 Implementations of Symmetric Filters

Corollary 16 provides us with an efficient implementation of S_r, since the DCT has an efficient implementation. This result is particularly useful when S has many filter coefficients. When S has few filter coefficients it may be just as well to compute the filter entry by entry, but also then we can reduce the number of arithmetic operations for symmetric filters. To see this write

$$
\begin{aligned}
(S\boldsymbol{x})_n &= \sum_{k=0}^{N-1} s_k x_{(n-k) \bmod N} \\
&= s_0 x_n + \sum_{k=1}^{(N-1)/2} s_k x_{(n-k) \bmod N} + \sum_{k=(N+1)/2}^{N-1} s_k x_{(n-k) \bmod N} \\
&= s_0 x_n + \sum_{k=1}^{(N-1)/2} s_k x_{(n-k) \bmod N} + \sum_{k=1}^{(N-1)/2} s_k x_{(n-(N-k)) \bmod N} \\
&= s_0 x_n + \sum_{k=1}^{(N-1)/2} s_k \big(x_{(n-k) \bmod N} + x_{(n+k) \bmod N} \big).
\end{aligned} \tag{3.21}
$$

If we compare the first and last expressions here, the same number of summations is needed, but the number of multiplications needed in the latter expression has been halved.

Observation 17. Reducing Arithmetic Operations for Symmetric Filters.

Assume that a symmetric filter has $2s+1$ filter coefficients. The filter applied to a vector of length N can then be implemented using $(s+1)N$ multiplications and $2sN$ additions. In comparison, for a non-symmetric filter with the same number of coefficients, a direct implementation requires $(2s+1)N$ multiplications and $2sN$ additions.

The conv function may not pick up this reduction in the number of multiplications, since it can't assume symmetry. We will still use the conv function in implementations, however.

At the end of Sect. 3.2 we saw that any filter could be factored as a product of filters with 3 filter coefficients. It turns out that any symmetric filter can be factored into a product of symmetric filters. To see how, note first that a real polynomial is symmetric if and only if $1/a$ is a root whenever a is. If we pair together the factors for the roots $a, 1/a$ when a is real we get a component in the frequency response of degree 2. If we pair the factors for the roots $a, 1/a, \overline{a}, \overline{1/a}$ when a is complex, we get a component in the frequency response of degree 4:

Idea 18. Factorizing Symmetric Filters.

Let S be a symmetric filter with real coefficients. There exist constants K, $a_1, \ldots, a_m$, $b_1, c_1, \ldots, b_n, c_n$ so that

$$\lambda_S(\omega) = K(a_1 e^{i\omega} + 1 + a_1 e^{-i\omega}) \ldots (a_m e^{i\omega} + 1 + a_m e^{-i\omega})$$
$$\times (b_1 e^{2i\omega} + c_1 e^{i\omega} + 1 + c_1 e^{-i\omega} + b_1 e^{-2i\omega}) \ldots$$
$$\times (b_n e^{2i\omega} + c_n e^{i\omega} + 1 + c_n e^{-i\omega} + b_n e^{-2i\omega}).$$

We can write $S = KA_1 \ldots A_m B_1 \ldots B_n$, *where* $A_i = \{a_i, \underline{1}, a_i\}$ *and* $B_i = \{b_i, c_i, \underline{1}, c_i, b_i\}$.

We see that the component filters have 3 and 5 filter coefficients.

Exercise 3.46: Direct Proof of the Orthogonality of the DCT Basis

The proof we gave that the DCT basis was orthonormal was constructive in that it built on the orthogonality of the DFT. In the literature many instead give a direct proof. Give a direct proof of the DCT basis using only trigonometric identities.

Exercise 3.47: Writing Down Lower Order S_r

Consider the averaging filter $S = \{\frac{1}{4}, \frac{1}{2}, \frac{1}{4}\}$. Write down the matrix S_r for the case when $N = 4$.

Exercise 3.48: Computing Eigenvalues

Consider the matrix

$$S = \frac{1}{3} \begin{pmatrix} 2 & 1 & 0 & 0 & 0 & 0 \\ 1 & 1 & 1 & 0 & 0 & 0 \\ 0 & 1 & 1 & 1 & 0 & 0 \\ 0 & 0 & 1 & 1 & 1 & 0 \\ 0 & 0 & 0 & 1 & 1 & 1 \\ 0 & 0 & 0 & 0 & 1 & 2 \end{pmatrix}$$

a) Compute the eigenvalues and eigenvectors of S using the results of this section. You should only need to perform one DFT or one DCT in order to achieve this.

b) Use a computer to find the eigenvectors and eigenvalues of S also. What are the differences from what you found in a)?

c) Find a filter T so that $S = T_r$. What kind of filter is T?

Exercise 3.49: Linear Phase Filters

For a symmetric $(2N)\times(2N)$-filter the filter coefficients have symmetry around $N-1/2$. Show that, if a filter instead has symmetry around d (where d may or may not be an integer), then its frequency response takes the form $\lambda_S(\omega) = Ce^{-i\omega d}$, with C a real number.

This means that the phase (argument) of the frequency response is $-d\omega$ or $\pi - d\omega$, depending on the sign of C. In other words, the phase is linear in ω. Such filters are therefore called *linear phase filters* (see also Exercise 3.44). One can show that linear phase filters also preserve symmetric extensions, but one has to re-define such extensions to have another point of symmetry. An example of linear phase filters which are not symmetric are low-pass filters where the coefficients are taken from odd rows in Pascal's triangle.

Exercise 3.50: Component Expressions for a Symmetric Filter

Assume that $S = t_{-L},\ldots,\underline{t_0},\ldots,t_L$ is a symmetric filter. Use Eq. (3.21) to show that $z_n = (S\boldsymbol{x})_n$ in this case can be split into the following different formulas, depending on n:

a) $0 \leq n < L$:

$$z_n = t_0 x_n + \sum_{k=1}^{n} t_k(x_{n+k} + x_{n-k}) + \sum_{k=n+1}^{L} t_k(x_{n+k} + x_{n-k+N}).$$

b) $L \leq n < N - L$:

$$z_n = t_0 x_n + \sum_{k=1}^{L} t_k(x_{n+k} + x_{n-k}).$$

c) $N - L \leq n < N$:

$$z_n = t_0 x_n + \sum_{k=1}^{N-1-n} t_k(x_{n+k} + x_{n-k}) + \sum_{k=N-1-n+1}^{L} t_k(x_{n+k-N} + x_{n-k}).$$

3.6 Relations to Signal Processing

In signal processing textbooks matrix notation for the DFT and for filters is often absent, probably because their component-wise formulas have sticked. In signal processing it is also not that usual to consider filters as finite dimensional operations, i.e. restrict to periodic vectors: there may no mention of the length of the input (i.e. the input to Eq. (3.9) can be infinite in either direction). The matrices occurring in the literature are thus infinite matrices with a Toeplitz structure (i.e. constant on diagonals), and with no circulation. This Toeplitz structure is usually not mentioned and analyzed. The presentation given here assumes finite-dimensional input, as in the classical linear algebra tradition. For another book which also views filters as circulant Toeplitz matrices, see [21].

In signal processing it is often not mentioned that the frequency response can be interpreted as eigenvalues, with corresponding eigenvectors being the Fourier basis vectors. As we have seen there are immediate consequences from the realization in terms of eigenvectors, such as the orthogonality of the Fourier basis. Also, the general recipe in signal processing for computing a filter by

1. first applying a DFT to the input (to go to the frequency domain),
2. then multiplying with the frequency response, and
3. finally apply the Inverse DFT (to go back to the time domain),

is clearly nothing but the three multiplications in the diagonalization $S = (F_N)^H D F_N$. We have proved that is the same as circular convolution, which is different from convolution. Signal processing literature often fails in mentioning that the three-step recipe for computing a filter gives an error (at the beginning and end of the signal) when compared to using the convolution formula.

Signal processing often defines convolution first, and then shows that it is frequency preserving. We rather define filters in frequency, before we use established mappings between time and frequency to deduce time-expressions for the filters. Convolution corresponds as we have seen to multiplication with a rectangular matrix, with the size of the output larger than that of the input. The advantage with periodic and symmetric extensions is that they turn filtering into multiplication with square matrices, diagonalizable with the DFT/DCT. In signal processing some also define *symmetric convolution* [45], which correspond to the matrices we deduced by specializing to symmetric filters. We have avoided using this concept.

Signal processing literature often starts with discrete-time systems, leaving Fourier series and continuous-time systems to later chapters. We have addressed that there can be a disadvantage with this: Although computations are usually carried out in discrete-time, there usually is some underlying continuous-time model. From this model one can infer things such as slow convergence of the Fourier series in case of a discontinuity, which implies bad discrete-time approximations. Circular convolution and convolution can be interpreted in terms of periodization and zero-padding, and both these create discontinuities in the input, so that they will suffer from the same precision problem.

Any trick to speed up the convergence of the Fourier series, such as symmetric extensions, propagates to a discrete model as well. Indeed, this argument makes the case for using the DCT instead of the DFT. Signal processing literature often does a bad job in motivating the DCT, in particular why it can be considered as more "precise" in some way. Much literature omits the DCT altogether, or merely mentions its existence. In this book we have tried to give it the place it deserves, motivating it in the right way, and addressing its widespread use in the processing of images, sound, and video.

In this book we have avoided using the Z-transform of a filter, which is defined as $\sum_k t_k z^{-k}$ for any complex z where the sum converges (i.e. it is an extension of the frequency response to outside the unit circle). Only once in Sect. 3.4 we substituted $z = e^{i\omega}$ in the frequency response to arrive at the Z-transform, in order to make it clear that we actually perform division of polynomials in one variable. The Z-transform is heavily used in signal processing, and is an important tool in *filter design*, which is a field on its own outside the scope of this book. In terms of a realization in terms of difference equations as in Sect. 3.4, the Z-transform takes the form $\dfrac{\sum_k b_k z^{-k}}{\sum_{0 \le k \le N} a_k z^{-k}}$. The zeros in the denominator are here called *poles*. Placement of poles and circles in the Z-transform is a much used design requirement for filters.

In practice IIR filters are very useful, and deserve more space than given here. We concentrate mostly on FIR filters since the filters we encounter later in terms of wavelets and subband coding are FIR. Engineering courses elaborate much more on IIR filters, and in particular on how an IIR filter can be factored by rewriting the Z-transform in various ways.

We have encountered three different transforms between time and frequency: Fourier series, the DFT, and the DTFT. The *Continuous-time Fourier Transform* (CTFT), which serves as the bridge between the time- and frequency domains for non-periodic functions, will only be handled briefly in Chap. 6. Together those four transforms cover discrete and continuous aspects, as well as periodic and non-periodic aspects, of Fourier analysis. Signal processing literature spends much more time on these transforms and their relationships and dualities than we do here.

In order to learn more about the different signal processing concepts, and a more thorough introduction to IIR filters, the reader can consult many excellent textbooks, such as [78, 59, 1, 54, 68]. While we restrict to applications towards sound, images, and video, these also contain much more on other applications.

3.7 Summary

We defined digital filters, which do the same job for digital sound as analog filters do for (continuous) sound. We proved several equivalent characterizations of digital filters, such as being time-invariant, being diagonalizable by the DFT, and having a matrix which is circulant Toeplitz. Digital filters are, just as continuous-time filters, characterized by their frequency response. We also went through several important examples of filters, corresponding to meaningful operations such as adjustment of bass and treble in sound, and adding echo. We also looked at the role of filters in the MP3 standard.

We saw that symmetric filters preserve symmetric extensions. Restricting a symmetric filter to symmetric extensions gave rise to mappings from $\mathbb{R}^N$ to $\mathbb{R}^N$ diagonalized by the DCT.

Filters used in practice can be very complex (in that they may have many nonzero coefficients), be highly specialized, and adapted to a particular class of signals. Finding good filters not only depends on a thorough frequency analysis of the class of signals in question, but also on how we perceive them. Examples of such classes (besides audio and images, which this book mainly concentrates on) are speech, sonar, radar, and seismic signals. In medicine one can also find important classes such as X-ray signals, and those arising in Magnetic Resonance (MR) imaging. Many signal processing textbooks specialize filter design to various such classes.

What You Should Have Learned in This Chapter

- The formal definition of a digital filter in terms of having the Fourier vectors as eigenvectors.
- The compact filter notation for filters.
- The equivalent characterizations of a filter: time-invariance, circulant Toeplitz matrix, diagonalizable by the DFT.
- Convolution and circular convolution.

- Convolution in time corresponds to multiplication in frequency.
- Properties of the frequency response.
- Low-pass- and high-pass filters, such as moving average filters, and filters arising from rows in Pascal's triangle.

Chapter 4
Motivation for Wavelets and Some Simple Examples

In the first part of the book the focus was on approximating functions or vectors with trigonometric counterparts. We saw that Fourier series and the Discrete Fourier transform could be used to obtain such approximations, and that the FFT provided an efficient algorithm. This approach was useful for analyzing and filtering data, but had some limitations. Firstly, the frequency content is fixed over time in a trigonometric representation. This is in contrast to most sound, where the characteristics change over time. Secondly, we have seen that even if a sound has a simple trigonometric representation on two different time intervals, the representation as a whole may not be simple. In particular this is the case if the function is nonzero only on a very small time interval.

In this chapter we will introduce an alternative to Fourier analysis called *wavelets*. Wavelets are bases of functions with different properties than the Fourier basis, so that they can be used to address some of the shortcomings mentioned above. Contrary to the Fourier basis, wavelet bases are not fixed: there exist a wide range of such bases, for different applications. We will start this chapter with a motivation for wavelets, before we continue by introducing the simplest wavelet bases. The first wavelet we present can be interpreted as an approximation scheme based on piecewise constant functions. The next one is similar, but using piecewise linear functions instead. Based on these examples a more general framework will be established. In the following chapters we will put this framework in the context of filters, and construct more interesting wavelets.

4.1 Motivation for Wavelets, and a Wavelet Based on Piecewise Constant Functions

The left image in Fig. 4.1 shows a view of the entire earth. The startup image in Google Earth^{TM}, a program for viewing satellite images, maps and other geographic information, is very similar to this. In the middle image we have zoomed in on the Gulf of Mexico, as marked with a rectangle in the left image. Similarly, in the right image we have further zoomed in on Cuba and a small portion of Florida, as marked with a rectangle in the middle image. Since one can zoom in further and obtain enough detail to differentiate between buildings and even trees or cars all over the earth, there is clearly

© Springer Nature Switzerland AG 2019
Ø. Ryan, *Linear Algebra, Signal Processing, and Wavelets - A Unified Approach*,
Springer Undergraduate Texts in Mathematics and Technology,
https://doi.org/10.1007/978-3-030-01812-2_4

an amazing amount of information available behind a program like this. When the earth is spinning in the opening screen of Google Earth^{TM}, all the earth's buildings appear to be spinning with it. If this was the case the earth would not be spinning on the screen, since there would just be so much information to process that the computer would not be able to display a rotating earth.

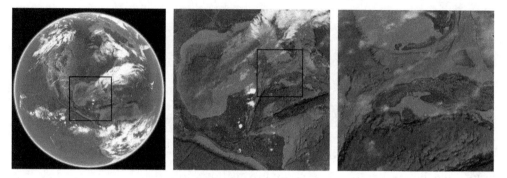

Fig. 4.1 A view of earth from space, together with versions of the image where we have zoomed in

There is a simple reason that the globe can be shown spinning in spite of the huge amounts of information that need to be handled. We are going to see later that a digital image is just a rectangular array of numbers that represent the color at a dense set of points. As an example, the images in Fig. 4.1 are made up of a grid of 1064×1064 points, which gives a total of 1,132,096 points. The color at a point is represented by three eight-bit integers, which means that the image files contain a total of 3,396,288 bytes each. So regardless of how close to the surface of the earth our viewpoint is, the resulting image always contains the same number of points. This means that when we are far away from the earth we can use a very coarse model of the geographic information that is being displayed, but as we zoom in, we need to display more details and therefore need a more accurate model.

A consequence of this observation is that for applications like Google Earth^{TM} we should use a mathematical model that makes it easy to switch between different levels of detail, or resolutions. Such models are called *multiresolution models*, and wavelets are prominent examples of such models. We will see that multiresolution models also can be used to approximate functions, just as Taylor series and Fourier series. This new approximation scheme differs from these in one important respect, however: When we approximate with Taylor series and Fourier series, the error must be computed at the same data points as well, so that the error contains just as much information as the approximating function and the function to be approximated. Multiresolution models on the other hand will be defined in such a way that the error and the approximation each contain half of the information in the function, i.e. their amount of data is reduced. This makes multiresolution models attractive for the problems at hand.

When we zoom in with Google Earth^{TM}, it seems that this is done continuously. The truth is probably that the program only has representations at some given resolutions (since each representation requires memory), and that one interpolates between these to give the impression of a continuous zoom. We will shortly look at how multiresolution models represent information at different resolutions.

We will now define wavelets formally, and construct the simplest wavelet. Its construction goes as follows: First we introduce what we call resolution spaces, and a corresponding scaling function. Then we introduce the detail spaces, and a corresponding mother wavelet. The scaling function and mother wavelet will in fact provide bases for the resolution spaces and detail spaces, and the Discrete Wavelet Transform will be defined as a change of coordinates between these bases. Let us start with the resolution spaces.

Definition 1. *The Resolution Space V_0.*
Let N be a natural number. The resolution space V_0 is defined as the space of functions defined on the interval $[0, N)$ that are constant on each subinterval $[n, n+1)$ for $n = 0, \ldots, N-1$.

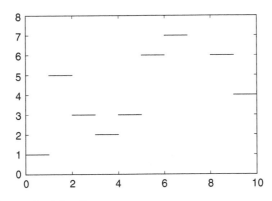

Fig. 4.2 A piecewise constant function

We will, just as we did in Fourier analysis, identify a function defined on $[0, N)$ with its (period N) periodic extension. An example of a function in V_0 for $N = 10$ is shown in Fig. 4.2.

V_0 is clearly a linear space. Let us first find its dimension and a corresponding basis.

Lemma 2. The Function ϕ.
Define the function $\phi(t)$ by

$$\phi(t) = \begin{cases} 1, & \text{if } 0 \leq t < 1; \\ 0, & \text{otherwise;} \end{cases} \tag{4.1}$$

The space V_0 has dimension N, and the N functions $\{\phi(t-n)\}_{n=0}^{N-1}$ form an orthonormal basis for V_0, denoted $\boldsymbol{\phi}_0$, with respect to the inner product

$$\langle f, g \rangle = \int_0^N f(t)g(t)\, dt. \tag{4.2}$$

Thus, any $f \in V_0$ can be represented as

$$f(t) = \sum_{n=0}^{N-1} c_n \phi(t - n) \tag{4.3}$$

for suitable coefficients $(c_n)_{n=0}^{N-1}$. The function $\phi(t-n)$ is referred to as the characteristic *function of the interval $[n, n+1)$.*

Note the small difference between the inner product we define here from the inner product we used in Fourier analysis: Here there is no scaling $1/T$ involved. Also, for wavelets we will only consider real functions, and the inner product will therefore not be defined for complex functions. Two examples of the basis functions defined in Lemma 2 are shown in Fig. 4.3.

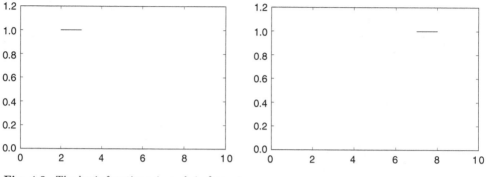

Fig. 4.3 The basis functions ϕ_2 and ϕ_7 from ϕ_0

Proof. Two functions ϕ_{n_1} and ϕ_{n_2} with $n_1 \neq n_2$ clearly satisfy $\int_0^N \phi_{n_1}(t)\phi_{n_2}(t)dt = 0$ since $\phi_{n_1}(t)\phi_{n_2}(t) = 0$ for all values of x. It is also easy to check that $\|\phi(t-n)\| = 1$ for all n. Finally, any function in V_0 can be written as a linear combination the functions $\phi_0, \phi_1, \ldots, \phi_{N-1}$, so the conclusion of the lemma follows. $\square$

Now, for f a given function, let us denote by $\mathrm{Supp}(f)$ the closure of the set of points where f is nonzero ($\mathrm{Supp}(f)$ is also called the *support* of f). Since ϕ is nonzero only on $[0, 1)$, we have that $\mathrm{Supp}(\phi) = [0, 1]$. If the support is a compact set, we will say that we have *compact support*. ϕ thus has compact support.

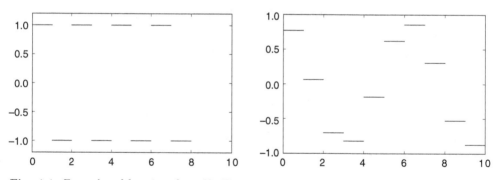

Fig. 4.4 Examples of functions from V_0. The square wave in V_0 (left), and an approximation to $\cos t$ from V_0 (right)

In our discussion of Fourier analysis, the starting point was the function $\sin(2\pi t)$. We can think of the space V_0 as being analogous to the space spanned by this function: The function $\sum_{n=0}^{N-1}(-1)^n\phi(t-n)$ is (part of the) square wave that we discussed in Chap. 1, and which also oscillates regularly like the sine function (left plot in Fig. 4.4). The difference is that we have more flexibility since we have a whole space at our disposal instead of just one function. The right plot in Fig. 4.4 shows another function in V_0.

In Fourier analysis we obtained a space of possible approximations by including sines of frequencies up to some maximum. We use a similar approach for constructing wavelets, but we double the frequency each time and label the spaces as $V_0, V_1, V_2, \ldots$

Definition 3. *Refined Resolution Spaces.*

The resolution space V_m is defined as the space of functions defined on the interval $[0, N)$ that are constant on each subinterval $[n/2^m, (n+1)/2^m)$ for $n = 0, 1, \ldots, 2^m N - 1$.

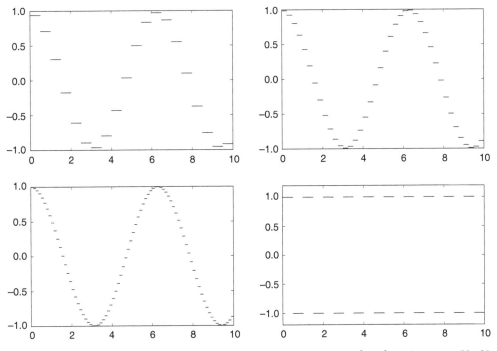

Fig. 4.5 Piecewise constant approximations to $\cos t$ on the interval $[0, 10]$ in the spaces V_1, V_2, and V_3. The lower right plot shows the square wave in V_2

Some examples of functions in the spaces V_1, V_2 and V_3 for the interval $[0, 10]$ are shown in Fig. 4.5. As m increases, we can represent smaller details. It is easy to find a basis for V_m, we just use the characteristic functions of each subinterval.

Lemma 4. Basis for V_m.

The dimension of V_m is $2^m N$, and the functions

$$\phi_{m,n}(t) = 2^{m/2}\phi(2^m t - n), \quad for \ n = 0, 1, \ldots, 2^m N - 1 \tag{4.4}$$

form an orthonormal basis for V_m. We will denote this basis by ϕ_m. Any function $f \in V_m$ can thus be represented uniquely as

$$f(t) = \sum_{n=0}^{2^m N - 1} c_{m,n}\phi_{m,n}(t).$$

Proof. The functions given by Eq. (4.4) are nonzero on the subintervals $[n/2^m, (n+1)/2^m)$, so that $\phi_{m,n_1}\phi_{m,n_2} = 0$ when $n_1 \neq n_2$, since these intervals are disjoint. The only mysterious thing may be the normalization factor $2^{m/2}$. This comes from the fact that

$$\int_0^N \phi(2^m t - n)^2 \, dt = \int_{n/2^m}^{(n+1)/2^m} \phi(2^m t - n)^2 \, dt = 2^{-m} \int_0^1 \phi(u)^2 \, du = 2^{-m}.$$

The normalization therefore ensures that $\|\phi_{m,n}\| = 1$ for all m. $\square$

In the following we will as above denote the coordinates in the basis $\boldsymbol{\phi}_m$ by $c_{m,n}$. Note that our definition restricts the dimensions of the spaces we study to be on the form $N2^m$. In Chap. 5 we will explain how this restriction can be dropped, but until then the dimensions will be assumed to be on this form. In the theory of wavelets, the function ϕ is also called a *scaling function*. The origin behind this name is that the scaled (and translated) functions $\phi_{m,n}$ of ϕ are used as basis functions for the resolution spaces. Later on we will find other scaling functions ϕ, and use their scaled versions $\phi_{m,n}$ similarly.

4.1.1 Function Approximation Property

Each time m is increased by 1, the dimension of V_m doubles, and the subinterval on which the functions are constant is halved in size. It therefore seems reasonable that, for most functions, we can find good approximations in V_m provided m is big enough. For continuous functions this is easy to show.

Theorem 5. Resolution Spaces and Approximation.
Let f be a given function that is continuous on the interval $[0, N]$. Given $\epsilon > 0$, there exists an integer $m \geq 0$ and a function $g \in V_m$ such that

$$\left| f(t) - g(t) \right| \leq \epsilon$$

for all t in $[0, N]$.

Proof. Since f is (uniformly) continuous on $[0, N]$, we can find an integer m so that $\left| f(t_1) - f(t_2) \right| \leq \epsilon$ for any two numbers t_1 and t_2 in $[0, N]$ with $|t_1 - t_2| \leq 2^{-m}$. Define the approximation g by

$$g(t) = \sum_{n=0}^{2^m N - 1} f\left(t_{m,n+1/2}\right) \phi_{m,n}(t),$$

where $t_{m,n+1/2}$ is the midpoint of the subinterval $\left[n2^{-m}, (n+1)2^{-m}\right)$,

$$t_{m,n+1/2} = (n + 1/2)2^{-m}.$$

For t in this subinterval we then obviously have $|f(t) - g(t)| \leq \epsilon$, and since these intervals cover $[0, N]$, the conclusion holds for all $t \in [0, N]$. $\square$

Theorem 5 does not tell us how to find the best approximation to f. Note that if we measure the error in the L^2-norm, we have

$$\|f - g\|^2 = \int_0^N \left| f(t) - g(t) \right|^2 \, dt \leq N\epsilon^2,$$

so $\|f - g\| \le \epsilon\sqrt{N}$. We therefore have the following corollary.

Corollary 6. Resolution Spaces and Approximation.
Let f be a given continuous function on the interval $[0, N]$. Then

$$\lim_{m \to \infty} \|f - proj_{V_m}(f)\| = 0.$$

Figure 4.6 illustrates how some of the approximations of the function $f(x) = x^2$ from the resolution spaces for the interval $[0, 1]$ improve with increasing m.

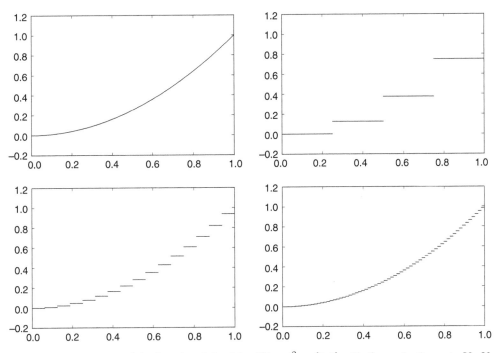

Fig. 4.6 Comparison of the function defined by $f(t) = t^2$ on $[0, 1]$ with the projection onto V_2, V_4, and V_6, respectively

4.1.2 Detail Spaces and Wavelets

So far we have described a family of function spaces that allow us to determine arbitrarily good approximations to a continuous function. The next step is to introduce the so-called detail spaces and the mother wavelet. We start by observing that since

$$[n, n + 1) = [2n/2, (2n + 1)/2) \cup [(2n + 1)/2, (2n + 2)/2),$$

we have

$$\phi_{0,n} = \frac{1}{\sqrt{2}}\phi_{1,2n} + \frac{1}{\sqrt{2}}\phi_{1,2n+1}. \tag{4.5}$$

This provides a formal proof of the intuitive observation that $V_0 \subset V_1$ (since all basis vectors of ϕ_0 are in V_1). It is easy to generalize this result.

Lemma 7. Resolution Spaces Are Nested.

The spaces V_0, V_1, ..., V_m, ... satisfy

$$V_0 \subset V_1 \subset V_2 \subset \cdots \subset V_m \cdots .$$

Proof. We have that

$$\phi_{m-1,n}(t) = 2^{(m-1)/2}\phi(2^{m-1}t - n) = 2^{(m-1)/2}\phi_{0,n}(2^{m-1}t)$$

$$= 2^{(m-1)/2}\frac{1}{\sqrt{2}}(\phi_{1,2n}(2^{m-1}t) + \phi_{1,2n+1}(2^{m-1}t))$$

$$= 2^{(m-1)/2}(\phi(2^m t - 2n) + \phi(2^m t - (2n+1)))$$

$$= \frac{1}{\sqrt{2}}(\phi_{m,2n}(t) + \phi_{m,2n+1}(t)),$$

so that

$$\phi_{m-1,n} = \frac{1}{\sqrt{2}}\phi_{m,2n} + \frac{1}{\sqrt{2}}\phi_{m,2n+1}. \tag{4.6}$$

It follows that $V_{m-1} \subset V_m$ for any integer $m \geq 1$.   □

Equation (4.6) is also called the *two-scale* equation. The next step is to characterize the projection from V_1 onto V_0, and onto the orthogonal complement of V_0 in V_1. Before we do this, let us make the following definitions.

Definition 8. *Detail Spaces.*

The orthogonal complement of V_{m-1} in V_m is denoted W_{m-1}. All the spaces W_k are also called *detail spaces*, or *error spaces*.

The name detail space is used since the projection from V_m onto V_{m-1} is considered as a (low-resolution) approximation, and the error, which lies in W_{m-1}, is the detail left out when we replace with this approximation. If $g_m \in V_m$, we can write $g_m = g_{m-1} + e_{m-1}$, with $g_{m-1} \in V_{m-1}$ and $e_{m-1} \in W_{m-1}$. In the context of our Google Earth[TM] example, in Fig. 4.1 you can interpret g_0 as the left image, the middle image as an excerpt of g_1, and e_0 as the additional details which are needed to reproduce the middle image from the left image.

Since V_0 and W_0 are mutually orthogonal spaces they are also linearly independent spaces. When U and V are two such linearly independent spaces, we will write $U \oplus V$ for the vector space consisting of all vectors of the form $\boldsymbol{u} + \boldsymbol{v}$, with $\boldsymbol{u} \in U$, $\boldsymbol{v} \in V$. $U \oplus V$ is also called the *direct sum* of U and V. This also makes sense if we have more than two vector spaces (such as $U \oplus V \oplus W$), and the direct sum clearly obeys the associate law $U \oplus (V \oplus W) = (U \oplus V) \oplus W$. Using the direct sum notation, we can first write

$$V_m = V_{m-1} \oplus W_{m-1}. \tag{4.7}$$

Since V_m has dimension $2^m N$, it follows that also W_m has dimension $2^m N$. We can continue the direct sum decomposition by also writing V_{m-1} as a direct sum, then V_{m-2} as a direct sum, and so on, and end up with

$$V_m = V_0 \oplus W_0 \oplus W_1 \oplus \cdots \oplus W_{m-1}, \tag{4.8}$$

where the spaces on the right hand side have dimension $N, N, 2N, \ldots, 2^{m-1}N$, respectively. This decomposition will be important for our purposes. It says that the resolution space V_m can be written as the sum of a lower order resolution space V_0, and m detail spaces $W_0, \ldots, W_{m-1}$.

It turns out that the following function will play the same role for the detail spaces as the function ϕ plays for the resolution spaces.

Definition 9. *The Function ψ.*

We define

$$\psi(t) = \frac{1}{\sqrt{2}}\phi_{1,0}(t) - \frac{1}{\sqrt{2}}\phi_{1,1}(t) = \phi(2t) - \phi(2t - 1), \tag{4.9}$$

and

$$\psi_{m,n}(t) = 2^{m/2}\psi(2^m t - n), \quad \text{for } n = 0, 1, \ldots, 2^m N - 1. \tag{4.10}$$

The functions ϕ and ψ are shown in Fig. 4.7.

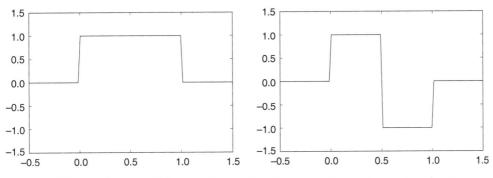

Fig. 4.7 The functions ϕ and ψ we used to analyze the space of piecewise constant functions

As in the proof for Eq. (4.6), it follows that

$$\psi_{m-1,n} = \frac{1}{\sqrt{2}}\phi_{m,2n} - \frac{1}{\sqrt{2}}\phi_{m,2n+1}, \tag{4.11}$$

Clearly $\text{Supp}(\psi) = [0, 1]$, and $\|\psi\| = 1$ on the support. From this it follows as for ϕ_0 that the $\{\psi_{0,n}\}_{n=0}^{N-1}$ are orthonormal. In the same way it follows that the $\{\psi_{m,n}\}_{n=0}^{2^m N-1}$ are orthonormal for any m, and we will write $\boldsymbol{\psi}_m$ for this basis. The next result motivates the definition of ψ, and states how we can project from V_1 onto V_0 and W_0, i.e. find the low-resolution approximation and the detail component of any function.

Lemma 10. Orthonormal Bases.

For $0 \leq n < N$ we have that

$$proj_{V_0}(\phi_{1,n}) = \begin{cases} \phi_{0,n/2}/\sqrt{2}, & \text{if } n \text{ is even;} \\ \phi_{0,(n-1)/2}/\sqrt{2}, & \text{if } n \text{ is odd.} \end{cases} \tag{4.12}$$

$$proj_{W_0}(\phi_{1,n}) = \begin{cases} \psi_{0,n/2}/\sqrt{2}, & \text{if } n \text{ is even;} \\ -\psi_{0,(n-1)/2}/\sqrt{2}, & \text{if } n \text{ is odd.} \end{cases} \tag{4.13}$$

In particular, $\boldsymbol{\psi}_0$ is an orthonormal basis for W_0.

Proof. We first observe that $\phi_{1,n}(t) \neq 0$ if and only if $n/2 \leq t < (n+1)/2$. Suppose that n is even. Then the intersection

$$\left[\frac{n}{2}, \frac{n+1}{2}\right) \cap [n_1, n_1 + 1) \qquad (4.14)$$

is nonempty only if $n_1 = \frac{n}{2}$. Using the orthogonal decomposition formula we get

$$\text{proj}_{V_0}(\phi_{1,n}) = \sum_{k=0}^{N-1} \langle \phi_{1,n}, \phi_{0,k} \rangle \phi_{0,k} = \langle \phi_{1,n}, \phi_{0,n_1} \rangle \phi_{0,n_1}$$

$$= \int_{n/2}^{(n+1)/2} \sqrt{2} \, dt \, \phi_{0,n/2} = \frac{1}{\sqrt{2}} \phi_{0,n/2}.$$

Using this we also get

$$\text{proj}_{W_0}(\phi_{1,n}) = \phi_{1,n} - \frac{1}{\sqrt{2}} \phi_{0,n/2} = \phi_{1,n} - \frac{1}{\sqrt{2}} \left(\frac{1}{\sqrt{2}} \phi_{1,n} + \frac{1}{\sqrt{2}} \phi_{1,n+1} \right)$$

$$= \frac{1}{2} \phi_{1,n} - \frac{1}{2} \phi_{1,n+1} = \psi_{0,n/2}/\sqrt{2}.$$

This proves the expressions for both projections when n is even. When n is odd, the intersection (4.14) is nonempty only if $n_1 = (n-1)/2$, which gives the expressions for both projections when n is odd in the same way. In particular we get

$$\text{proj}_{W_0}(\phi_{1,n}) = \phi_{1,n} - \frac{\phi_{0,(n-1)/2}}{\sqrt{2}} = \phi_{1,n} - \frac{1}{\sqrt{2}} \left(\frac{1}{\sqrt{2}} \phi_{1,n-1} + \frac{1}{\sqrt{2}} \phi_{1,n} \right)$$

$$= \frac{1}{2} \phi_{1,n} - \frac{1}{2} \phi_{1,n-1} = -\psi_{0,(n-1)/2}/\sqrt{2}.$$

$\boldsymbol{\psi}_0$ must be an orthonormal basis for W_0 since $\boldsymbol{\psi}_0$ is contained in W_0, and both have dimension N. $\quad\square$

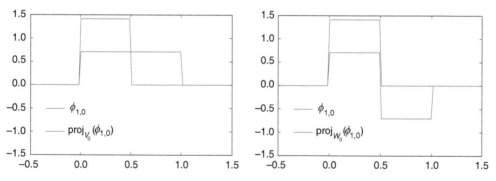

Fig. 4.8 The projection of $\phi_{1,0} \in V_1$ onto V_0 and W_0

In Fig. 4.8 we have used Lemma 10 to plot the projections of $\phi_{1,0} \in V_1$ onto V_0 and W_0. In the same way as in Lemma 10, it is possible to show that

$$\mathrm{proj}_{W_{m-1}}(\phi_{m,n}) = \begin{cases} \psi_{m-1,n/2}/\sqrt{2}, & \text{if } n \text{ is even;} \\ -\psi_{m-1,(n-1)/2}/\sqrt{2}, & \text{if } n \text{ is odd.} \end{cases} \tag{4.15}$$

From this it follows as before that $\boldsymbol{\psi}_m$ is an orthonormal basis for W_m. The function ψ thus has the property that its dilations and translations together span the detail components. Later we will encounter other functions, which also will be denoted by ψ, and have similar properties. In the theory of wavelets, such ψ are called *mother wavelets*.

From the above it also follows that both $\boldsymbol{\phi}_m$ and

$$\mathcal{C}_m = \{\phi_{m-1,0}, \psi_{m-1,0}, \phi_{m-1,1}, \psi_{m-1,1}, \cdots, \phi_{m-1,2^{m-1}N-1}, \psi_{m-1,2^{m-1}N-1}\} \tag{4.16}$$

(i.e. the basis vectors from $\boldsymbol{\phi}_{m-1}$ and $\boldsymbol{\psi}_{m-1}$ are put in alternating order) are bases for V_m. There is a good reason why the basis vectors in $\mathcal{C}_m$ are in alternating order. $\phi_{m-1,n}$ and $\psi_{m-1,n}$ are both supported on $[n2^{-m+1}, (n+1)2^{-m+1})$, so that this alternation will ensure that they have the correct order with respect to time. The index in the $\mathcal{C}_m$ basis thus corresponds to time, which is very natural. Let us make the following definition.

Definition 11. *Kernel Transformations.*
    The matrices

$$H = P_{\mathcal{C}_m \leftarrow \phi_m} \text{ and } G = P_{\phi_m \leftarrow \mathcal{C}_m}$$

are called *kernel transformations*.

The index in the basis $\boldsymbol{\phi}_m$ also corresponds to time (since $\phi_{m,n}$ is supported on $[n2^{-m}, (n+1)2^{-m})$). As a consequence kernel transformations can be computed sequentially in time, similarly to a filter. The kernel transformations are in fact very similar to filters, as we will return to in the next chapter.

If $\boldsymbol{c}_m$ and $\boldsymbol{w}_m$ are the coordinates in the bases $\boldsymbol{\phi}_m$ and $\boldsymbol{\psi}_m$, respectively, then

$$\begin{pmatrix} c_{m-1,0} \\ w_{m-1,0} \\ c_{m-1,1} \\ w_{m-1,1} \\ \vdots \end{pmatrix} = H \begin{pmatrix} c_{m,0} \\ c_{m,1} \\ c_{m,2} \\ c_{m,3} \\ \vdots \end{pmatrix} \text{ and } \begin{pmatrix} c_{m,0} \\ c_{m,1} \\ c_{m,2} \\ c_{m,3} \\ \vdots \end{pmatrix} = G \begin{pmatrix} c_{m-1,0} \\ w_{m-1,0} \\ c_{m-1,1} \\ w_{m-1,1} \\ \vdots \end{pmatrix}.$$

It is useful to interpret m as frequency, n as time, and $w_{m,n}$ as the contribution at frequency m and time n. In this sense our bases provide a *time-frequency representation* of functions, contrary to Fourier analysis which only provides frequency representations. From Eqs. (4.6) and (4.11) the following is apparent.

Theorem 12. Expressions for the Kernel Transformations H and G.
    G and H both equal the matrix where

$$\begin{pmatrix} \frac{1}{\sqrt{2}} & \frac{1}{\sqrt{2}} \\ \frac{1}{\sqrt{2}} & -\frac{1}{\sqrt{2}} \end{pmatrix} \tag{4.17}$$

is repeated along the main diagonal $2^{m-1}N$ times, i.e.

$$H = G = \frac{1}{\sqrt{2}} \begin{pmatrix} 1 & 1 & 0 & 0 & \cdots\cdots & 0 & 0 \\ 1 & -1 & 0 & 0 & \cdots\cdots & 0 & 0 \\ 0 & 0 & 1 & 1 & \cdots\cdots & 0 & 0 \\ 0 & 0 & 1 & -1 & \cdots\cdots & 0 & 0 \\ \vdots & \vdots & \vdots & \vdots & \vdots & \vdots & \vdots \\ 0 & 0 & 0 & 0 \cdots & \cdots & 1 & 1 \\ 0 & 0 & 0 & 0 \cdots & \cdots & 1 & -1 \end{pmatrix} \tag{4.18}$$

The only additional thing we need to address above is that the matrix $\begin{pmatrix} \frac{1}{\sqrt{2}} & \frac{1}{\sqrt{2}} \\ \frac{1}{\sqrt{2}} & -\frac{1}{\sqrt{2}} \end{pmatrix}$ equals its own inverse. A matrix as the one above is called a *block diagonal matrix* (see Appendix A). The matrices H and G are very sparse, with only two nonzero elements in each row/column. Multiplication with them can thus be performed efficiently, and there is no need to allocate the full matrix, just as for filters.

In addition to $\mathcal{C}_m$, it is customary to use a basis where the coordinates in ϕ_m are included before the coordinates in ψ_m. This means that the low-resolution coordinates are listed first, followed by detail coordinates of increasing resolution. The lower resolutions are most important for our overall impression of the object, with the higher resolutions important for the detail. Thus, this order would be natural for the data storage in Google Earth^{TM}: When images are downloaded to a web browser, the browser can very early show a low-resolution of the image, while waiting for the rest of the details in the image to be downloaded. If $\{\mathcal{B}_i\}_{i=1}^n$ are mutually independent bases, we will in the following write $(\mathcal{B}_1, \mathcal{B}_2, \ldots, \mathcal{B}_n)$ for the basis where vectors from $\mathcal{B}_i$ are included before $\mathcal{B}_j$ when $i < j$. With this notation, all of ϕ_m, $\mathcal{C}_m$ (ϕ_{m-1}, ψ_{m-1}) are bases for V_m, and by iteration we see that

$$(\phi_0, \psi_0, \psi_1, \cdots, \psi_{m-1})$$

also is a basis for V_m. We now have all the tools needed to define the Discrete Wavelet Transform.

Definition 13. *Discrete Wavelet Transform.*

The Discrete Wavelet Transform, or DWT, is defined as the change of coordinates from ϕ_1 to (ϕ_0, ψ_0). More generally, the m-level DWT is defined as the change of coordinates from ϕ_m to $(\phi_0, \psi_0, \psi_1, \cdots, \psi_{m-1})$. In an m-level DWT, the change of coordinates from

$$(\phi_{m-k+1}, \psi_{m-k+1}, \psi_{m-k+2}, \cdots, \psi_{m-1}) \text{ to } (\phi_{m-k}, \psi_{m-k}, \psi_{m-k+1}, \cdots, \psi_{m-1}) \tag{4.19}$$

is also called the k'th stage. The (m-level) IDWT (Inverse Discrete Wavelet Transform) is defined as the change of coordinates the opposite way.

The DWT thus corresponds to splitting into a sum of a low-resolution approximation, and detail components from increasing resolutions. Note that

$$\text{DWT} = P_{(\phi_{m-1}, \psi_{m-1}) \leftarrow \phi_m} = P_{(\phi_{m-1}, \psi_{m-1}) \leftarrow \mathcal{C}_m} P_{\mathcal{C}_m \leftarrow \phi_m} = P_{(\phi_{m-1}, \psi_{m-1}) \leftarrow \mathcal{C}_m} H,$$

and similarly

$$\text{IDWT} = G P_{\mathcal{C}_m \leftarrow (\phi_{m-1}, \psi_{m-1})}.$$

The DWT and the IDWT can thus be computed in terms of the kernel transformations by applying permutation matrices, since

- $P_{(\phi_{m-1},\psi_{m-1})\leftarrow\mathcal{C}_m}$ is a permutation matrix which groups the even elements first, then the odd elements,
- $P_{\mathcal{C}_m\leftarrow(\phi_{m-1},\psi_{m-1})}$ is a permutation matrix which places the first half at the even indices, the last half at the odd indices.

At each level in a DWT, V_k is split into one low-resolution component from V_{k-1}, and one detail component from W_{k-1}. We have illustrated this in Fig. 4.9.

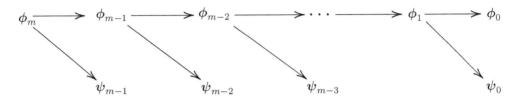

Fig. 4.9 Illustration of a wavelet transform

The detail component from W_{k-1} is not subject to further transformation. This is seen in the figure since ψ_{k-1} is a leaf node, i.e. there are no arrows going out from ψ_{m-1}. In a similar illustration for the IDWT, the arrows would go the opposite way.

The Discrete Wavelet Transform is the analog for wavelets to the Discrete Fourier transform. When applying the DFT to a vector of length N, one starts by viewing this vector as coordinates relative to the standard basis. When applying the DWT to a vector of length N, one instead views the vector as coordinates relative to the basis ϕ_m. This makes sense in light of Exercise 4.1.

Exercise 4.1: The Vector of Samples Is the Coordinate Vector

Show that the coordinate vector for $f \in V_0$ in the basis $\{\phi_{0,0}, \phi_{0,1}, \ldots, \phi_{0,N-1}\}$ is $(f(0), f(1), \ldots, f(N-1))$. This shows that, for $f \in V_m$, there is no loss in working with the samples of f rather than f itself.

Exercise 4.2: Realizing the DWT in Terms of Averages and Differences

Let $f(t) \in V_1$, and let $f_{n,1}$ be the value f attains on $[n, n+1/2)$, and $f_{n,2}$ the value f attains on $[n+1/2, n+1)$. Show that

- $\text{proj}_{V_0}(f)$ is the function in V_0 which equals $(f_{n,1}+f_{n,2})/2$ on the interval $[n, n+1)$.
- $\text{proj}_{W_0}(f)$ is the function in W_0 which is $(f_{n,1}-f_{n,2})/2$ on $[n, n+1/2)$, and $-(f_{n,1}-f_{n,2})/2$ on $[n+1/2, n+1)$.

Hint

Apply Lemma 10.

Exercise 4.3: Computing Projections

a) Use Lemma 10 to write down the matrix of $\text{proj}_{V_0} : V_1 \to V_0$ relative to ϕ_1 and ϕ_0, and the matrix of $\text{proj}_{W_0} : V_1 \to W_0$ relative to ϕ_1 and ψ_0.

b) Write down the matrix of $\text{proj}_{V_0} : V_1 \to V_0$ relative to (ϕ_0, ψ_0) and ϕ_0, and the matrix of $\text{proj}_{W_0} : V_1 \to W_0$ relative to (ϕ_0, ψ_0) and ψ_0.

c) Generalize the points above to the projections from V_{m+1} onto V_m and W_m.

Exercise 4.4: Finding the Least Squares Error

Show that

$$\|\text{proj}_{V_0} f - f\|^2 = \langle f, f \rangle - \sum_n \left(\int_n^{n+1} f(t) dt \right)^2.$$

Exercise 4.5: Direct Sums

Let $C_1, C_2 \ldots, C_n$ be independent vector spaces, and let $T_i : C_i \to C_i$ be linear transformations. The direct sum of $T_1, T_2, \ldots, T_n$, written as $T_1 \oplus T_2 \oplus \ldots \oplus T_n$, denotes the linear transformation from $C_1 \oplus C_2 \oplus \cdots \oplus C_n$ to itself defined by

$$T_1 \oplus T_2 \oplus \ldots \oplus T_n(c_1 + c_2 + \cdots + c_n) = T_1(c_1) + T_2(c_2) + \cdots + T_n(c_n)$$

when $c_1 \in C_1$, $c_2 \in C_2$, $\ldots$, $c_n \in C_n$. Also, when $A_1, A_2, \ldots, A_n$ are square matrices, $\text{diag}(A_1, A_2, \cdots, A_n)$ is defined as the block matrix where the blocks along the diagonal are $A_1, A_2, \ldots, A_n$, and all other blocks are 0 (see Appendix A). Show that, if $\mathcal{B}_i$ is a basis for C_i then

$$[T_1 \oplus T_2 \oplus \ldots \oplus T_n]_{(\mathcal{B}_1, \mathcal{B}_2, \ldots, \mathcal{B}_n)} = \text{diag}([T_1]_{\mathcal{B}_1}, [T_2]_{\mathcal{B}_2}, \ldots, [T_n]_{\mathcal{B}_n}).$$

Exercise 4.6: Properties of Block Diagonal Matrices

Assume that A, B, C, and D are square matrices of the same dimensions.

a) Assume that the eigenvalues of A are equal to those of B. What are the eigenvalues of $\text{diag}(A, B)$? Can you express the eigenvectors of $\text{diag}(A, B)$ in terms of those of A and B?

b) Assume that A and B also are non-singular. Show that $\mathrm{diag}(A, B)$ is non-singular, and that $(\mathrm{diag}(A, B))^{-1} = \mathrm{diag}(A^{-1}, B^{-1})$.

c) Show that $\mathrm{diag}(A, B)\mathrm{diag}(C, D) = \mathrm{diag}(AC, BD)$.

Exercise 4.7: The DWT When N Is Odd

When N is odd, the (first stage in a) DWT is defined as the change of coordinates from $(\phi_{1,0}, \phi_{1,1}, \ldots, \phi_{1,N-1})$ to

$$(\phi_{0,0}, \psi_{0,0}, \phi_{0,1}, \psi_{0,1}, \ldots, \phi_{0,(N-3)/2}, \psi_{0,(N-3)/2}, \phi_{0,(N-1)/2}).$$

Since all functions are assumed to have period N, we have that

$$\phi_{0,(N-1)/2} = \frac{1}{\sqrt{2}}(\phi_{1,N-1} + \phi_{1,N}) = \frac{1}{\sqrt{2}}(\phi_{1,0} + \phi_{1,N-1}).$$

This says that the last column in the change of coordinate matrix from $\mathcal{C}_1$ to ϕ_1 (i.e. the IDWT matrix) equals $\frac{1}{\sqrt{2}}(1, 0, \ldots, 0, 1)$. In particular, for $N = 3$, the IDWT matrix equals

$$\frac{1}{\sqrt{2}}\begin{pmatrix} 1 & 1 & 1 \\ 1 & -1 & 0 \\ 0 & 0 & 1 \end{pmatrix},$$

The DWT matrix, i.e. the inverse of this, is

$$\frac{1}{\sqrt{2}}\begin{pmatrix} 1 & 1 & -1 \\ 1 & -1 & -1 \\ 0 & 0 & 2 \end{pmatrix}.$$

a) Explain from this that, when N is odd, the DWT matrix can be constructed from the DWT matrix with $N - 1$ columns by adding $\frac{1}{\sqrt{2}}(-1, -1, 0, \ldots, 0, 2)^T$ as the last columns, and adding zeros in the last row.

b) Explain that the DWT matrix is orthogonal if and only if N is even. Also explain that it is only the last column which spoils the orthogonality.

Exercise 4.8: General Kernel Transformations

In the next sections we will define other functions ϕ and ψ. For these we will also define $\phi_{m,n}(t) = 2^{m/2}\phi(2^m t - n)$ and $\psi_{m,n}(t) = 2^{m/2}\psi(2^m t - n)$, and we will define resolution- and detail spaces using these instead, and find similar properties to those we found in this section. These new functions give rise to new kernel transformations G and H. From this exercise it will be clear that, also in this more general setting, it causes no confusion to suppress the resolution m in the notation for the kernel transformations (i.e. H and G), as we have done.

a) Show that if

$$\phi(t) = \sum_n a_n \phi_{1,n}(t) \qquad\qquad \psi(t) = \sum_n b_n \phi_{1,n}(t),$$

then also

$$\phi_{m,k}(t) = \sum_n a_n \phi_{m+1,2k+n}(t) \qquad \psi_{m,k}(t) = \sum_n b_n \phi_{m+1,2k+n}(t).$$

b) Show that if

$$\phi_{1,0}(t) = \sum_n \left(c_n \phi_{0,n}(t) + d_n \psi_{0,n}(t) \right)$$

$$\phi_{1,1}(t) = \sum_n \left(e_n \phi_{0,n}(t) + f_n \psi_{0,n}(t) \right),$$

then also

$$\phi_{m+1,2k}(t) = \sum_n \left(c_n \phi_{m,n+k}(t) + d_n \psi_{m,n+k}(t) \right)$$

$$\phi_{m+1,2k+1}(t) = \sum_n \left(e_n \phi_{m,n+k}(t) + f_n \psi_{m,n+k}(t) \right).$$

4.2 Implementation of the DWT

The DWT is straightforward to implement: Simply iterate the *kernel transformation* (4.18) for $m, m-1, \ldots, 1$ by calling a *kernel function*. We will consider the following implementation of this:

```
function x = dwt_kernel_haar(x, bd_mode)
    x = x/sqrt(2);
    N = size(x, 1);
    for k = 1:2:(N-1)
        x(k:(k+1), :) = [x(k, :) + x(k+1, :); x(k, :) - x(k+1, :)];
    end
```

For simplicity this first version of the code assumes that N is even. See Exercises 4.7 and 4.18 for the case when N is odd. The name `dwt_kernel_haar` is inspired by Alfred Haar, often considered as the inventor of the wavelet based on piecewise constant functions, which is also called the *Haar wavelet*. The code above, just as `fft_impl`, accepts higher dimensional data, and can thus be applied to all channels in a sound simultaneously. The reason for using a kernel function will be apparent later, when we construct other types of wavelets. It is not meant that you call `dwt_kernel_haar` directly yourself. Instead call the function

```
dwt_impl(x, wave_name, m, bd_mode, prefilter_mode, ...
         dims, dual, transpose, data_layout)
```

where

- `x` is the input to the DWT,
- `wave_name` is a name identifying the wavelet (for the current wavelet this is `"Haar"`),
- `m` is the number of levels.

The parameter m is optional. So are also the remaining parameters, which will be addressed later. The optional parameters have meaningful default values.[1] We do not list the entire code for `dwt_impl` here (as well as for some other functions we now go through), only go through the main steps, and encourage you to go through the remaining parts on your own.

When `dwt_impl` is called, the `wave_name` parameter is passed to a function called `find_wav_props`, which computes quantities needed for computation with the wavelet in question. These quantities are then passed to a function `find_kernel`, which returns a kernel function which use these quantities. When "Haar" is input to `dwt_impl`, `find_kernel` will return a kernel very similar to the `dwt_kernel_haar` listed above. Finally the returned kernel is passed to a function `dwt1_impl_internal`, which computes the actual DWT (the `1` in `dwt1_impl_internal` reflects that a one-dimensional DWT is computed. We will later consider extensions to several dimensions, for which other versions are optimized).

In what follows you will be encouraged to write kernel functions for other wavelets on your own. You can then by-pass `dwt_impl` and use your own kernel by simply writing

```
x = dwt1_impl_internal(x, @dwt_kernel_haar);
```

Computing the DWT this way is preferred for another reason as well: `dwt_impl` can be applied to many types of wavelets, of which many require some pre-computation to find a kernel function. `dwt_impl` will make those pre-computations every time. If you pass a kernel as above to `dwt1_impl_internal`, however, you can reuse a kernel which encapsulates these pre-computations, so that computations are not done every time.[2] Such kernels can also be stored in files for reuse, using the built-in persistence functions in MATLAB.

`dwt1_impl_internal` invokes the kernel function one time for each resolution. Note that the coordinates from ϕ_m end up at indices $k2^m$ apart after m iterations of the kernel. `dwt1_impl_internal` calls a function which sorts the data according to the `data_layout` parameter, which is optional and dictates how the output should be sorted:

- If `data_layout` is `"resolution"`, the output is sorted by resolution after a change of coordinates to $(\phi_0, \psi_0, \psi_1, \cdots, \psi_{m-1})$.
- If `data_layout` is `"time"`, nothing is done, as kernels are time-preserving.

`dwt1_impl_internal` is non-recursive in that a for-loop runs through the different DWT levels. The first levels require the most operations, since the latter levels leave an increasing part of the coordinates unchanged.

There is a similar setup for the IDWT:

```
idwt_impl(x, wave_name, m, bd_mode, prefilter_mode, ...
          dims, dual, transpose, data_layout)
```

Also this function will use `find_kernel` to find the kernel function corresponding to the name `wave_name`. This kernel is then sent as input to a function `idwt1_impl_internal`, which computes the actual IDWT. The `data_layout` parameter again dictates how `idwt1_impl_internal` sorts the data.

Let us round off this section with some important examples using these functions.

---

[1] See the documentation and implementation in the library for what these default values are.

[2] Consult the documentation of `dwt_impl` in the library for a full code example.

Example 4.9: Computing the DWT by Hand

In some cases, the DWT can be computed by hand, keeping in mind its definition as a change of coordinates. Consider the simple vector x of length $2^{10} = 1024$ defined by

$$x_n = \begin{cases} 1 & \text{for } 0 \le n < 512 \\ 0 & \text{for } 512 \le n < 1024, \end{cases}$$

and let us compute the 10-level DWT of x by first visualizing the function with these coordinates. Since $m = 10$, x are the coordinates in the basis ϕ_{10} of a function $f \in V_{10}$. More precisely, $f(t) = \sum_{n=0}^{511} \phi_{10,n}(t)$, and since $\phi_{10,n}$ is supported on $[2^{-10}n, 2^{-10}(n+1))$, the support of f has width $512 \times 2^{-10} = 1/2$ (512 translates, each of width 2^{-10}). Moreover, since $\phi_{10,n}$ is $2^{10/2} = 2^5 = 32$ on $[2^{-10}n, 2^{-10}(n+1))$ and 0 elsewhere, it is clear that

$$f(t) = \begin{cases} 32 & \text{for } 0 \le t < 1/2 \\ 0 & \text{for } t \ge 1/2. \end{cases}$$

This is by definition a function in V_1: f must in fact be a multiple of $\phi_{1,0}$, since this also is supported on $[0, 1/2)$. We can thus write $f(t) = c\phi_{1,0}(t)$ for some c. We can find c by setting $t = 0$. This gives that $32 = 2^{1/2}c$ (since $f(0) = 32$, $\phi_{1,0}(0) = 2^{1/2}$), so that $c = 32/\sqrt{2}$. This means that $f(t) = \frac{32}{\sqrt{2}}\phi_{1,0}(t)$, f is in V_1, and with coordinates $(32/\sqrt{2}, 0, \ldots, 0)$ in ϕ_1.

When we run a 10-level DWT we make a change of coordinates from ϕ_{10} to $(\phi_0, \psi_0, \cdots, \psi_9)$. The first 9 levels give us the coordinates in $(\phi_1, \psi_1, \psi_2, \ldots, \psi_9)$, and these are $(32/\sqrt{2}, 0, \ldots, 0)$ from what we showed. It remains thus only to perform the last level in the DWT, i.e. perform the change of coordinates from ϕ_1 to (ϕ_0, ψ_0). Since $\phi_{1,0}(t) = \frac{1}{\sqrt{2}}(\phi_{0,0}(t) + \psi_{0,0}(t))$ we get

$$f(t) = \frac{32}{\sqrt{2}}\phi_{1,0}(t) = \frac{32}{\sqrt{2}}\frac{1}{\sqrt{2}}(\phi_{0,0}(t) + \psi_{0,0}(t)) = 16\phi_{0,0}(t) + 16\psi_{0,0}(t).$$

From this we see that the coordinate vector of f in $(\phi_0, \psi_0, \cdots, \psi_9)$, i.e. the 10-level DWT of x, is $(16, 16, 0, 0, \ldots, 0)$. Note that here V_0 and W_0 are both 1-dimensional, since V_{10} was assumed to be of dimension 2^{10} (in particular, $N = 1$).

It is straightforward to verify what we just found as follows:

```
x= dwt_impl([ones(512,1); zeros(512,1)], 'Haar', 10);
```

The reason why the method from this example worked was that the vector we started with had a simple representation in the wavelet basis, actually it equaled the coordinates of a basis function in ϕ_1. Usually this is not the case, and our only possibility is to run the DWT on a computer.

Example 4.10: DWT on Sound

Let us plot the samples of our audio sample file, and compare them with the first order DWT. Both are shown in Fig. 4.10.

The first part of the DWT plot represents the low resolution part, the second the detail. Since $\phi(2^m t - n) \in V_m$ oscillates more quickly than $\phi(t - n) \in V_0$, one is lead to

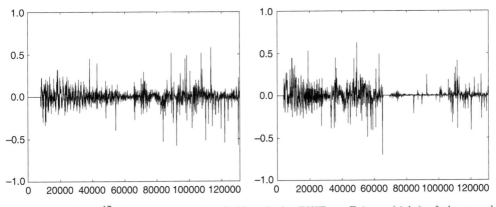

Fig. 4.10   The 2^{17} first sound samples (left) and the DWT coefficients (right) of the sound castanets.wav

believe that coefficients from lower order resolution spaces correspond to lower frequencies. The functions $\phi_{m,n}$ do not correspond to pure tones in the setting of wavelets, however, but let us nevertheless listen to sound from the different resolution spaces. The library includes a function forw_comp_rev_dwt1 which runs an m-level DWT on the first samples of the audio sample file, extracts the detail or the low-resolution approximation, and runs an IDWT to reconstruct the sound. Since the returned values may lie outside the legal range $[-1, 1]$, the values are normalized at the end. To listen to the low-resolution approximation, write

```
[x, fs] = forw_comp_rev_dwt1(m, 'Haar');
playerobj = audioplayer(x, fs);
playblocking(playerobj)
```

For $m = 2$ we clearly hear a degradation in the sound. For $m = 4$ and above most of the sound is unrecognizable, as too much of the detail is omitted. To be more precise, when listening to the sound by throwing away detail from W_0, W_1,...,W_{m-1}, we are left with a 2^{-m} share of the data.

For $m = 1$ and $m = 2$ the detail can be played as follows

```
[x, fs] = forw_comp_rev_dwt1(1, 'Haar', 0);
playerobj = audioplayer(x, fs);
playblocking(playerobj)
```

```
[x, fs] = forw_comp_rev_dwt1(2, 'Haar', 0);
playerobj = audioplayer(x, fs);
playblocking(playerobj)
```

They are plotted in Fig. 4.11.

It is seen that the detail is larger in the part of the sound where there are bigger variations. The detail is clearly a very degraded version of the sound, but if you increase m, the sound will improve in quality. The reason is that more and more information is contained in the detail components for large m, which can be seen when we compare the plot for $m = 1$ and $m = 2$.

The previous example illustrates that wavelets as well may be used to perform operations on sound. In this book the main application for wavelets will be images, however, where they have found a more important role. Images typically display variations which are less abrupt than the ones found in sound. The main motivation behind wavelets

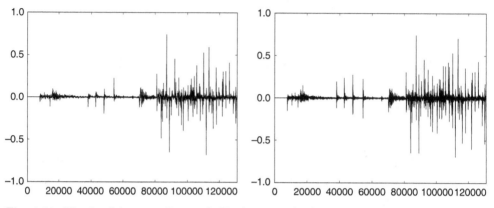

Fig. 4.11 The detail in our audio sample file, for $m = 1$ (left) and $m = 2$ (right)

comes from the fact that the detail components often are very small, and in less abrupt data such as images, the detail components will be even smaller. After a DWT one is therefore often left with a couple of significant coefficients, while most of the detail is small. This is a very good starting point for compression methods. When we later look at how wavelets are applied to images, we will need to handle one final hurdle, namely that images are two-dimensional.

Example 4.11: DWT on the Samples of a Mathematical Function

Above we plotted the DWT coefficients of a sound, as well as the detail/error. We can also experiment with samples generated from a mathematical function. Figure 4.12 plots the error for different functions, with $N = 1024$.

In these cases, we see that we require large m before the detail/error becomes significant. We see also that there is no error for the square wave. The reason is that the square wave is a piecewise constant function, so that it can be represented exactly by the basis functions. For the other functions this is not the case, however, so we get an error.

Example 4.12: Computing the Wavelet Coefficients

In the previous example, `dwt_impl` was used to compute the error. For some functions it is also possible to compute this by hand. You will be asked to do this in Exercise 4.16. To exemplify the general procedure for this, consider the function $f(t) = 1 - t/N$. This decreases linearly from 1 to 0 on $[0, N]$, so that it is not piecewise constant, and does not lie in any of the spaces V_m. We can instead consider $\text{proj}_{V_m} f \in V_m$, and apply the

DWT to this. Let us compute the ψ_m-coordinates $w_{m,n}$ of $\mathrm{proj}_{V_m} f$ in the orthonormal basis $(\phi_0, \psi_0, \psi_1, \ldots, \psi_{m-1})$. The orthogonal decomposition theorem says that

$$w_{m,n} = \langle f, \psi_{m,n} \rangle = \int_0^N f(t)\psi_{m,n}(t)dt = \int_0^N (1 - t/N)\psi_{m,n}(t)dt.$$

Using the definition of $\psi_{m,n}$ we see that this can also be written as

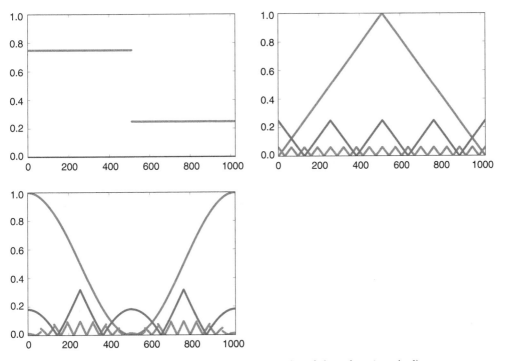

Fig. 4.12 The error in a 1024-point DWT of the samples of three functions (red): a square wave, the linear function $f(t) = 1 - 2|1/2 - t/N|$, and the trigonometric function $f(t) = 1/2 + \cos(2\pi t/N)/2$. The detail is indicated for $m = 6$ (green) and $m = 8$ (blue)

$$2^{m/2} \int_0^N (1 - t/N)\psi(2^m t - n)dt$$

$$= 2^{m/2} \left(\int_0^N \psi(2^m t - n)dt - \int_0^N \frac{t}{N}\psi(2^m t - n)dt \right).$$

Clearly $\int_0^N \psi(t) = 0$, so that $\int_0^N \psi(2^m t - n)dt = 0$, so that the first term above vanishes. Moreover, $\psi_{m,n}$ is nonzero only on $[2^{-m}n, 2^{-m}(n+1))$, and is 1 on $[2^{-m}n, 2^{-m}(n+1/2))$, and -1 on $[2^{-m}(n+1/2), 2^{-m}(n+1))$. We therefore get

$$w_{m,n} = -2^{m/2} \left(\int_{2^{-m}n}^{2^{-m}(n+1/2)} \frac{t}{N}dt - \int_{2^{-m}(n+1/2)}^{2^{-m}(n+1)} \frac{t}{N}dt \right)$$

$$= -2^{m/2} \left(\left[\frac{t^2}{2N} \right]_{2^{-m}n}^{2^{-m}(n+1/2)} - \left[\frac{t^2}{2N} \right]_{2^{-m}(n+1/2)}^{2^{-m}(n+1)} \right)$$

$$= -2^{m/2} \left(\frac{2^{-2m}(n+1/2)^2}{2N} - \frac{2^{-2m}n^2}{2N} \right)$$

$$+ 2^{m/2} \left(\frac{2^{-2m}(n+1)^2}{2N} - \frac{2^{-2m}(n+1/2)^2}{2N} \right)$$

$$= -2^{m/2} \left(-\frac{2^{-2m}n^2}{2N} + \frac{2^{-2m}(n+1/2)^2}{N} - \frac{2^{-2m}(n+1)^2}{2N} \right)$$

$$= -\frac{2^{-3m/2}}{2N} \left(-n^2 + 2(n+1/2)^2 - (n+1)^2 \right) = \frac{1}{N2^{2+3m/2}}.$$

We see in particular that $w_{m,n} \to 0$ when $m \to \infty$. Also, all coordinates were equal, i.e. $w_{m,0} = w_{m,1} = w_{m,2} = \cdots$. It is not too hard to see that this will always happen when f is linear. We see also that there were a lot of computations even in this very simple example. For most functions we therefore usually do not compute $w_{m,n}$ symbolically, but instead run an implementation.

Exercise 4.13: Finding N

Assume that you run an m-level DWT on a vector of length r. What value of N does this correspond to?

Exercise 4.14: Different DWT's for Similar Vectors

In Fig. 4.13 we have plotted the DWT's of two vectors $\boldsymbol{x}_1$ and $\boldsymbol{x}_2$. In both vectors we have 16 ones followed by 16 zeros, and this pattern repeats cyclically so that the length of both vectors is 256. The only difference is that the second vector is obtained by delaying the first vector with one element.

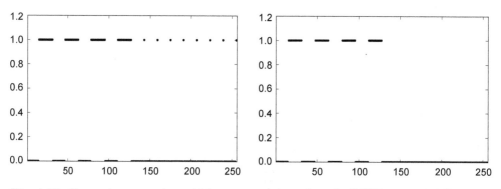

Fig. 4.13 Two vectors $\boldsymbol{x}_1$ and $\boldsymbol{x}_2$ which seem equal, but where the DWT's are very different

You see that the two DWT's are very different: For the first vector we see that there is much detail present (the second part of the plot), while for the second vector there is no detail present. Attempt to explain why this is the case. Based on your answer, also attempt to explain what can happen if you change the point of discontinuity for the piecewise constant function to something else.

Exercise 4.15: Construct a Sound

Construct a (nonzero) sound where the low resolution approximations equal the sound itself for $m = 1$, $m = 2$.

Exercise 4.16: Exact Computation of Wavelet Coefficients

For the following functions compute the detail coefficients analytically, i.e. compute the quantities $w_{m,n} = \int_0^N f(t)\psi_{m,n}(t)dt$ similarly to how this was done in Example 4.12:

a) The functions in Example 4.11.

b) The functions $f(t) = \left(\frac{t}{N}\right)^k$.

Exercise 4.17: Computing the DWT of a Simple Vector

Suppose that we have the vector x with length $2^{10} = 1024$, defined by $x_n = 1$ for n even, $x_n = -1$ for n odd. What will be the result if you run a 10-level DWT on x? Use the function dwt_impl to verify what you have found.

Hint

We defined ψ by $\psi(t) = (\phi_{1,0}(t) - \phi_{1,1}(t))/\sqrt{2}$. From this connection it follows that $\psi_{9,n} = (\phi_{10,2n} - \phi_{10,2n+1})/\sqrt{2}$, and thus $\phi_{10,2n} - \phi_{10,2n+1} = \sqrt{2}\psi_{9,n}$. Try to couple this identity with the alternating sign you see in x.

Exercise 4.18: Implementing the DWT When N Is Odd

Use the results from Exercise 4.7 to rewrite the implementations dwt_kernel_haar and idwt_kernel_haar so that they also work in the case when N is odd.

Exercise 4.19: In-Place DWT and Partial Bit-Reversal

The kernel transformations in this book can be computed in-place. A DWT is also required to reorder coordinates as in the $(\phi_0, \psi_0, \ldots, \psi_{m-1})$-basis, however, and this exercise addresses how this reordering also can be computed in-place. As a result the entire DWT can be computed in-place.

a) Show that the coordinates in ϕ_0 after m iterations of a kernel transformation end up at indices $k2^m$, $k = 0, 1, 2, \ldots$, and that the coordinates in ψ_0 end up at indices $2^{m-1} + k2^m$, $k = 0, 1, 2, \ldots$.

b) After m iterations of a kernel transformation, show that if the indices are rearranged so that the last m bits are placed in front in reverse order, then coordinates from ϕ_0 are placed first, while coordinates from ψ_i are placed before those from $\{\psi_j\}_{j>i}$. This is also called a *partial bit-reversal*. When an m-level DWT on a vector of length 2^m is computed, the partial bit-reversal is actually a full bit-reversal. After a partial bit-reversal, what can you say about the internal ordering in $\{\psi_j\}_{j>i}$?

c) Write a function `partial_bit_reversal(x, m)` which computes a partial bit-reversal of the vector x in-place. Assume that x has length 2^n with $n \geq m$.

Hint

Partial bit-reversal can be split in two: First a full bit-reversal, and then reversing the last $n - m$ bits again. It turns out that reversal of the last bits can be computed easily using bit-reversal of $n - m$ bits. Due to this we can easily implement partial bit-reversal by adapting our previous function for bit-reversal.

d) Contrary to full bit-reversal, partial bit-reversal is not its own inverse. It is straightforward, however, to compute the inverse following the same lines as in c). Extend the function `partial_bit_reversal` so that it takes a third parameter `forward`, which indicates whether a forward- or reverse partial bit-reversal should be computed.

4.3 A Wavelet Based on Piecewise Linear Functions

Unfortunately, piecewise constant functions are too simple to provide good approximations. In this section we will make a similar construction, instead using *piecewise linear functions*. The advantage is that piecewise linear functions are better for approximating smooth functions and data, and our hope is that this translates into smaller components (errors) in the detail spaces in many practical situations. In the setting of piecewise linear functions it turns out that we loose the orthonormal bases we previously had. On the other hand, we will see that the new scaling functions and mother wavelets are symmetric functions, and we will later see implementations of the DWT and IDWT can be adapted to this. Our experience from deriving the Haar wavelet will guide us in our new construction. Let us first define new resolution spaces.

Definition 14. *Resolution Spaces of Piecewise Linear Functions.*
    The space V_m is the subspace of continuous functions on $\mathbb{R}$ which are periodic with period N, and linear on each subinterval of the form $[n2^{-m}, (n+1)2^{-m})$.

    The left plot in Fig. 4.14 shows an example of a piecewise linear function in V_0 on the interval $[0, 10]$. The functions in V_0 that are 1 at one integer and 0 at all others are particularly simple, see the right plot in Fig. 4.14. These simple functions are all translates of each other and therefore gives rise to a new kind of scaling function.

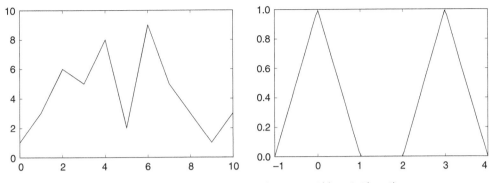

Fig. 4.14 A piecewise linear function and the two functions $\phi(t)$ and $\phi(t-3)$

Lemma 15. The Function ϕ.

Let the function ϕ be defined by

$$\phi(t) = \begin{cases} 1 - |t|, & \text{if } -1 \le t \le 1; \\ 0, & \text{otherwise;} \end{cases} \tag{4.20}$$

and for any $m \ge 0$ set

$$\phi_{m,n}(t) = 2^{m/2}\phi(2^m t - n)$$

and $\boldsymbol{\phi}_m = \{\phi_{m,n}\}_{n=0}^{2^m N - 1}$. $\boldsymbol{\phi}_m$ is a basis for V_m.

Since we now assume that all functions are periodic, ϕ would split into two non-zero parts if we plotted it over $[0, N]$: one close to 0, and one close to N. Plotting over $[-N/2, N/2]$ and using the periodicity is preferred, since the graph then appear as one connected part.

Proof. We have that

$$\phi_{m,n'}(n2^{-m}) = 2^{m/2}\phi(2^m(2^{-m}n) - n') = 2^{m/2}\phi(n - n') = \begin{cases} 2^{m/2} & n' = n \\ 0 & n' \ne n. \end{cases}$$

If $\sum_{n'=0}^{2^m N - 1} \alpha_{n'}\phi_{m,n'} = 0$, we have in particular that

$$\sum_{n'=0}^{2^m N - 1} \alpha_{n'}\phi_{m,n'}(n2^{-m}) = \alpha_n\phi_{m,n}(n2^{-m}) = \alpha_n 2^{m/2} = 0,$$

so that $\alpha_n = 0$ for all n. It follows that $\boldsymbol{\phi}_m$ is a linearly independent set of functions in V_m. Also, any function in V_m is uniquely defined from its values in the $n2^{-m}$, so that the dimension of V_m is $2^m N$. Since $\boldsymbol{\phi}_m$ has the same number of functions, it follows that $\boldsymbol{\phi}_m$ is a basis for V_m. $\square$

Clearly also $\phi_{0,n}(t)$ is the function in V_0 with smallest support that is nonzero at $t = n$, since any such function in V_0 must be nonzero on $(n-1, n+1)$. The function ϕ and its translates and dilates are often referred to as *hat functions* for obvious reasons. Also in the piecewise linear case coordinates correspond to samples:

Lemma 16. Writing in Terms of the Samples.

A function $f \in V_m$ may be written as

$$f(t) = \sum_{n=0}^{2^m N - 1} f(n/2^m) 2^{-m/2} \phi_{m,n}(t). \tag{4.21}$$

Also here the spaces are nested.

Lemma 17. Resolution Spaces Are Nested.

The piecewise linear resolution spaces are nested, i.e.

$$V_0 \subset V_1 \subset \cdots \subset V_m \subset \cdots.$$

Proof. We only need to prove that $V_0 \subset V_1$ since the other inclusions are similar. But this is immediate since any function in V_0 is continuous, and linear on any subinterval in the form $[n/2, (n+1)/2)$. □

In the piecewise constant case, we saw in Lemma 2 that the basis functions were orthogonal since their supports did not overlap. This is not the case in the linear case, but we could orthogonalize the basis ϕ_m with the Gram-Schmidt process from linear algebra. The disadvantage is that we loose the nice local behavior of the scaling functions and end up with basis functions that are nonzero over all of $[0, N]$. And for many applications, orthogonality is not essential; we just need a basis. The next step is to find formulas that let us express a function given in the basis ϕ_0 for V_0 in terms of the basis ϕ_1 for V_1.

Lemma 18. The Two-Scale Equation.

The function ϕ satisfy the relation

$$\phi = \frac{1}{\sqrt{2}} \left(\frac{1}{2}\phi_{1,-1} + \phi_{1,0} + \frac{1}{2}\phi_{1,1} \right). \tag{4.22}$$

This equation has a similar interpretation as Eq. (4.6) for the Haar wavelet, so that it is also called the two-scale equation.

Proof. Since $\phi \in V_0$ it may be expressed in the basis ϕ_1 with formula (4.21),

$$\phi(t) = 2^{-1/2} \sum_{k=0}^{2N-1} \phi(k/2)\phi_{1,k}(t).$$

The relation (4.22) now follows since

$$\phi(-1/2) = \phi(1/2) = 1/2, \quad \phi(0) = 1,$$

and $\phi(k/2) = 0$ for all other values of k. □

The relationship given by Eq. (4.22) is shown in Fig. 4.15.

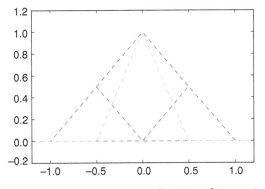

Fig. 4.15 How ϕ can be decomposed as a linear combination of $\phi_{1,-1}$, $\phi_{1,0}$, and $\phi_{1,1}$

4.3.1 Detail Spaces and Wavelets

The next step is the to define new detail spaces. We need to determine a space W_0 that is linearly independent from V_0, and so that $V_1 = V_0 \oplus W_0$. For piecewise constant functions we simply defined W_0 as an orthogonal complement. This strategy is less appealing in the case of piecewise linear functions since our new basis functions are not orthogonal anymore (see Exercise 4.23), so that the orthogonal decomposition theorem fails to give us projections in a simple way. Instead of using projections to find low-resolution approximations, and orthogonal complements to find the detail, we will attempt the following simple approximation method:

Definition 19. *Alternative Projection.*
Let g_1 be a function in V_1. The approximation $g_0 = P(g_1)$ in V_0 is defined as the unique function in V_0 which has the same values as g_1 at the integers, i.e.

$$g_0(n) = g_1(n), \quad n = 0, 1, \ldots, N-1. \tag{4.23}$$

It is easy to show that $P(g_1)$ actually is different from the projection of g_1 onto V_0: If $g_1 = \phi_{1,1}$, then g_1 is zero at the integers, and then clearly $P(g_1) = 0$. But in Exercise 4.25 you will be asked to compute the projection onto V_0 using different means than the orthogonal decomposition theorem, and the result will be seen to be nonzero. It is also very easy to see that the coordinates of g_0 in ϕ_0 can be obtained by dropping every second coordinate of g_0 in ϕ_1. To be more precise, the following holds:

Lemma 20. Expression for the Alternative Projection.
We have that

$$P(\phi_{1,n}) = \begin{cases} \sqrt{2}\phi_{0,n/2}, & \textit{if } n \textit{ is an even integer;} \\ 0, & \textit{otherwise.} \end{cases}$$

With this approximation method we can define simple candidates for the detail spaces, and corresponding bases.

Lemma 21. Resolution Spaces.
Define

$$W_0 = \{ f \in V_1 \mid f(n) = 0, \quad for\ n = 0,\ 1,\ \ldots,\ N-1 \},$$

and

$$\psi(t) = \frac{1}{\sqrt{2}} \phi_{1,1}(t) \qquad\qquad \psi_{m,n}(t) = 2^{m/2} \psi(2^m t - n). \qquad (4.24)$$

Suppose that $g_1 \in V_1$ and that $g_0 = P(g_1)$. Then

- *the error $e_0 = g_1 - g_0$ lies in W_0,*
- *$\boldsymbol{\psi}_0 = \{\psi_{0,n}\}_{n=0}^{N-1}$ is a basis for W_0.*
- *V_0 and W_0 are linearly independent, and $V_1 = V_0 \oplus W_0$.*

Proof. Since $g_0(n) = g_1(n)$ for all integers n, $e_0(n) = (g_1 - g_0)(n) = 0$, so that $e_0 \in W_0$. This proves the first statement.

For the second statement, note first that

$$\psi_{0,n}(t) = \psi(t-n) = \frac{1}{\sqrt{2}} \phi_{1,1}(t-n) = \phi(2(t-n) - 1)$$

$$= \phi(2t - (2n+1)) = \frac{1}{\sqrt{2}} \phi_{1,2n+1}(t). \qquad (4.25)$$

$\boldsymbol{\psi}_0$ is thus a linearly independent set of dimension N, since it corresponds to a subset of $\boldsymbol{\phi}_1$. Since $\phi_{1,2n+1}$ is nonzero only on $(n, n+1)$, it follows that all of $\boldsymbol{\psi}_0$ lies in W_0. Clearly then $\boldsymbol{\psi}_0$ is also a basis for W_0, since W_0 also has dimension N (since any function in W_0 is uniquely defined from its values in $1/2, 3/2, \ldots$).

Consider finally a linear combination from $\boldsymbol{\phi}_0$ and $\boldsymbol{\psi}_0$ which gives zero:

$$\sum_{n=0}^{N-1} a_n \phi_{0,n} + \sum_{n=0}^{N-1} b_n \psi_{0,n} = 0.$$

If we evaluate this at $t = k$, we see that $\psi_{0,n}(k) = 0$, $\phi_{0,n}(k) = 0$ when $n \neq k$, and $\phi_{0,k}(k) = 1$. When we evaluate at k we thus get a_k, which must be zero. If we then evaluate at $t = k + 1/2$ we get in a similar way that all $b_n = 0$, and it follows that V_0 and W_0 are linearly independent. That $V_1 = V_0 \oplus W_0$ follows from the fact that V_1 has dimension $2N$, and V_0 and W_0 both have dimension N. $\square$

We can define W_m in a similar way for $m > 0$, and generalize this lemma to W_m:

Theorem 22. Decomposing V_m.
We have that $V_m = V_{m-1} \oplus W_{m-1}$ where

$$W_{m-1} = \{ f \in V_m \mid f(n/2^{m-1}) = 0, \quad for\ n = 0,\ 1,\ \ldots,\ 2^{m-1}N - 1 \}.$$

W_m has the basis $\boldsymbol{\psi}_m = \{\psi_{m,n}\}_{n=0}^{2^m N - 1}$, and V_m has the two bases

$$\boldsymbol{\phi}_m = \{\phi_{m,n}\}_{n=0}^{2^m N - 1}$$

$$(\boldsymbol{\phi}_{m-1}, \boldsymbol{\psi}_{m-1}) = \left(\{\phi_{m-1,n}\}_{n=0}^{2^{m-1}N - 1}, \{\psi_{m-1,n}\}_{n=0}^{2^{m-1}N - 1} \right).$$

With this result we can define the DWT and the IDWT with their stages as before, but the matrices are different. Equations (4.22) and (4.24) said that

$$\phi = \frac{1}{\sqrt{2}}\left(\frac{1}{2}\phi_{1,-1} + \phi_{1,0} + \frac{1}{2}\phi_{1,1}\right)$$

$$\psi = \frac{1}{\sqrt{2}}\phi_{1,1}. \tag{4.26}$$

Using Exercise 4.8 we see that

$$G = P_{\phi_m \leftarrow C_m} = \frac{1}{\sqrt{2}}\begin{pmatrix} 1 & 0 & 0 & 0 & \cdots & 0 & 0 & 0 \\ 1/2 & 1 & 1/2 & 0 & \cdots & 0 & 0 & 0 \\ 0 & 0 & 1 & 0 & \cdots & 0 & 0 & 0 \\ \vdots & \vdots & \vdots & \vdots & \vdots & \vdots & \vdots & \vdots \\ 0 & 0 & 0 & 0 & \cdots & 0 & 1 & 0 \\ 1/2 & 0 & 0 & 0 & \cdots & 0 & 1/2 & 1 \end{pmatrix} \tag{4.27}$$

In general we will call a matrix on the form

$$\begin{pmatrix} 1 & 0 & 0 & 0 & \cdots & 0 & 0 & 0 \\ \lambda & 1 & \lambda & 0 & \cdots & 0 & 0 & 0 \\ 0 & 0 & 1 & 0 & \cdots & 0 & 0 & 0 \\ \vdots & \vdots & \vdots & \vdots & \vdots & \vdots & \vdots & \vdots \\ 0 & 0 & 0 & 0 & \cdots & 0 & 1 & 0 \\ \lambda & 0 & 0 & 0 & \cdots & 0 & \lambda & 1 \end{pmatrix} \tag{4.28}$$

an *elementary lifting matrix of odd type*, and denote it by B_λ. This results from the identity matrix by adding λ times the preceding and succeeding rows to the odd-indexed rows. Using this notation we see that $G = P_{\phi_m \leftarrow C_m} = \frac{1}{\sqrt{2}}B_{1/2}$.

Since the even-indexed rows are left untouched when we multiply with B_λ, its inverse is clearly obtained by subtracting λ times the preceding and succeeding rows to the odd-indexed rows, i.e. $(B_\lambda)^{-1} = B_{-\lambda}$. This means that the matrix for the DWT can also be easily found:

$$H = P_{C_m \leftarrow \phi_m} = \sqrt{2}B_{-1/2} = \sqrt{2}\begin{pmatrix} 1 & 0 & 0 & 0 & \cdots & 0 & 0 & 0 \\ -1/2 & 1 & -1/2 & 0 & \cdots & 0 & 0 & 0 \\ 0 & 0 & 1 & 0 & \cdots & 0 & 0 & 0 \\ \vdots & \vdots & \vdots & \vdots & \vdots & \vdots & \vdots & \vdots \\ 0 & 0 & 0 & 0 & \cdots & 0 & 1 & 0 \\ -1/2 & 0 & 0 & 0 & \cdots & 0 & -1/2 & 1 \end{pmatrix}$$

Due to its very simple structure (every second element is unchanged, and the computations of the remaining elements do not conflict with each other), it is actually possible to multiply with B_λ in such a way that the output can be stored in the same memory as the input. In Sect. 2.3 this was called in-place computation. In the exercises you will be asked to implement a function `lifting_odd_symm` which computes multiplication with B_λ in-place. Using this function the new DWT kernel transformation can be implemented as follows.

```
x = x*sqrt(2);
x = lifting_odd_symm(-0.5, x, 'per');
```

The IDWT kernel transformation can be computed similarly. While the name `"Haar"` identified the wavelet for piecewise constant functions, the name `"pwl0"` will identify this new wavelet for piecewise linear functions (`"pwl"` stands for piecewise linear. The `"0"` will be explained later). The library implements the kernel for this wavelet in this way, but does it a more general way so that several such `lifting_odd_symm`-steps can be combined. We will later see that this is useful in a more general context for wavelets. The library version of `lifting_odd_symm` can be found in `find_kernel.m`.

In the two examples on wavelets considered so far, we also chose a mother wavelet ψ. The Haar wavelet exemplified this by projecting onto the orthogonal complement. For the piecewise linear wavelet, ψ was chosen more at random, and there is no reason to believe that this initial guess is a good one—in fact the listening experiments below turn out to be not very convincing. We will see in Sect. 4.4 how we may modify an initial guess on ψ, in order to obtain a mother wavelet with more desirable properties.

4.3.2 Multiresolution Analysis: A Generalization

We have now seen two examples of resolution spaces V_m. They shared the following properties.

- The spaces were nested, i.e. $V_m \subset V_{m+1}$ for all m.
- Continuous functions could be approximated arbitrarily well from the spaces.
- The spaces were defined in terms of translates and dilations of a prototype function ϕ.

These properties are compatible with the following concept.

Definition 23. *Multiresolution Analysis.*
A Multiresolution analysis, or MRA, is a nested sequence of function spaces

$$V_0 \subset V_1 \subset V_2 \subset \cdots \subset V_m \subset \cdots, \qquad (4.29)$$

called resolution spaces, where all functions are defined on $(-\infty, \infty)$, and so that

- Any function from $L^2(\mathbb{R})$ can be approximated arbitrarily well from the V_m in $\| \cdot \|$,
- $f(t) \in V_0$ if and only if $f(2^m t) \in V_m$,
- $f(t) \in V_0$ if and only if $f(t - n) \in V_0$ for all n.
- There is a function $\phi \in L^2(\mathbb{R})$, called a *scaling function*, so that $\boldsymbol{\phi}_0 = \{\phi(t-n)\}_{0 \leq n < N}$ is a basis for V_0.

When $\boldsymbol{\phi}_0$ is an orthonormal basis we say that the MRA is *orthonormal*.

It is not hard to show that the functions ϕ we already have defined gives rise to MRA's. The wavelet of piecewise constant functions was an orthonormal MRA, while the wavelet for piecewise linear functions was not. The reason why restrict to square integrable functions is that this is the case one can prove the existence of MRA's in the literature.

Also in the setting of an MRA we attempt to find a function ψ (which again is called a mother wavelet) so that $\boldsymbol{\psi}_0 = \{\psi(t-n)\}_{0 \leq n < N}$ and $\boldsymbol{\phi}_0$ together span V_1. For orthonormal wavelets this can be done by projecting onto the orthogonal complement of V_0 in V_1, We then define $\phi_{m,n} = 2^{m/2}\phi(2^t - n)$, $\psi_{m,n} = 2^{m/2}\phi(2^t - n)$, $\boldsymbol{\phi}_m$, $\boldsymbol{\psi}_m$, and W_m as before, so that $V_m = V_{m-1} \oplus W_{m-1}$. With all these definitions in place, the Discrete Wavelet Transform can be defined as before. The differences from the previous constructions are that the bases are infinite, and that the functions are defined on the whole real line instead of being periodic with period N. The following proposition connects the concept of multiresolution analysis with our previous finite-dimensional constructions.

Proposition 24. Finite-Dimensional Bases from an MRA.
Assume that ϕ gives rise to a multiresolution analysis V_m with mother wavelet ψ. The functions

$$\phi_{m,n}^{per}(t) = \sum_{k=-\infty}^{\infty} \phi_{m,n}(t-kN) \qquad \psi_{m,n}^{per}(t) = \sum_{k=-\infty}^{\infty} \psi_{m,n}(t-kN).$$

are periodic with period N, are in $L^2[0,N]$, and the sets

$$\boldsymbol{\phi}_m^{per} = \{\phi_{m,n}^{per}\}_{n=0}^{2^m N} \qquad \boldsymbol{\psi}_m^{per} = \{\psi_{m,n}^{per}\}_{n=0}^{2^m N},$$

are finite-dimensional bases for spaces V_m^{per} and W_m^{per}, respectively. Any function $f \in L^2[0,N]$ can be approximated arbitrarily well from the V_m^{per} in $\|\cdot\|$.

The functions $\phi_{0,n}^{per}$ and $\psi_{0,n}^{per}$ to $[0,N]$ are the functions we started with at the beginning of the chapter. Thus, from an MRA on the whole real line, we can construct finite-dimensional resolution spaces which can be used to approximate any $f \in L^2[0,N]$. The spaces V_m are different in the two settings, however and that the basis functions $\phi_{m,n}, \psi_{m,n}$ and $\phi_{m,n}^{per}, \psi_{m,n}^{per}$ are different. Also a function in $L^2[0,N]$ can have a different expansion in the finite-dimensional and the infinite-dimensional setting.

Proof. If $f \in L^2[0,N]$, let

$$\check{f}(t) = \begin{cases} f(t) & t \in [0,N) \\ 0 & \text{other } t. \end{cases}$$

be the extension of f to $(-\infty, \infty)$. $\check{f}$ is clearly in $L^2(\mathbb{R})$, so that we can find linear combinations so that

$$\sum_{n=-\infty}^{\infty} c_{0,n}\phi_{0,n} + \sum_{m=0}^{\infty} \sum_{n=-\infty}^{\infty} w_{m,n}\psi_{m,n} \overset{L^2(\mathbb{R})}{\to} \check{f}$$

In particular, this converges to f in $L^2[0,N]$. If we periodically extend this restriction to $(-\infty, \infty)$ we get

$$\sum_{k=-\infty}^{\infty} \left(\sum_{n=-\infty}^{\infty} c_{0,n}\phi_{0,n}(t-kN) + \sum_{m=0}^{\infty} \sum_{n=-\infty}^{\infty} w_{m,n}\psi_{m,n}(t-kN) \right)$$

It is straightforward to show that this converges to f in $L^2[0, N]$ (use the convergence to $\check{f}$ in $L^2(\mathbb{R})$). This can be rewritten as

$$\sum_{n=-\infty}^{\infty} c_{0,n} \phi_{0,n}^{\mathrm{per}} + \sum_{m=0}^{\infty} \sum_{n=-\infty}^{\infty} w_{m,n} \psi_{m,n}^{\mathrm{per}} \stackrel{L^2[0,N]}{\longrightarrow} f.$$

□

Example 4.20: DWT on Sound

With the new wavelet, let us plot and listen to the new low resolution approximations and detail, as in Example 4.10. First we listen to the low-resolution approximation.

```
[x, fs] = forw_comp_rev_dwt1(m, 'pwl0');
playerobj = audioplayer(x, fs);
playblocking(playerobj)
```

There is a new and undesired effect when we increase m here: The castanet sound seems to grow strange. The sounds from the castanets are perhaps the sound with the highest frequencies.

For $m = 1$ and $m = 2$ the detail can be played as follows

```
[x, fs] = forw_comp_rev_dwt1(1, 'pwl0', 0);
playerobj = audioplayer(x, fs);
playblocking(playerobj)
```

```
[x, fs] = forw_comp_rev_dwt1(2, 'pwl0', 0);
playerobj = audioplayer(x, fs);
playblocking(playerobj)
```

The errors are shown in Fig. 4.16.

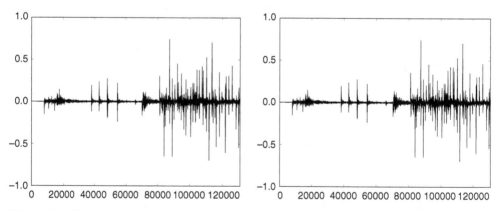

Fig. 4.16 The detail in our audio sample file for the piecewise linear wavelet, for $m = 1$ (left) and $m = 2$ (right)

When comparing with Example 4.10 we see much of the same, but it seems here that the error is bigger than before. In the next section we will try to explain why this is the case, and attempt to modify the definition of ψ to remedy this.

Example 4.21: DWT on the Samples of a Mathematical Function

Let us also repeat Example 4.11, where we plotted the detail/error at different resolutions for the samples of a mathematical function.

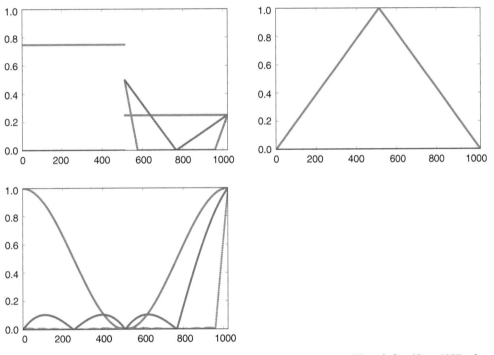

Fig. 4.17 The error (i.e. the contribution from $W_0 \oplus W_1 \oplus \cdots \oplus W_{m-1}$) for $N = 1025$ when f is a square wave, the linear function $f(t) = 1 - 2|1/2 - t/N|$, and the trigonometric function $f(t) = 1/2 + \cos(2\pi t/N)/2$, respectively. The detail is indicated for $m = 6$ and $m = 8$

Figure 4.17 shows the new plot. With the square wave we see now that there is an error. The reason is that a piecewise constant function can not be represented exactly by piecewise linear functions, due to discontinuity. For the second function we see that there is no error, since this function is piecewise linear, so there is no error when we represent the function from the space V_0. With the third function we see an error as before: a trigonometric function can not be represented exactly by piecewise constant nor piecewise linear functions.

Exercise 4.22: The Vector of Samples Is the Coordinate Vector

Show that, for $f \in V_0$ we have that $[f]_{\phi_0} = (f(0), f(1), \ldots, f(N-1))$. This shows that, also for the piecewise linear wavelet, there is no loss in working with the samples of f rather than f itself.

Exercise 4.23: The Piecewise Linear Wavelet Is Not Orthogonal

Show that

$$\langle \phi_{0,n}, \phi_{0,n} \rangle = \frac{2}{3} \qquad \langle \phi_{0,n}, \phi_{0,n\pm1} \rangle = \frac{1}{6} \qquad \langle \phi_{0,n}, \phi_{0,n\pm k} \rangle = 0 \text{ for } k > 1.$$

As a consequence, the $\{\phi_{0,n}\}_n$ are neither orthogonal, nor have norm 1.

Exercise 4.24: Implementation of Elementary Lifting of Odd Type

Write a function

```
lifting_odd_symm(lambda, x, bd_mode)
```

which applies an elementary lifting matrix of odd type (Eq. (4.28)) to $\boldsymbol{x}$. Assume that N is even. The parameter bd_mode should do nothing, as we will return to this parameter later. The function should not allocate the full matrix, and apply as few multiplications as possible. How many local variables are needed in your code? Compare with the corresponding number you found for filters in Exercise 3.26.

Exercise 4.25: Computing Projections

In this exercise we will show how the projection of $\phi_{1,1}$ onto V_0 can be computed. We will see from this that it is nonzero, and that its support is the entire $[0, N]$. Let $f = \text{proj}_{V_0} \phi_{1,1}$, and let $x_n = f(n)$ for $0 \le n < N$. This means that, on $(n, n+1)$, $f(t) = x_n + (x_{n+1} - x_n)(t - n)$.

a) Show that $\int_n^{n+1} f(t)^2 dt = (x_n^2 + x_n x_{n+1} + x_{n+1}^2)/3$.

b) Show that

$$\int_0^{1/2} (x_0 + (x_1 - x_0)t)\phi_{1,1}(t)dt = 2\sqrt{2}\left(\frac{1}{12}x_0 + \frac{1}{24}x_1\right)$$

$$\int_{1/2}^1 (x_0 + (x_1 - x_0)t)\phi_{1,1}(t)dt = 2\sqrt{2}\left(\frac{1}{24}x_0 + \frac{1}{12}x_1\right).$$

c) Use the fact that

$$\int_0^N (\phi_{1,1}(t) - \sum_{n=0}^{N-1} x_n\phi_{0,n}(t))^2 dt$$

$$= \int_0^1 \phi_{1,1}(t)^2 dt - 2\int_0^{1/2}(x_0 + (x_1 - x_0)t)\phi_{1,1}(t)dt - 2\int_{1/2}^1 (x_0 + (x_1 - x_0)t)\phi_{1,1}(t)dt$$

$$+ \sum_{n=0}^{N-1}\int_n^{n+1}(x_n + (x_{n-1} - x_n)t)^2 dt$$

and a) and b) to find an expression for $\|\phi_{1,1}(t) - \sum_{n=0}^{N-1} x_n\phi_{0,n}(t)\|^2$.

d) To find the minimum least squares error, we can set the gradient of the expression in c) to zero, and thus find the expression for the projection of $\phi_{1,1}$ onto V_0. Show that the values $\{x_n\}_{n=0}^{N-1}$ can be found by solving the equation $S\boldsymbol{x} = \boldsymbol{b}$, where $S = \frac{1}{3}\{1, \underline{4}, 1\}$ is an $N \times N$ symmetric filter, and $\boldsymbol{b}$ is the vector with components $b_0 = b_1 = \sqrt{2}/2$, and $b_k = 0$ for $k \geq 2$.

e) Solve the system in d. for some values of N to verify that the projection of $\phi_{1,1}$ onto V_0 is nonzero, and that its support covers the entire $[0, N]$.

Exercise 4.26: Convolution Relation Between the Piecewise Constant- and Piecewise Linear Scaling Functions

The *convolution* of two functions defined on $(-\infty, \infty)$ is defined by

$$(f * g)(x) = \int_{-\infty}^{\infty} f(t)g(x-t)dt.$$

Show that we can obtain the piecewise linear ϕ we have defined as $\phi = \chi_{[-1/2,1/2)} * \chi_{[-1/2,1/2)}$ (recall that $\chi_{[-1/2,1/2)}$ is the function which is 1 on $[-1/2, 1/2)$ and 0, see Eq. (1.16)). This gives us a nice connection between the piecewise constant scaling function (which is similar to $\chi_{[-1/2,1/2)}$) and the piecewise linear scaling function in terms of convolution.

4.4 Alternative Wavelet Based on Piecewise Linear Functions

The scaling function we used for piecewise linear functions did not give rise to an orthogonal basis ϕ_0, contrary to the case for piecewise constant functions. We were still able to construct resolution spaces and detail spaces. In our listening experiments we identified some shortcomings when comparing to the wavelet for piecewise constant functions, however. It turns out that this can be seen in connection with the following concept:

Definition 25. *Vanishing Moments.*
A function ψ is said to have k vanishing moments if $\int t^l \psi(t)dt = 0$ for all $0 \leq l \leq k-1$.

The Haar wavelet clearly has at least one vanishing moment. It is a simple exercise to see that it has only one vanishing moment, i.e. $\int_0^N t\psi(t)dt \neq 0$. To see why vanishing moments are desirable, we will prove the following theorem. The theorem is stated only in the case of orthonormal wavelets, so that it applies for the Haar wavelet, but not necessarily for the piecewise linear wavelet. In Chap. 6 we will see how the result can be generalized to wavelet bases which are not orthonormal.

Theorem 26. Vanishing Moments, Regularity, and the Decay of Wavelet Coefficients.
If ϕ_m and ψ_m is an orthonormal wavelet basis, then the following hold:

- *If ψ has one vanishing moment and $|f(x) - f(t)| \leq C|t - x|$ for $t = n2^{-m}$ and all x close to t (C is some constant), then there exist a constant D so that $|w_{m',n2^{m'-m}}| \leq D2^{-3m'/2}$ for all $m' \geq m$.*
- *If ψ has k vanishing moments, then*

  1. *If f is k times continuously differentiable, then there exists a constant D so that $|w_{m,n}| \leq D2^{-m/2-mk}$ for all m.*
  2. *If f is a polynomial of degree less than or equal to $k-1$ then $w_{m,n} = 0$ for all m.*

Thus, due to the factor 2^{-mk}, the wavelet coefficients $w_{m,n}$ go quickly to zero when the function is regular and we have many vanishing moments. This makes vanishing moments desirable, since the wavelet bases then can approximate many functions well (i.e. with relatively few $\psi_{m,n}$, we can create good approximations). Note that we do not assume that $f \in V_m$. The first statement is a local result and states that, if f is regular close to t, the wavelet coefficients $w_{m,n}$ with $n2^{-m}$ close to t will be small, even if f is not regular in other points. We will use this property in Chap. 5.

Proof. The statement on one vanishing moment says that $f(t) = f(n2^{-m}) + Q_0(t)$, where Q_0 satisfies $|Q_0(t)| \leq C|t - n2^{-m}|$. If f also is k times continuously differentiable, it can be written as $f(t) = P_{k-1}(t) + Q_{k-1}(t)$, where P_{k-1} is the Taylor polynomial of degree $k-1$ around $n2^{-m}$, and Q_{k-1} is the remainder, which must satisfy $Q_{k-1}(t) \leq C|t - n2^{-m}|^k$ for some constant C. This means that the statement on one vanishing moment can be treated in the same way as 1 time continuously differentiable at $t = n2^{-m}$. We have that

$$w_{m,n} = \int f(t)\psi_{m,n}(t)dt = \int P_{k-1}(t)\psi_{m,n}(t)dt + \int Q_{k-1}(t)\psi_{m,n}(t)dt.$$

Now, if we apply the change of variables $u = 2^m t - n$ (so that $t = 2^{-m}(u + n)$), we obtain that $\int t^l \psi_{m,n}(t)dt = 2^{-m-ml} \int (t+n)^l \psi(t)dt$. The first integral is thus zero since ψ has k vanishing moments. Making the same substitution in the second integral, it can be bounded as follows:

$$\left| \int Q_{k-1}(t)\psi_{m,n}(t)dt \right| \le C \int |t - n2^{-m}|^k |\psi_{m,n}(t)| dt$$

$$= 2^{m/2-m}C \int |2^{-m}(u+n) - n2^{-m}|^k |\psi(u)| du$$

$$= 2^{-m/2-mk} \int |t^k \psi(t)| dt$$

Setting $D = \int |t^k \psi(t)| dt$ gives the desired conclusion. The last statement follows by only using the Taylor polynomial at $t = n2^{-m}$. □

For the Haar wavelet clearly $\int_0^N \psi(t)dt = 0$. This can also be seen directly from the plot in Fig. 4.7, since the areas above and below the x-axis are equal. It is straightforward to check that the Haar wavelet does not have more vanishing moments that one. The piecewise linear wavelet, however, has no vanishing moments, since $\psi \ge 0$ (the name for this wavelet was "pwl0", with "0" standing for 0 vanishing moments). Therefore, this is not a very good choice of mother wavelet. We will attempt the following adjustment strategy to construct an alternative mother wavelet $\hat{\psi}$ which has two vanishing moments, i.e. one more than the Haar wavelet.

Idea 27. Adjusting the Wavelet Construction.
  Adjust the wavelet construction in Theorem 22 to

$$\hat{\psi} = \psi - \alpha\phi_{0,0} - \beta\phi_{0,1} \tag{4.30}$$

and choose α, β so that

$$\int_0^N \hat{\psi}(t)\, dt = \int_0^N t\hat{\psi}(t)\, dt = 0, \tag{4.31}$$

and define $\psi_m = \{\hat{\psi}_{m,n}\}_{n=0}^{N2^m-1}$, and W_m as the space spanned by ψ_m.

We have two free variables α, β in Eq. (4.30) to enforce the two conditions in Eq. (4.31). In Exercise 4.30 you are taken through solving these equations to arrive at the following result:

Lemma 28. The New Function $\hat{\psi}$.
  The function

$$\hat{\psi} = \psi - \frac{1}{4}(\phi_{0,0} + \phi_{0,1}) \tag{4.32}$$

satisfies the conditions (4.31).

Using Eq. (4.22), which stated that

$$\phi = \frac{1}{\sqrt{2}}\left(\frac{1}{2}\phi_{1,-1} + \phi_{1,0} + \frac{1}{2}\phi_{1,1} \right), \tag{4.33}$$

we get

$$
\begin{aligned}
\hat{\psi} &= \psi - \frac{1}{4}\left(\phi_{0,0} + \phi_{0,1}\right) \\
&= \frac{1}{\sqrt{2}}\phi_{1,1} - \frac{1}{4}\frac{1}{\sqrt{2}}\left(\frac{1}{2}\phi_{1,-1} + \phi_{1,2n} + \frac{1}{2}\phi_{1,1}\right) \\
&\quad - \frac{1}{4}\frac{1}{\sqrt{2}}\left(\frac{1}{2}\phi_{1,1} + \phi_{1,2} + \frac{1}{2}\phi_{1,3}\right) \\
&= \frac{1}{\sqrt{2}}\left(-\frac{1}{8}\phi_{1,-1} - \frac{1}{4}\phi_{1,0} + \frac{3}{4}\phi_{1,1} - \frac{1}{4}\phi_{1,2} - \frac{1}{8}\phi_{1,3}\right)
\end{aligned}
\tag{4.34}
$$

In summary we have

$$
\phi = \frac{1}{\sqrt{2}}\left(\frac{1}{2}\phi_{1,-1} + \phi_{1,0} + \frac{1}{2}\phi_{1,1}\right)
$$
$$
\hat{\psi} = \frac{1}{\sqrt{2}}\left(-\frac{1}{8}\phi_{1,-1} - \frac{1}{4}\phi_{1,0} + \frac{3}{4}\phi_{1,1} - \frac{1}{4}\phi_{1,2} - \frac{1}{8}\phi_{1,3}\right),
\tag{4.35}
$$

The new function $\hat{\psi}$ is plotted in Fig. 4.18. We see that $\mathrm{Supp}(\hat{\psi}) = [-1,2]$, and consist of four linear segments glued together. This is in contrast to the old ψ, which was simpler with the shorter support $[0,1]$, and only two linear segments.

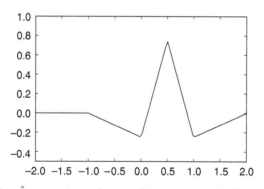

Fig. 4.18 The function $\hat{\psi}$ we constructed as an alternative wavelet for piecewise linear functions

The DWT in this new setting is the change of coordinates from ϕ_m to

$$
\hat{\mathcal{C}}_m = \{\phi_{m-1,0}, \hat{\psi}_{m-1,0}, \phi_{m-1,1}, \hat{\psi}_{m-1,1}, \cdots, \phi_{m-1,2^{m-1}N-1}, \hat{\psi}_{m-1,2^{m-1}N-1}\}.
$$

Equation (4.32) states that

$$
P_{\mathcal{C}_m \leftarrow \hat{\mathcal{C}}_m} =
\begin{pmatrix}
1 & -1/4 & 0 & 0 & \cdots & 0 & 0 & -1/4 \\
0 & 1 & 0 & 0 & \cdots & 0 & 0 & 0 \\
0 & -1/4 & 1 & -1/4 & \cdots & 0 & 0 & 0 \\
\vdots & \vdots & \vdots & \vdots & \vdots & \vdots & \vdots & \vdots \\
0 & 0 & 0 & 0 & \cdots & -1/4 & 1 & -1/4 \\
0 & 0 & 0 & 0 & \cdots & 0 & 0 & 1
\end{pmatrix}
$$

(Column j for j even equals e_j, since the basis functions $\phi_{0,n}$ are not altered). In general we will call a matrix on the form

$$
\begin{pmatrix}
1 & \lambda & 0 & 0 & \cdots & 0 & 0 & \lambda \\
0 & 1 & 0 & 0 & \cdots & 0 & 0 & 0 \\
0 & \lambda & 1 & \lambda & \cdots & 0 & 0 & 0 \\
\vdots & \vdots & \vdots & \vdots & \vdots & \vdots & \vdots & \vdots \\
0 & 0 & 0 & 0 & \cdots & \lambda & 1 & \lambda \\
0 & 0 & 0 & 0 & \cdots & 0 & 0 & 1
\end{pmatrix}
\tag{4.36}
$$

an *elementary lifting matrix of even type*, and denote it by A_λ. Using Eq. (4.27) we can write

$$
G = P_{\phi_m \leftarrow \hat{\mathcal{C}}_m} = P_{\phi_m \leftarrow \mathcal{C}_m} P_{\mathcal{C}_m \leftarrow \hat{\mathcal{C}}_m} = \frac{1}{\sqrt{2}} B_{1/2} A_{-1/4}
$$

This gives us a factorization of the IDWT in terms of elementary lifting matrices. In general such factorizations are called *lifting factorizations*, and will be the focus in Chap. 7. The inverse of elementary lifting matrices of even type can be found similarly to how we found the inverse of elementary lifting matrices of odd type, i.e. $(A_\lambda)^{-1} = A_{-\lambda}$. This means that the matrix for the DWT is easily found also in this case, and

$$
H = G^{-1} = \sqrt{2} A_{1/4} B_{-1/2}.
\tag{4.37}
$$

Note that Eqs. (4.35) also gives the matrix G, but we will rather use lifting factorizations since they turn out to be attractive in terms of few arithmetic operations, and in terms of ease of implementation. In the exercises you will be asked to implement a function `lifting_even_symm` which applies A_λ to a vector. Using this the DWT kernel transformation for the alternative piecewise linear wavelet can be applied as follows.

```
x = x*sqrt(2);
x = lifting_odd_symm(-0.5, x, 'per');
x = lifting_even_symm(0.25, x, 'per');
```

and similarly for the IDWT. The name "pwl2" will be used to identify the piecewise linear wavelet with 2 vanishing moments.

Example 4.27: DWT on Sound

Using the new kernels, let us again listen to the low resolution approximations and the detail. First the low-resolution approximation:

```
[x, fs] = forw_comp_rev_dwt1(m, 'pwl2');
playerobj = audioplayer(x, fs);
playblocking(playerobj)
```

The undesired effect in the castanets from Example 4.20 seems to be gone. The detail for $m = 1$ and $m = 2$ can be played as follows

```
[x, fs] = forw_comp_rev_dwt1(1, 'pwl2', 0);
playerobj = audioplayer(x, fs);
playblocking(playerobj)
```

```
[x, fs] = forw_comp_rev_dwt1(2, 'pwl2', 0);
playerobj = audioplayer(x, fs);
playblocking(playerobj)
```

The errors are shown in Fig. 4.19.

Again, when comparing with Example 4.10 we see much of the same. It is difficult to see an improvement from this figure. However, this figure also clearly shows a smaller error than the piecewise linear wavelet. A partial explanation is that the wavelet we now have constructed has two vanishing moments, while the other had not.

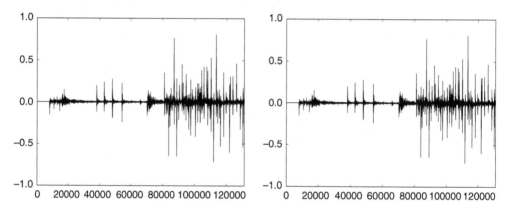

Fig. 4.19 The detail in our audio sample file for the alternative piecewise linear wavelet, for $m = 1$ (left) and $m = 2$ (right)

Example 4.28: DWT on the Samples of a Mathematical Function

Let us also repeat Example 4.11 for our alternative wavelet, where we plotted the detail/error at different resolutions for the samples of a mathematical function.

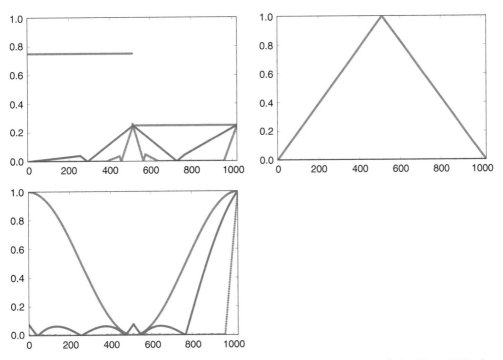

Fig. 4.20 The error (i.e. the contribution from $W_0 \oplus W_1 \oplus \cdots \oplus W_{m-1}$) for $N = 1025$ when f is a square wave, the linear function $f(t) = 1 - 2|1/2 - t/N|$, and the trigonometric function $f(t) = 1/2 + \cos(2\pi t/N)/2$, respectively. The detail is indicated for $m = 6$ and $m = 8$

Figure 4.20 shows the new plot. Again for the square wave there is an error, which seems to be slightly lower than for the previous wavelet. For the second function we see that there is no error, as before. The reason is the same as before, since the function is piecewise linear. With the third function there is an error. The error seems to be slightly lower than for the previous wavelet, which fits well with the increased number of vanishing moments.

Exercise 4.29: Implementation of Elementary Lifting of Even Type

Write a function

```
lifting_even_symm(lambda, x, bd_mode)
```

which applies an elementary lifting matrix of even type (Eq. (4.36)) to x. As before, assume that N is even, and that the parameter `bd_mode` does nothing.

Exercise 4.30: Two Vanishing Moments

In this exercise we will show that there is a unique function on the form given by Eq. (4.30) which has two vanishing moments.

a) Show that, when $\hat{\psi}$ is defined by Eq. (4.30), we have that

$$
\hat{\psi}(t) = \begin{cases}
-\alpha t - \alpha & \text{for } -1 \le t < 0 \\
(2 + \alpha - \beta)t - \alpha & \text{for } 0 \le t < 1/2 \\
(\alpha - \beta - 2)t - \alpha + 2 & \text{for } 1/2 \le t < 1 \\
\beta t - 2\beta & \text{for } 1 \le t < 2 \\
0 & \text{for all other } t
\end{cases}
$$

b) Show that

$$
\int_0^N \hat{\psi}(t)dt = \frac{1}{2} - \alpha - \beta, \qquad \int_0^N t\hat{\psi}(t)dt = \frac{1}{4} - \beta.
$$

c) Explain why there is a unique function on the form given by Eq. (4.30) which has two vanishing moments, and that this function is given by Eq. (4.32).

Exercise 4.31: More than Two Vanishing Moments

In the previous exercise we ended up with a lot of calculations to find α, β in Eq. (4.30). Let us try to make a program which does this for us, and which also makes us able to generalize the result.

a) Define

$$
a_k = \int_{-1}^{1} t^k (1 - |t|)dt, \quad b_k = \int_0^2 t^k(1 - |t - 1|)dt, \quad e_k = \int_0^1 t^k(1 - 2|t - 1/2|)dt,
$$

for $k \ge 0$. Explain why finding α, β so that we have two vanishing moments in Eq. (4.30) is equivalent to solving the following equation:

$$
\begin{pmatrix} a_0 & b_0 \\ a_1 & b_1 \end{pmatrix} \begin{pmatrix} \alpha \\ \beta \end{pmatrix} = \begin{pmatrix} e_0 \\ e_1 \end{pmatrix}
$$

Write a program which sets up and solves this system of equations, and use this program to verify the values for α, β we previously have found.

Hint

you can integrate functions in MATLAB with the function `quad`. As an example, the function $\phi(t)$, which is nonzero only on $[-1, 1]$, can be integrated as follows:

```
quad(@(t)t.^k.*(1-abs(t)),-1,1)
```

b) The procedure where we set up a matrix equation in a) allows for generalization to more vanishing moments. Define

$$\hat{\psi} = \psi_{0,0} - \alpha\phi_{0,0} - \beta\phi_{0,1} - \gamma\phi_{0,-1} - \delta\phi_{0,2}.$$

We would like to choose $\alpha, \beta, \gamma, \delta$ so that we have 4 vanishing moments. Define also

$$g_k = \int_{-2}^{0} t^k (1 - |t + 1|)dt, \qquad\qquad d_k = \int_{1}^{3} t^k (1 - |t - 2|)dt$$

for $k \geq 0$. Show that $\alpha, \beta, \gamma, \delta$ must solve the equation

$$\begin{pmatrix} a_0 & b_0 & g_0 & d_0 \\ a_1 & b_1 & g_1 & d_1 \\ a_2 & b_2 & g_2 & d_2 \\ a_3 & b_3 & g_3 & d_3 \end{pmatrix} \begin{pmatrix} \alpha \\ \beta \\ \gamma \\ \delta \end{pmatrix} = \begin{pmatrix} e_0 \\ e_1 \\ e_2 \\ e_3 \end{pmatrix},$$

and solve this with your computer.

c) Plot the function defined by the equation from b).

Hint

If t is the vector of t-values, and you write

(t >= 0).*(t <= 1).*(1-2*abs(t-0.5))

you get the points $\phi_{1,1}(t)$.

d) Explain why the coordinate vector of $\hat{\psi}$ in the basis (ϕ_0, ψ_0) is

$$[\hat{\psi}]_{(\phi_0, \psi_0)} = (-\alpha, -\beta, -\delta, 0, \ldots, 0 - \gamma) \oplus (1, 0, \ldots, 0).$$

Hint

The placement of $-\gamma$ may seem a bit strange here, and has to do with that $\phi_{0,-1}$ is not one of the basis functions $\{\phi_{0,n}\}_{n=0}^{N-1}$. However, we have that $\phi_{0,-1} = \phi_{0,N-1}$, i.e. $\phi(t+1) = \phi(t-N+1)$, since we always assume that the functions we work with have period N.

e) Compute the coordinates of $\hat{\psi}$ in the basis ϕ_1 (i.e. $[\hat{\psi}]_{\phi_1}$) with $N = 8$, i.e. compute the IDWT of

$$[\hat{\psi}]_{(\phi_0, \psi_0)} = (-\alpha, -\beta, -\delta, 0, 0, 0, 0, -\gamma) \oplus (1, 0, 0, 0, 0, 0, 0, 0),$$

which is the coordinate vector you computed in d). For this, you should use the function idwt_impl, with the kernel of the piecewise linear wavelet without symmetric extension as input.

f) Sketch a more general procedure than the one you found in b), which can be used to find wavelet bases where we have even more vanishing moments.

Exercise 4.32: Two Vanishing Moments for the Haar Wavelet

Let ϕ be the scaling function of the Haar wavelet.

a) Compute $\text{proj}_{V_0} f$, where $f(t) = t^2$, and where f is defined on $[0, N)$.

b) Find constants α, β so that $\hat{\psi} = \psi - \alpha\phi_{0,0} - \beta\phi_{0,1}$ has two vanishing moments, i.e. so that $\langle \hat{\psi}, 1 \rangle = \langle \hat{\psi}, t \rangle = 0$. Plot also the function $\hat{\psi}$.

Hint

Start with computing the integrals $\int \psi(t)dt$, $\int t\psi(t)dt$, $\int \phi_{0,0}(t)dt$, $\int \phi_{0,1}(t)dt$, and $\int t\phi_{0,0}(t)dt$, $\int t\phi_{0,1}(t)dt$.

c) Express ϕ and $\hat{\psi}$ with the help of functions from ϕ_1, and use this to write down the change of coordinate matrix from $(\phi_0, \hat{\psi}_0)$ to ϕ_1.

Exercise 4.33: More than Two Vanishing Moments for the Haar Wavelet

It is also possible to add more vanishing moments to the Haar wavelet. Define

$$\hat{\psi} = \psi_{0,0} - a_0\phi_{0,0} - \cdots - a_{k-1}\phi_{0,k-1}.$$

Define also $c_{r,l} = \int_l^{l+1} t^r dt$, and $e_r = \int_0^1 t^r \psi(t) dt$.

a) Show that $\hat{\psi}$ has k vanishing moments if and only if $a_0, \ldots, a_{k-1}$ solves the system

$$\begin{pmatrix} c_{0,0} & c_{0,1} & \cdots & c_{0,k-1} \\ c_{1,0} & c_{1,1} & \cdots & c_{1,k-1} \\ \vdots & \vdots & \vdots & \vdots \\ c_{k-1,0} & c_{k-1,1} & \cdots & c_{k-1,k-1} \end{pmatrix} \begin{pmatrix} a_0 \\ a_1 \\ \vdots \\ a_{k-1} \end{pmatrix} = \begin{pmatrix} e_0 \\ e_1 \\ \vdots \\ e_{k-1} \end{pmatrix}$$

b) Write a function `vanishing_moms_haar` which takes k as input, solves the system in a), and returns the vector $\boldsymbol{a} = (a_0, a_1, \ldots, a_{k-1})$.

Exercise 4.34: Listening Experiments

Run the function `forw_comp_rev_dwt1` for different m for the Haar wavelet, the piecewise linear wavelet, and the alternative piecewise linear wavelet, but listen to the detail components instead. Describe the sounds you hear for different m, and try to explain why the sound seems to get louder when you increase m.

4.5 Summary

We started this chapter by motivating wavelets as a function approximation scheme which solves some of the shortcomings of Fourier series. While one approximates functions with trigonometric functions in Fourier analysis, wavelet-based analysis is split in stages, each stage capturing information at a given resolution. Bases at all resolutions were constructed from a prototype called the scaling function. The bases are localized in time, contrary to the Fourier basis, making them suitable for time-frequency representations of signals. The wavelet-based scheme can represent an image at different resolutions in a scalable way, so that passing from one resolution to another simply amounts to adding some detail information to the lower resolution version. This also made wavelets useful for compression, since images at different resolutions can serve as compressed versions of the image.

We defined the simplest wavelets and deduced their properties: First the Haar wavelet, which is based on piecewise constant functions, and then wavelets based on piecewise linear functions. The latter basis functions were not orthonormal, contrary to the Haar wavelet. We had some freedom in defining the mother wavelet and the detail spaces: By adding what we called vanishing moments, the mother wavelet could be made more suitable for approximating functions. From piecewise linear and constant functions we made a generalization to what we called a multiresolution analysis (MRA), and in the next chapters we will construct more general wavelets within this framework.

We defined the Discrete Wavelet Transform (DWT) as a change of coordinates between the corresponding function bases. This is the computational component for wavelets, and is a crucial object to study when it comes to more general wavelets as well.

Wavelets have been recognized for quite some time, and the Haar wavelet can be traced all the way back to 1910. It was first in the 80's that one obtained a solid mathematical framework for them in the context of multiresolution analysis, through the pioneering work of Mallat [39] and Meyer [47]. Mallat [40] goes through these developments in detail. Wavelets have also found many new applications in later years. The main application we will focus on in later chapters is image compression.

So far we have focused on the simpler aspects of wavelets. The next chapters will be more advanced, but we will not go into all the details of the theory: Understanding all foundations of wavelet theory requires a lot of mathematics. Several books attempt to introduce the reader to wavelets in various ways, requiring more or less background mathematics. Boggess and Narcowich [3] and Frazier [21] may suit (advanced) undergraduates, and also promote a linear algebra background. They may not go as far as this book when it comes to the connection to signal processing and the computational perspective, however. Kaiser [29] may suit graduates better.

What You Should Have Learned in This Chapter

- Definition of resolution spaces (V_m), detail spaces (W_m), scaling function (ϕ), and mother wavelet (ψ) for the wavelets based on piecewise constant and linear functions.
- The nesting of resolution spaces, and how one can project from one resolution space onto a lower order resolution space, and onto its orthogonal complement.
- The definition of the Discrete Wavelet Transform as a change of coordinates, and how this can be written down from relations between basis functions.

- Implementation of the Haar wavelet transform and its inverse, and using this implementation to experiment with sound.
- How one alters the mother wavelet for piecewise linear functions, in order to add a vanishing moment.
- Definition of a multiresolution analysis.

Chapter 5
The Filter Representation of Wavelets

We saw in the previous chapter that the wavelet kernels G and H had some repeating structure in the rows and columns, similar to the circular Toeplitz structure in filters. Clearly the matrices are not filters, however. Nevertheless, we will now prove that wavelet kernels can be implemented easily in terms of filters. but that several filters are needed in the computation. Each of these filters will have an interpretation in terms of how the wavelet transform treats different frequencies. Much has been done in establishing efficient implementations of filters, and by expressing a wavelet transform in terms of filters we can take advantage of this.

In this chapter we will also see that expressing the DWT in terms of filters paves the ground for defining more general transforms, where even more filters are used. In this setting it is fruitful to think about each filter as concentrating on a particular frequency range, and that these transforms split the input into different frequency bands. Such transforms have important applications to the processing and compression of sound, and we will show that the MP3 standard applies such a transform.

5.1 DWT and IDWT in Terms of Filters

In the cases we have considered we have seen that every second row/column in the kernel transformations

$$G = P_{\phi_m \leftarrow \mathcal{C}_m} \qquad\qquad H = P_{\mathcal{C}_m \leftarrow \phi_m}$$

repeat, as in a circulant Toeplitz matrix. Since there are two different columns repeating, they are not circulant Toeplitz, however, and hence not filters. To establish relations with filters, let us start by giving names to our new matrices.

Definition 1. *MRA-Matrices.*
An $N \times N$-matrix T, with N even, is called an *MRA-matrix* if the columns are translates of the first two columns in alternating order.

© Springer Nature Switzerland AG 2019
Ø. Ryan, *Linear Algebra, Signal Processing, and Wavelets - A Unified Approach*,
Springer Undergraduate Texts in Mathematics and Technology,
https://doi.org/10.1007/978-3-030-01812-2_5

5 The Filter Representation of Wavelets

Clearly G and H will always be MRA-matrices, once we have a scaling function ϕ for an MRA and a mother wavelet ψ. Note also that every second column repeats if and only if every second row repeats. To relate an MRA-matrix to filters we will separate the low resolution and detail in a wavelet transform. We have that

$$\begin{pmatrix} c_{m-1,0} \\ w_{m-1,0} \\ c_{m-1,1} \\ w_{m-1,1} \\ \vdots \end{pmatrix} = H \begin{pmatrix} c_{m,0} \\ c_{m,1} \\ c_{m,2} \\ c_{m,3} \\ \vdots \end{pmatrix} \quad \text{and} \quad \begin{pmatrix} c_{m,0} \\ c_{m,1} \\ c_{m,2} \\ c_{m,3} \\ \vdots \end{pmatrix} = G \begin{pmatrix} c_{m-1,0} \\ w_{m-1,0} \\ c_{m-1,1} \\ w_{m-1,1} \\ \vdots \end{pmatrix}.$$

For the first equation, since every second row in H repeats, the even-indexed rows coincide with those from a unique filter H_0, the odd-indexed rows with those from a unique filter H_1. This means that we can write

$$\begin{pmatrix} c_{m-1,0} \\ w_{m-1,0} \\ c_{m-1,1} \\ w_{m-1,1} \\ \vdots \end{pmatrix} = \begin{pmatrix} (H_0)_{0,:} \\ (H_1)_{1,:} \\ (H_0)_{2,:} \\ (H_1)_{3,:} \\ \vdots \end{pmatrix} \begin{pmatrix} c_{m,0} \\ c_{m,1} \\ c_{m,2} \\ c_{m,3} \\ \vdots \end{pmatrix},$$

where $(H_j)_{i,:}$ denotes row i of H_j. We can now separate the detail and low resolution coordinates to obtain

$$\begin{pmatrix} c_{m-1,0} \\ x_1 \\ c_{m-1,1} \\ x_3 \\ \vdots \end{pmatrix} = H_0 \begin{pmatrix} c_{m,0} \\ c_{m,1} \\ c_{m,2} \\ c_{m,3} \\ \vdots \end{pmatrix} \qquad \begin{pmatrix} x_0 \\ w_{m-1,0} \\ x_2 \\ w_{m-1,1} \\ \vdots \end{pmatrix} = H_1 \begin{pmatrix} c_{m,0} \\ c_{m,1} \\ c_{m,2} \\ c_{m,3} \\ \vdots \end{pmatrix},$$

where the x_i are numbers we are not interested in. For the second equation we can separate the detail and low resolution coordinates by writing

$$G \begin{pmatrix} c_{m-1,0} \\ w_{m-1,0} \\ c_{m-1,1} \\ w_{m-1,1} \\ \vdots \end{pmatrix} = G \left(\begin{pmatrix} c_{m-1,0} \\ 0 \\ c_{m-1,1} \\ 0 \\ \vdots \end{pmatrix} + \begin{pmatrix} 0 \\ w_{m-1,0} \\ 0 \\ w_{m-1,1} \\ \vdots \end{pmatrix} \right)$$

$$= G \begin{pmatrix} c_{m-1,0} \\ 0 \\ c_{m-1,1} \\ 0 \\ \vdots \end{pmatrix} + G \begin{pmatrix} 0 \\ w_{m-1,0} \\ 0 \\ w_{m-1,1} \\ \vdots \end{pmatrix}$$

$$= G_0 \begin{pmatrix} c_{m-1,0} \\ 0 \\ c_{m-1,1} \\ 0 \\ \vdots \end{pmatrix} + G_1 \begin{pmatrix} 0 \\ w_{m-1,0} \\ 0 \\ w_{m-1,1} \\ \vdots \end{pmatrix},$$

where we split the vector into its even-indexed and odd-indexed components. In the last equation, we replaced G with G_0, G_1, which are the unique filters so that the even-indexed/odd-indexed columns coincide with those of G: Since every second element is zero in the vector, it does not matter if we replace the even-index/odd-indexed columns of G with those from G_0/G_1. We have thus proved the following important connection between wavelets and filters.

Theorem 2. The DWT and IDWT in Terms of Filters.
We denote by

- H_0/H_1 *the unique filters with the same even-indexed/odd-indexed rows as H.*
- G_0/G_1 *the unique filters with the same even-indexed/odd-indexed columns as G.*

A stage in the DWT can be computed as follows:

- *Compute $H_0\boldsymbol{c}_m$. The even-indexed entries in the result are the coordinates $\boldsymbol{c}_{m-1}$,*
- *Compute $H_1\boldsymbol{c}_m$. The odd-indexed entries in the result are the coordinates $\boldsymbol{w}_{m-1}$.*

A stage in the IDWT can be computed as follows

$$\boldsymbol{c}_m = G_0 \begin{pmatrix} c_{m-1,0} \\ 0 \\ c_{m-1,1} \\ 0 \\ \vdots \end{pmatrix} + G_1 \begin{pmatrix} 0 \\ w_{m-1,0} \\ 0 \\ w_{m-1,1} \\ \vdots \end{pmatrix}. \tag{5.1}$$

Note that the filters G_0, G_1 were defined in terms of the columns of G, while the filters H_0, H_1 were defined in terms of the rows of H. It is easy to mix up rows and columns here. We see that the difference comes from how we separated the low resolution and detail components in the proof.

H_0 and H_1 are called the *DWT filter components*, and G_0 and G_1 the *IDWT filter components*. Keeping only every second coordinate (as in the application of H_0 and H_1) is called *downsampling*. The downsampling means that one should avoid computing the full output of the filters. Making a new vector with zeroes inserted, as done here for G_0 and G_1, is called *upsampling*. When H and G are defined in terms of filters this way, they are also called *forward-* and *reverse filter bank transforms*.

In Figs. 5.1 and 5.2 we have complemented Fig. 4.9 for the DWT and IDWT by giving names to the arrows. Figure 5.3 additionally indicates the downsampling and upsampling steps from Theorem 2. They are denoted by $\downarrow_2$ and $\uparrow_2$, respectively. $\oplus$ represents summing the elements which point inwards. The left side represents the DWT, the right side the IDWT. Wavelet transforms are more often illustrated in this way in the literature, and such diagrams open for generalization, as we will shortly see.

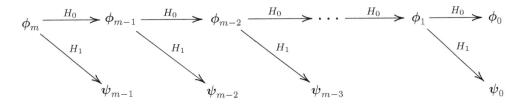

Fig. 5.1 Detailed illustration of a DWT

In Chap. 7 we will go through how forward and reverse filter bank transform can be implemented. The filter coefficients in the four filters can be found from the relations between the bases ϕ_1 and (ϕ_0, ψ_0). Filters can also be constructed from outside a

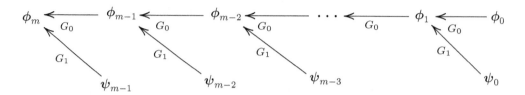

Fig. 5.2 Detailed illustration of an IDWT

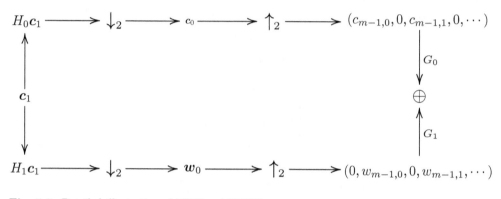

Fig. 5.3 Detailed illustration of DWT and IDWT

wavelet setting, so that they at the start do not originate from change of coordinates between wavelet bases. The important point is that the matrices invert each other, but in a signal processing setting it may also be meaningful to allow for the reverse transform to not invert the forward transform exactly. This will give some loss of information when we reconstruct the original signal with the reverse transform. A small such loss can, as we will see at the end of this chapter, be acceptable.

In Example 4.10 we argued that the elements in V_{m-1} correspond to frequencies at lower frequencies than those in V_m, since $V_0 = \mathrm{Span}(\{\phi_{0,n}\}_n)$ should be interpreted as content of lower frequency than the $\phi_{1,n}$, with $W_0 = \mathrm{Span}(\{\psi_{0,n}\}_n)$ the remaining high frequency detail. To elaborate more on this, note that

$$\phi(t) = \sum_{n=0}^{2N-1} (G_0)_{n,0} \phi_{1,n}(t) \tag{5.2}$$

$$\psi(t) = \sum_{n=0}^{2N-1} (G_1)_{n-1,1} \phi_{1,n}(t)., \tag{5.3}$$

where $(G_k)_{i,j}$ are the entries in the matrix G_k. Similar equations hold for $\phi(t-k), \psi(t-k)$. Due to Eq. (5.2), the filter G_0 should have low-pass characteristics since it extracts contents of lower frequency, and G_1 should have high-pass characteristics due to Eq. (5.3) since it extracts high frequency detail.

5.1.1 Difference Between MRA Matrices and Filters

MRA matrices are not filters, unless the two filters defining them are equal. As a result, the output for some frequency must contain a contribution from some other frequency. This effect is called *aliasing*. We will here show that only one other frequency besides the input frequency contributes in the output of a filter bank transform. Thus, MRA-matrices are not that different from filters—a frequency is not spread out across many frequencies.

Theorem 3. Aliasing in Filter Bank Transforms.
We have that

$$H\phi_r = \frac{1}{2}(\lambda_{H_0,r} + \lambda_{H_1,r})\phi_r + \frac{1}{2}(\lambda_{H_0,r} - \lambda_{H_1,r})\phi_{r+N/2} \tag{5.4}$$

$$G\phi_r = \frac{1}{2}(\lambda_{G_0,r} + \lambda_{G_1,r})\phi_r + \frac{1}{2}(\lambda_{G_0,r+N/2} - \lambda_{G_1,r+N/2})\phi_{r+N/2}, \tag{5.5}$$

where ϕ_r is a Fourier basis vector.

Thus, in the output for angular frequency ω, only the frequencies ω and $\omega + \pi$ contribute.

Proof. Since H_0 and H_1 are filters with frequency responses λ_{H_0} and λ_{H_1}, we have that

$$(H(e^{2\pi irk/N}))_k = \begin{cases} \lambda_{H_0,r}e^{2\pi irk/N} & k \text{ even} \\ \lambda_{H_1,r}e^{2\pi irk/N} & k \text{ odd.} \end{cases}$$

Component n in the DFT of $H(e^{2\pi irk/N})$ is

$$\sum_{k=0}^{N/2-1} \lambda_{H_0,r}e^{2\pi ir(2k)/N}e^{-2\pi i(2k)n/N}$$

$$+ \sum_{k=0}^{N/2-1} \lambda_{H_1,r}e^{2\pi ir(2k+1)/N}e^{-2\pi i(2k+1)n/N}$$

$$= \sum_{k=0}^{N/2-1} \lambda_{H_0,r}e^{2\pi i(r-n)(2k)/N} + \sum_{k=0}^{N/2-1} \lambda_{H_1,r}e^{2\pi i(r-n)(2k+1)/N}$$

$$= (\lambda_{H_0,r} + \lambda_{H_1,r}e^{2\pi i(r-n)/N}) \sum_{k=0}^{N/2-1} e^{2\pi i(r-n)k/(N/2)}.$$

Clearly, $\sum_{k=0}^{N/2-1} e^{2\pi i(r-n)k/(N/2)} = N/2$ if $n = r$ or $n = r + N/2$, and 0 else. It follows that

$$\text{DFT}_N(H(e^{2\pi irk/N})) = \frac{N}{2}(\lambda_{H_0,r} + \lambda_{H_1,r})e_r + \frac{N}{2}(\lambda_{H_0,r} - \lambda_{H_1,r})e_{r+N/2}.$$

Taking an IDFT we obtain

$$H(e^{2\pi irk/N}) = \frac{1}{2}(\lambda_{H_0,r} + \lambda_{H_1,r})e^{2\pi irk/N} + \frac{1}{2}(\lambda_{H_0,r} - \lambda_{H_1,r})e^{2\pi i(r+N/2)k/N}.$$

The first equation now follows by substituting with ϕ_r.

For the reverse filter bank transform, in Exercise 2.18a) we proved that

$$(e^{2\pi ir\cdot 0/N}, 0, e^{2\pi ir\cdot 2/N}, 0, \ldots, e^{2\pi ir(N-2)/N}, 0) = \frac{1}{2}(e^{2\pi irk/N} + e^{2\pi i(r+N/2)k/N})$$

$$(0, e^{2\pi ir\cdot 1/N}, 0, e^{2\pi ir\cdot 3/N}, \ldots, 0, e^{2\pi ir(N-1)/N}) = \frac{1}{2}(e^{2\pi irk/N} - e^{2\pi i(r+N/2)k/N})$$

Using these we have that

$$G(e^{2\pi irk/N}) = G_0\left(\frac{1}{2}\left(e^{2\pi irk/N} + e^{2\pi i(r+N/2)k/N}\right)\right)$$

$$+ G_1\left(\frac{1}{2}\left(e^{2\pi irk/N} - e^{2\pi i(r+N/2)k/N}\right)\right)$$

$$= \frac{1}{2}(\lambda_{G_0,r}e^{2\pi irk/N} + \lambda_{G_0,r+N/2}e^{2\pi i(r+N/2)k/N})$$

$$+ \frac{1}{2}(\lambda_{G_1,r}e^{2\pi irk/N} - \lambda_{G_1,r+N/2}e^{2\pi i(r+N/2)k/N})$$

$$= \frac{1}{2}(\lambda_{G_0,r} + \lambda_{G_1,r})e^{2\pi irk/N}$$

$$+ \frac{1}{2}(\lambda_{G_0,r+N/2} - \lambda_{G_1,r+N/2})e^{2\pi i(r+N/2)k/N},$$

which gives the second equation. $\square$

If H and G invert each other we say that (H,G) gives *perfect reconstruction*. In a wavelet setting perfect reconstruction is always the case by construction. Perfect reconstruction makes sense also outside a wavelet setting, i.e. when one assumes only knowledge of the filters. Perfect reconstruction requires in particular that the aliasing introduced by H must be canceled by the aliasing introduced by G. This is called *alias cancellation*. When we have alias cancellation GH is a filter, and we say that there is no phase distortion if the frequency response of GH also is real. In Exercise 5.9 we will find necessary conditions on the filters (irrespective of whether they appear in a wavelet setting or not) for such alias cancellation and perfect reconstruction.

It may be easier to construct filters so that G only approximately inverts H, in which case we speak of *near-perfect reconstruction*. Near-perfect reconstruction systems have been around long before wavelets, and exist in more general variants, as we will shortly see. In signal processing, one also says that we have perfect- or near-perfect reconstruction if GH equals, or is close to, E_d (i.e. the overall result is a delay). A delay occurs naturally in many forward/reverse transforms. In particular, in real-time processing there is a delay in computing the output, since that computation requires that enough input is available. In terms of reconstructing the input from the output, an additional delay is unproblematic.

Example 5.1: The Haar Wavelet

For the Haar wavelet we saw that, in G, the matrix

$$\begin{pmatrix} \frac{1}{\sqrt{2}} & \frac{1}{\sqrt{2}} \\ \frac{1}{\sqrt{2}} & -\frac{1}{\sqrt{2}} \end{pmatrix}$$

repeated along the diagonal. The filters G_0 and G_1 can be found directly from this as

$$G_0 = \{1/\sqrt{2}, 1/\sqrt{2}\}$$
$$G_1 = \{1/\sqrt{2}, \underline{-1/\sqrt{2}}\}.$$

We have seen these filters previously: G_0 is a moving average filter of two elements (up to multiplication with a constant), i.e. a low-pass filter, and G_1 is the corresponding high-pass filter (up to a delay), which also can be seen as an approximation to the derivative. The frequency responses are

$$\lambda_{G_0}(\omega) = \frac{1}{\sqrt{2}} + \frac{1}{\sqrt{2}}e^{-i\omega} = \sqrt{2}e^{-i\omega/2}\cos(\omega/2)$$

$$\lambda_{G_1}(\omega) = \frac{1}{\sqrt{2}}e^{i\omega} - \frac{1}{\sqrt{2}} = \sqrt{2}ie^{i\omega/2}\sin(\omega/2).$$

By considering filters which repeat with the same rows, it is clear that

$$H_0 = \{1/\sqrt{2}, \underline{1/\sqrt{2}}\}$$
$$H_1 = \{-1/\sqrt{2}, \underline{1/\sqrt{2}}\},$$

and these frequency responses have the same low-pass/high-pass characteristics. The frequency responses can be plotted as follows.

```
plot(omega, abs(1+exp(-i*omega))/sqrt(2))
plot(omega, abs(exp(i*omega)-1)/sqrt(2))
```

They are shown in Fig. 5.4.

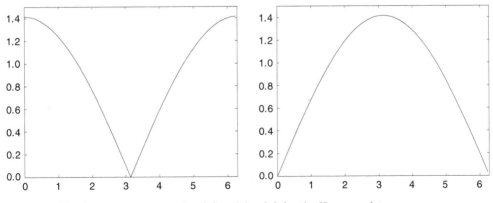

Fig. 5.4 The frequency responses $\lambda_{G_0}(\omega)$ and $\lambda_{G_1}(\omega)$ for the Haar wavelet

Example 5.2: The Piecewise Linear Wavelet

For the wavelet for piecewise linear functions, Eq. (4.27) gives that

$$G_0 = \frac{1}{\sqrt{2}}\{1/2, \underline{1}, 1/2\}$$

$$G_1 = \frac{1}{\sqrt{2}}\{\underline{1}\}.$$

G_0 is again a filter we have seen before: Up to multiplication with a constant, it is the low-pass filter from row 2 of Pascal's triangle. We see something different here when compared to the Haar wavelet, in that the filter G_1 is not the high-pass filter corresponding to G_0. The frequency responses are now

$$\lambda_{G_0}(\omega) = \frac{1}{2\sqrt{2}}e^{i\omega} + \frac{1}{\sqrt{2}} + \frac{1}{2\sqrt{2}}e^{-i\omega} = \frac{1}{\sqrt{2}}(\cos\omega + 1)$$

$$\lambda_{G_1}(\omega) = \frac{1}{\sqrt{2}}.$$

$\lambda_{G_1}(\omega)$ thus has magnitude $\frac{1}{\sqrt{2}}$ at all points. The frequency response of G_0 is shown in the left part of Fig. 5.5. Also here the frequency response is zero at π, but seems to be flatter around π. For the DWT we have that

$$H_0 = \sqrt{2}\{\underline{1}\}$$

$$H_1 = \sqrt{2}\{-1/2, \underline{1}, -1/2\}.$$

Even though G_1 was not the high-pass filter corresponding to G_0, we see that (up to a constant), H_1 is, since it too is taken from row 2 of Pascal's triangle (with an alternating sign added).

Example 5.3: The Alternative Piecewise Linear Wavelet

Previously we found the first two columns in $P_{\phi_m \leftarrow C_m}$ for the alternative piecewise linear wavelet. This gives us that the filters G_0 and G_1 are

$$G_0 = \frac{1}{\sqrt{2}}\{1/2, \underline{1}, 1/2\}$$

$$G_1 = \frac{1}{\sqrt{2}}\{-1/8, -1/4, \underline{3/4}, -1/4, -1/8\}.$$

Here G_0 is as for the wavelet of piecewise linear functions since we use the same scaling function. G_1 is different, however. It turns out that G_1 now has high-pass characteristics.

We can perform the matrix multiplication in Eq. (4.37) to compute the DWT filter components:

$$H_0 = \sqrt{2}\{-1/8, 1/4, \underline{3/4}, 1/4, -1/8\}$$

$$H_1 = \sqrt{2}\{-1/2, \underline{1}, -1/2\}.$$

The frequency responses can be plotted as follows

```
plot(omega, abs(1+cos(omega))/sqrt(2))
plot(omega, abs(3/4 - cos(omega)/2 - cos(2*omega)/4)/sqrt(2))
```

and are shown in Fig. 5.5. Low-pass/high-pass characteristics are clearly seen here as well.

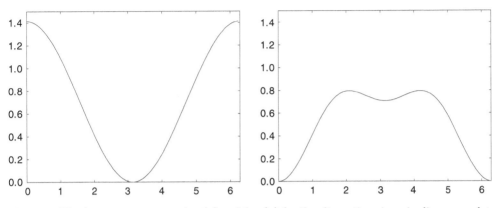

Fig. 5.5   The frequency responses $\lambda_{G_0}(\omega)$ and $\lambda_{G_1}(\omega)$ for the alternative piecewise linear wavelet

The filters G_0, G_1, H_0, H_1 are particularly important in applications: Apart from the scaling factors $1/\sqrt{2}$, $\sqrt{2}$ in front, we see that the filter coefficients are all dyadic fractions, i.e. they are on the form $\beta/2^j$. Arithmetic operations with dyadic fractions can be carried out exactly on a computer, due to representations as binary numbers. These filters are thus important in applications, since they can be used for lossless coding. The same argument can be made for the Haar wavelet, but this wavelet had one less vanishing moment.

Note that the role of H_1 as the high-pass filter corresponding to G_0 is the case in both previous examples. We will prove in the next chapter that this is a much more general result.

Exercise 5.4: Symmetry of MRA Matrices vs. Symmetry of Filters

a) Find two symmetric filters, so that the corresponding MRA-matrix is not symmetric.

b) Assume that an MRA-matrix is symmetric. Are the corresponding filters H_0, H_1, G_0, G_1 also symmetric? If not, find a counterexample.

Exercise 5.5: Passing Between the MRA-Matrices and the Filters

a) Assume that we have the forward filter bank transform

$$H = \begin{pmatrix} 1/5 & 1/5 & 1/5 & 0 & 0 & 0 & \cdots & 0 & 1/5 & 1/5 \\ -1/3 & 1/3 & -1/3 & 0 & 0 & 0 & \cdots & 0 & 0 & 0 \\ 1/5 & 1/5 & 1/5 & 1/5 & 1/5 & 0 & \cdots & 0 & 0 & 0 \\ 0 & 0 & -1/3 & 1/3 & -1/3 & 0 & \cdots & 0 & 0 & 0 \\ \vdots & \vdots & \vdots & \vdots & \vdots & \vdots & \vdots & \vdots & \vdots & \vdots \end{pmatrix}$$

Write down the corresponding filters H_0, H_1, and compute and plot the frequency responses. Are the filters symmetric?

b) Assume that we have the reverse filter bank transform

$$G = \begin{pmatrix} 1/2 & -1/4 & 0 & 0 & \cdots \\ 1/4 & 3/8 & 1/4 & 1/16 & \cdots \\ 0 & -1/4 & 1/2 & -1/4 & \cdots \\ 0 & 1/16 & 1/4 & 3/8 & \cdots \\ 0 & 0 & 0 & -1/4 & \cdots \\ 0 & 0 & 0 & 1/16 & \cdots \\ 0 & 0 & 0 & 0 & \cdots \\ \vdots & \vdots & \vdots & \vdots & \vdots \\ 0 & 0 & 0 & 0 & \cdots \\ 1/4 & 1/16 & 0 & 0 & \cdots \end{pmatrix}$$

Write down the filters G_0, G_1, and compute and plot the frequency responses. Are the filters symmetric?

c) Assume that $H_0 = \{1/16, 1/4, 3/8, 1/4, 1/16\}$, and $H_1 = \{-1/4, 1/2, -1/4\}$. Plot the frequency responses of H_0 and $\overline{H_1}$, and verify that H_0 is a low-pass filter, and that H_1 is a high-pass filter. Also write down the corresponding forward filter bank transform H.

d) Assume that $G_0 = \frac{1}{3}\{1, \underline{1}, 1\}$, and $G_1 = \frac{1}{5}\{1, -1, \underline{1}, -1, 1\}$. Plot the frequency responses of G_0 and G_1, and verify that G_0 is a low-pass filter, and that G_1 is a high-pass filter. Also write down the corresponding reverse filter bank transform G.

Exercise 5.6: Computing by Hand

In Exercise 4.14 we computed the DWT of two very simple vectors x_1 and x_2, using the Haar wavelet.

a) Compute $H_0 x_1$, $H_1 x_1$, $H_0 x_2$, and $H_1 x_2$, where H_0 and H_1 are the filters used by the Haar wavelet.

b) Compare the odd-indexed elements in $H_1 x_1$ with the odd-indexed elements in $H_1 x_2$. From this comparison, attempt to find an explanation to why the two vectors have very different detail components.

Exercise 5.7: Code for Applying Wavelets to Sound

Suppose that we run the following algorithm on the sound represented by the vector x:

```
N=size(x,1);
c = (x(1:2:N, :) + x(2:2:N, :))/sqrt(2);
w = (x(1:2:N, :) - x(2:2:N, :))/sqrt(2);

newx = [c; w];
newx = newx/max(abs(newx));
playerobj=audioplayer(newx,44100);
playblocking(playerobj)
```

a) Comment the code and explain what happens. Which wavelet is used? What do the vectors c and w represent? Describe the sound you believe you will hear.

b) Assume that we add lines in the code above which sets the elements in the vector w to 0 before we compute the inverse operation. What will you hear if you play this new sound?

Exercise 5.8: Computing Filters and Frequency Responses

a) Write down the filter coefficients for the corresponding filter G_1 obtained in Exercise 4.31, and plot its frequency response.

b) Write down the corresponding filters G_0 and G_1 for Exercise 4.32. Plot their frequency responses, and characterize the filters as low-pass- or high-pass.

c) Repeat b) for the Haar wavelet as in Exercise 4.33, and plot the corresponding frequency responses for $k = 2, 4, 6$.

Exercise 5.9: Perfect Reconstruction for Forward and Reverse Filter Bank Transforms

Combine Eqs. (5.4) and (5.5) in Theorem 3 to show that

$$GH\phi_r = \frac{1}{2}(\lambda_{H_0,r}\lambda_{G_0,r} + \lambda_{H_1,r}\lambda_{G_1,r})\phi_r$$
$$+ \frac{1}{2}(\lambda_{H_0,r}\lambda_{G_0,r+N/2} - \lambda_{H_1,r}\lambda_{G_1,r+N/2})\phi_{r+N/2}.$$

Conclude that, in terms of continuous frequency responses, we have alias cancellation if and only if

$$\lambda_{H_0}(\omega)\lambda_{G_0}(\omega + \pi) = \lambda_{H_1}(\omega)\lambda_{G_1}(\omega + \pi),$$

and perfect reconstruction if and only if in addition

$$\lambda_{H_0}(\omega)\lambda_{G_0}(\omega) + \lambda_{H_1}(\omega)\lambda_{G_1}(\omega) = 2.$$

These are also called the *alias cancellation-* and *perfect reconstruction conditions*, respectively.

Exercise 5.10: Conditions for Perfect Reconstruction

In this exercise we will prove that, when G_0 and H_0 are given filters which satisfy a certain property, we can define unique (up to a constant) filters H_1 and G_1 so that all four filters together give perfect reconstruction. In Sect. 7.1 we will also prove that, as long as all filters are FIR, these conditions are also necessary.

a) Show that, if there exist $\alpha \in \mathbb{R}$ and $d \in \mathbb{Z}$ so that

$$(H_1)_n = (-1)^n \alpha^{-1} (G_0)_{n-2d}$$
$$(G_1)_n = (-1)^n \alpha (H_0)_{n+2d}$$
$$2 = \lambda_{H_0,n} \lambda_{G_0,n} + \lambda_{H_0,n+N/2} \lambda_{G_0,n+N/2}$$

then we also have that

$$\lambda_{H_1}(\omega) = \alpha^{-1} e^{-2id\omega} \lambda_{G_0}(\omega + \pi)$$
$$\lambda_{G_1}(\omega) = \alpha e^{2id\omega} \lambda_{H_0}(\omega + \pi)$$
$$2 = \lambda_{H_0}(\omega) \lambda_{G_0}(\omega) + \lambda_{H_0}(\omega + \pi) \lambda_{G_0}(\omega + \pi)$$

b) Show that the first two conditions in a) implies the alias cancellation condition from Exercise 5.9, and that the last condition implies the perfect reconstruction condition from Exercise 5.9.

c) Show that if the conditions in a) are met, and the H_0, H_1, G_0, G_1 are non-trivial symmetric FIR filters, then necessarily $d = 0$. It follows that the conditions take the form

$$\lambda_{H_1}(\omega) = \alpha^{-1} \lambda_{G_0}(\omega + \pi)$$
$$\lambda_{G_1}(\omega) = \alpha \lambda_{H_0}(\omega + \pi)$$
$$2 = \lambda_{H_0}(\omega) \lambda_{G_0}(\omega) + \lambda_{H_0}(\omega + \pi) \lambda_{G_0}(\omega + \pi),$$

Exercise 5.11: Finding the DWT Filter Components from the IDWT Filter Components

Assume that H and G give perfect reconstruction. It may be that only one of H and G is known, and that we would like to find an expression for the other. To this end we will assume that the conditions in Exercise 5.10a) are met for some d and α.

a) Assume that G_0, G_1 are known and symmetric (so that $d = 0$). Deduce that

$$\alpha = \sum_n (-1)^n (G_0)_n (G_1)_n.$$

Deduce a similar formula for the case when H_0 and H_1 are known.

b) Use a) to verify the expressions for the filters H_0 and H_1 of the alternative piecewise linear wavelet from Example 5.3.

Exercise 5.12: Classical QMF Filter Bank Transforms

A classical QMF filter bank is one where $G_0 = H_0$ and $G_1 = H_1$ (i.e. the filters in the forward and reverse transforms are equal), and

$$\lambda_{H_1}(\omega) = \lambda_{H_0}(\omega + \pi).$$

QMF is short for *Quadrature Mirror Filter*.

a) Show that, for a classical QMF filter bank transform, the two alias cancellation conditions in Exercise 5.10a) are satisfied with $\alpha = 1, d = 0$.

b) Show that, for classical QMF filter bank transforms, the perfect reconstruction condition in Exercise 5.10a) can be written as

$$\lambda_{H_0}(\omega)^2 + \lambda_{H_0}(\omega + \pi)^2 = 2.$$

c) Show that it is impossible to find classical QMF filter banks with non-trivial FIR-filters, and which give perfect reconstruction.

Despite this negative result, classical QMF filter banks are still useful, as one can construct such filter banks which give near-perfect reconstruction. We shall see in Sect. 7.3 that the MP3 standard take use of such filters, and this explains our previous observation that the MP3 standard does not give perfect reconstruction.

Exercise 5.13: Alternative QMF Filter Bank Transforms

An alternative QMF filter bank is one where $G_0 = (H_0)^T$ and $G_1 = (H_1)^T$ (i.e. the filter coefficients in the forward and reverse transforms are reverse of one another), and

$$\lambda_{H_1}(\omega) = \overline{\lambda_{H_0}(\omega + \pi)}.$$

There is some confusion in the literature between classical and alternative QMF filter banks.

a) Show that, for an alternative QMF filter bank, the two alias cancellation conditions in Exercise 5.10a) are satisfied with $\alpha = 1, d = 0$.

b) Show that the perfect reconstruction condition for alternative QMF filter banks can be written as

$$|\lambda_{H_0}(\omega)|^2 + |\lambda_{H_0}(\omega + \pi)|^2 = 2.$$

We see that the perfect reconstruction conditions of the two definitions of QMF filter banks only differ in that the latter takes absolute values. It turns out that the latter also has many interesting solutions, as we will see in Chap. 6.

Exercise 5.14: Orthogonal Filter Bank Transforms

Show that H is orthogonal (i.e. $G = H^T$) if and only if the filters satisfy $G_0 = (H_0)^T, G_1 = (H_1)^T$,

In the literature, a wavelet is called *orthonormal* if the filters satisfy $G_0 = (H_0)^T, G_1 = (H_1)^T$. Thus, alternative QMF filter banks with perfect reconstruction are examples of orthonormal wavelets.

Exercise 5.15: The Haar Wavelet as an Alternative QMF Filter Bank Transform

Show that the Haar wavelet satisfies $\lambda_{H_1}(\omega) = -\overline{\lambda_{H_0}(\omega + \pi)}$, and $G_0 = (H_0)^T$, $G_1 = (H_1)^T$. The Haar wavelet thus differs from an alternative QMF filter bank only in a sign. Despite this sign difference, the Haar wavelet is often called an alternative QMF filter bank transform in the literature. The additional sign leads to an orthonormal wavelet with $\alpha = -1, d = 0$ instead.

Exercise 5.16: The Support of the Scaling Function and the Mother Wavelet

The scaling functions and mother wavelets we encounter will always turn out to be functions with compact support (for a general proof of this, see [9]). An interesting consequence of Eqs. (5.2) and (5.3) is that the corresponding filters then are FIR, and that one can find a connection between the supports of the scaling function and mother wavelet, and the number of filter coefficients. In the following we will say that the *support of a filter* is $[E, F]$ if $t_E, \ldots, t_F$ are the only nonzero filter coefficients.

a) Assume that G_0 has support $[M_0, M_1]$, G_1 has support $[N_0, N_1]$. Show that $\mathrm{Supp}(\phi) = [M_0, M_1]$, and $\mathrm{Supp}(\psi) = [(M_0 + N_0 + 1)/2, (M_1 + N_1 + 1)/2]$.

b) Assume that all the filters are symmetric. Show that $\mathrm{Supp}(\phi)$ is on the form $[-M_1, M_1]$ (i.e. symmetric around 0), and that $\mathrm{Supp}(\psi)$ is on the form $1/2 + [-(M_1 + N_1)/2, (M_1 + N_1)/2]$, i.e. symmetric around 1/2. Verify that the alternative piecewise linear wavelet has $\mathrm{Supp}(\phi) = [-1, 1]$, and $\mathrm{Supp}(\psi) = [-1, 2]$ (as we already know from Fig. 4.18), and that the piecewise linear wavelet has $\mathrm{Supp}(\psi) = [0, 1]$.

c) The wavelet with symmetric filters with most filter coefficients we will consider has 7 and 9 filter coefficients, respectively. Explain why plotting over $[-4, 4]$ captures the entire support of both the scaling function and the mother wavelet when there are at most this number of filter coefficients. This explains why $[-4, 4]$ has been used as the plotting interval in many of the plots.

d) Verify that, for the Haar wavelet, G_0 has filter coefficients evenly distributed around 1/2, G_1 has equally many, and evenly distributed around $-1/2$, and the supports of both ϕ and ψ are symmetric around 1/2. The Haar wavelet is the simplest orthonormal wavelet, and these properties will be used as a template for the other orthonormal wavelets.

e) We will only consider orthonormal wavelets with at most 8 filter coefficients. Explain again why $[-4, 4]$ as plotting interval is desirable.

5.2 Dual Filter Bank Transform and Dual Wavelets

Since the reverse transform inverts the forward transform for wavelets, $GH = I$. If we transpose this expression we get that $H^T G^T = I$. Clearly H^T is a reverse filter bank transform with filters $(H_0)^T, (H_1)^T$, and G^T is a forward filter bank transform with filters $(G_0)^T, (G_1)^T$. Due to their usefulness, these transforms have their own name:

Definition 4. *Dual Filter Bank Transforms.*
     Assume that H_0, H_1 are the filters of a forward filter bank transform, and that G_0, G_1 are the filters of a reverse filter bank transform. By the *dual transforms* we mean the forward filter bank transform with filters $(G_0)^T, (G_1)^T$, and the reverse filter bank transform with filters $(H_0)^T, (H_1)^T$.

From Exercise 5.14 it follows that H is orthogonal if and only if the transform and the dual transform are equal. In the setting of wavelets we called this an orthonormal wavelet.
     The DWT and IDWT functions in the library accept a parameter called `dual`. This indicates whether the filter bank transforms should replace the filter bank transform with the dual filter bank transform. In particular, we can differ between the DWT, IDWT, and their duals as follows:

```
dwt_impl(x, wave_name, m, 'symm', 'none', 0, 0);  % DWT
idwt_impl(x, wave_name, m, 'symm', 'none', 0, 0); % IDWT
dwt_impl(x, wave_name, m, 'symm', 'none', 0, 1);  % Dual DWT
idwt_impl(x, wave_name, m, 'symm', 'none', 0, 1); % Dual IDWT
```

In Chap. 6 we will see that there exist functions $\tilde{\phi}$ and $\tilde{\psi}$, called the *dual scaling function* and *dual mother wavelet*, which also give rise to an MRA, and so that the associated kernel transformations are the dual transforms. Contrary to the ϕ and ψ, which are given at the start, these are given indirectly in terms of filters. So, how can we plot these dual functions?
     It turns out that we can easily get good approximations to function values of the scaling functions and mother wavelets (dual or not), once we have the filters. The starting point is that

- The coordinates of ϕ in $(\boldsymbol{\phi}_0, \boldsymbol{\psi}_0, \boldsymbol{\psi}_1, \ldots, \boldsymbol{\psi}_{m-1})$ is $(1, 0, \ldots, 0)$, where there are $2^m N - 1$ zeros.
- The coordinates of ψ in $(\boldsymbol{\phi}_0, \boldsymbol{\psi}_0, \boldsymbol{\psi}_1, \ldots, \boldsymbol{\psi}_{m-1})$ is $(0, \ldots, 0, 1, 0, \ldots, 0)$, where there are N zeros at the beginning.

Thus, there is only one non-zero coordinate, and the coordinates of ϕ and ψ in $\boldsymbol{\phi}_m$ can be found with an m-level IDWT. In Exercises 4.1 and 4.22 we saw that, for our first two wavelets, the samples of $f \in V_m$ corresponded directly to the coordinates in $\boldsymbol{\phi}_m$. In general the coordinate vector of $f \in V_0$ may not equal the samples $(f(0), f(1), \ldots)$, however. Our next result states that, even if the coordinates of f in $\boldsymbol{\phi}_m$ are something different than the samples of f, they still can be used to approximate these samples. The result can thus be used not only to make an approximate plot of the dual functions, but also of ϕ and ψ, in case they also are indirectly given.

Theorem 5. Relation Between Samples and Wavelet Coefficients.

Assume that $\tilde{\phi}$ has compact support, and that $\int_0^N |\tilde{\phi}(t)|\, dt < \infty$. Assume also that f is continuous and has coordinates $c_{m,n}$ in $\boldsymbol{\phi}_m$, i.e.

$$\sum_n c_{m,n}\phi_{m,n} \overset{L^2(\mathbb{R})}{\to} f$$

Then we have that

$$\lim_{m\to\infty} 2^{m/2} c_{m,n2^{m-m'}} = f(n/2^{m'}) \int_0^N \tilde{\phi}(t)\, dt.$$

Proof. We will only prove this result in the case of orthonormal wavelets, i.e. when $\phi = \tilde{\phi}$. We will comment on the general case in Chap. 6.

Since ϕ has compact support, $\phi_{m,n2^{m-m'}}$ will be supported on a small interval close to $n/2^{m'}$ for large m. Since f is continuous, given $\epsilon > 0$ we can choose m so large that $f(t) = f(n/2^{m'}) + r(t)$, where $|r(t)| < \epsilon$ on this interval. But then

$$c_{m,n2^{m-m'}} = \int_0^N f(t)\phi_{m,n2^{m-m'}}(t)\, dt$$

$$= f(n/2^{m'}) \int_0^N \phi_{m,n2^{m-m'}}(t)\, dt + \int_0^N r(t)\phi_{m,n2^{m-m'}}(t)\, dt$$

$$\leq 2^{-m/2} f(n/2^{m'}) \int_0^N \phi(t)\, dt + \epsilon \int_0^N |\phi_{m,n2^{m-m'}}(t)|\, dt$$

$$= 2^{-m/2} f(n/2^{m'}) \int_0^N \phi(t)\, dt + 2^{-m/2}\epsilon \int_0^N |\phi(t)|\, dt.$$

From this it follows that $\lim_{m\to\infty} 2^{m/2} c_{m,n2^{m-m'}} = f(n/2^{m'}) \int_0^N \phi(t)\, dt$, since ϵ was arbitrary, and since we assumed that $\int_0^N |\phi(t)|\, dt < \infty$. $\qquad\square$

Now, let us apply an m-stage IDWT to obtain the coordinates of ϕ and ψ in $\boldsymbol{\phi}_m$. If ϕ and ψ are continuous, Theorem 5 states that, if we scale the resulting coordinates with $2^{m/2}$, we obtain good estimates to the samples $\phi(n/2^{m'})$, $\psi(n/2^{m'})$. Even if we do not know the value of the integral $\int_0^N \tilde{\phi}(t)\, dt$, the resulting values can be used to make an accurate plot of ϕ and ψ. This algorithm is also called the cascade algorithm.

Definition 6. *The Cascade Algorithm.*

The *cascade algorithm* applies a change of coordinates for the functions ϕ, ψ from the basis $(\boldsymbol{\phi}_0, \boldsymbol{\psi}_0, \boldsymbol{\psi}_1 \ldots, \boldsymbol{\psi}_{m-1})$ to the basis $\boldsymbol{\phi}_m$, and uses the new coordinates as an approximation to the values of these functions.

Let us see how we can implement the cascade algorithm. The input to our implementation will be

- the number of levels m,
- integers a and b so that the interval $[a, b]$ contains the supports of ϕ, ψ (Exercise 5.16 explains how we can find these supports). This enables us to plot all of ϕ and ψ. With $N = b - a$, V_m is of dimension $N2^m$, and the cascade algorithm needs a coordinate vector of this size as starting point, which we previously described,
- the name of the wavelet,

- a variable stating whether to plot ϕ or ψ,
- a variable stating whether to plot the dual function or not.

The algorithm can be implemented as follows.

```
function cascade_alg(m, a, b, wave_name, scaling, dual)
    coords = zeros((b-a)*2^m, 1);
    if scaling
        coords(1) = 1;
    else
        coords(b - a + 1) = 1;
    end
    t = linspace(a, b, (b-a)*2^m);
    coords = idwt_impl(coords, wave_name, m, 'per', 'none', 1, dual);

    coords = [ coords((b*2^m+1):((b-a)*2^m)); coords(1:(b*2^m)) ];
    figure()
    plot(t, 2^(m/2)*coords, 'k-')
end
```

In Fig. 5.6 we have used this function to plot the dual scaling functions and mother wavelets for both piecewise linear wavelets we have considered. We see that these functions look very irregular. Also, they are very different from the original scaling function and mother wavelet. We will later argue that this is bad, it would be much better if $\phi \approx \tilde{\phi}$ and $\psi \approx \tilde{\psi}$. To see why the dual functions for the piecewise linear wavelets look like these spikes, note that they must satisfy

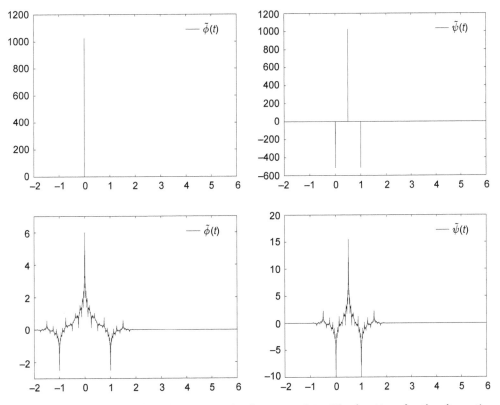

Fig. 5.6 Dual functions for the two piecewise linear wavelets. The functions for the alternative piecewise linear wavelet is shown at the bottom

$$\tilde{\phi} = \sqrt{2}\tilde{\phi}_{1,0} \qquad\qquad \tilde{\psi} = \sqrt{2}\left(-\frac{1}{2}\tilde{\phi}_{1,0} + \tilde{\phi}_{1,1} - \frac{1}{2}\tilde{\phi}_{1,2}\right).$$

The first equation gives a contradiction unless $\tilde{\phi}$ is a "spike" at zero. The second equation then says that $\tilde{\psi}$ must be the sum of three spikes, at 0, 1/2, and 1. The middle spike peaks at a positive value, the other two at the same negative value, being half of the positive value. All this can be verified in the plot.

In many books, one computes the DWT of a vector of sample function values, so that the underlying function is $\sum_n f(n2^{-m})\phi_{m,n}$. Since the sample values are different from the coordinates in ϕ_m in general, this is different from f, and we compute something completely different than we want. This wrong procedure for computing the DWT is also called *the first crime of wavelets*. The next result states that the error committed in a wavelet crime, i.e. when we replace $\sum_n f(n/2^m)\phi_{m,n}(t)$ with f, will be small if we increase m.

Theorem 7. Using the Samples.

If f is continuous and ϕ has compact support, for all t with a finite bit expansion (i.e. $t = n/2^{m'}$ for some integers n and m') we have that

$$\lim_{m\to\infty} 2^{-m/2} \sum_{s=0}^{2^m N-1} f(s/2^m)\phi_{m,s}(t) = f(t)\sum_s \phi(s).$$

This says that, up to the constant factor $c = \sum_s \phi(s)$, the functions $f_m \in V_m$ with coordinates $2^{-m/2}(f(0/2^m), f(1/2^m), \ldots)$ in ϕ_m converge pointwise to f as $m \to \infty$ (even though the samples of f_m may not equal those of f).

Proof. With $t = n/2^{m'}$, for $m > m'$ we have that

$$\phi_{m,s}(t) = \phi_{m,s}(n2^{m-m'}/2^m) = 2^{m/2}\phi(2^m n2^{m-m'}/2^m - s) = 2^{m/2}\phi(n2^{m-m'} - s).$$

We thus have that

$$\sum_{s=0}^{2^m N-1} 2^{-m/2}f(s/2^m)\phi_{m,s}(t) = \sum_{s=0}^{2^m N-1} f(s/2^m)\phi(n2^{m-m'} - s).$$

In the sum finitely many s close to $n2^{m-m'}$ contribute (due to finite support), and then $s/2^m \approx t$. Due to continuity of f this sum thus converges to $f(t)\sum_s \phi(s)$, and the proof is done. $\square$

Even if the error committed in a wavelet crime can be made small, the problem is that one usually does not address what would be acceptable values for m. If the samples $f(n2^{-m})$ are known, a better approach would be to find a linear combination $\sum_n c_{m,n}\phi_{m,n}(t)$ which interpolates these. This amounts to solving the equations

$$\sum_{n'} c_{m,n'}\phi_{m,n'}(n2^{-m}) = f(n2^{-m}),$$

so that

$$\sum_{n'} c_{m,n'}\phi(n - n') = 2^{-m/2}f(n2^{-m}).$$

This can be summarized as follows

Proposition 8. Finding Coordinates from the Samples.
Assume that $f(t) = \sum_n c_{m,n}\phi_{m,n}(t)$. The $c_{m,n}$ can be found from the samples $f(n2^{-m})$ by solving the system $Sc_m = b$, where S is the matrix with entries $S_{kn} = \phi(k-n)$, and b is the vector with entries $b_n = 2^{-m/2}f(n2^{-m})$.

S and its inverse are (circulant) Toeplitz matrices, so that finding c_m amounts to applying a filtering. Also, the sparsity of S is proportional to the support size of ϕ. Efficient DFT-based methods can thus be used to find c_m. Similarly, if we already have the coordinates $c_{m,n}$, and we want to know the sample values, we only need to filter c_m with S.

Observation 9. Pre-filtering the DWT.
If f is a function with known sample values, it is preferred to pre-filter the sample values with the filter S^{-1}, before applying the DWT.

A final remark should be made regarding the filter S: We may not know the values $\phi(n)$ exactly. These values can be approximated using the cascade algorithm again, however.

In a wavelet toolbox, the filters may not be stated explicitly, and we may have no way of knowing how the DWT and the IDWT are computed. Nevertheless, it is easy to find the filters from the DWT and IDWT implementations: Since the filter coefficients of G_0 and G_1 can be found in the columns of G, all we need to do is apply the IDWT kernel to e_0 (to find G_0) and e_1 (to find G_1). Finally the frequency responses can by computing a DFT of the filter coefficients. This is exemplified in the following algorithm.

```
function freqresp_alg(wave_name, lowpass, dual)
    N = 128;
    n = (0:(N-1))';
    omega = 2*pi*n/N;

    g = zeros(N, 1);
    if lowpass
        g(1) = 1;
    else
        g(2) = 1;
    end

    g = idwt_impl(g, wave_name, 1, 'per', 'none', 1, dual, 0, 'time');
    figure();
    plot(omega, abs(fft(g)), 'k-')
end
```

If the parameter `dual` is set to true, the dual filters $(H_0)^T$ and $(H_1)^T$ are plotted instead. If the filters have real coefficients, $|\lambda_{H_i^T}(\omega)| = |\lambda_{H_i}(\omega)|$, so the correct frequency responses are shown.

Example 5.17: Plotting Scaling Functions and Mother Wavelets

It turns out that the interval $[-2, 6]$ contains the supports of all scaling functions and mother wavelets we have considered up to now. We can therefore use $a = -2$ and $b = 6$ as parameters to `cascade_alg`, in order to plot all these functions. We will also use $m = 10$ levels. The following code then runs the cascade algorithm for the three wavelets we have considered, to reproduce all previous scaling functions and mother wavelets.

```
cascade_alg(10, -2, 6, 'Haar', 1, 0)
cascade_alg(10, -2, 6, 'Haar', 0, 0)


cascade_alg(10, -2, 6, 'pwl0', 1, 0)
cascade_alg(10, -2, 6, 'pwl0', 0, 0)


cascade_alg(10, -2, 6, 'pwl2', 1, 0)
cascade_alg(10, -2, 6, 'pwl2', 0, 0)
```

Example 5.18: Plotting Frequency Responses

In order to verify the low-pass/high-pass characteristics of G_0 and G_1, let us plot the frequency responses of the wavelets we have considered. To plot $\lambda_{G_0}(\omega)$ and $\lambda_{G_1}(\omega)$ for the Haar wavelet, we can write

```
freqresp_alg('Haar', 1, 0)
freqresp_alg('Haar', 0, 0)
```

To plot the same frequency response for the alternative piecewise linear wavelet, we can write

```
freqresp_alg('pwl2', 1, 0)
freqresp_alg('pwl2', 0, 0)
```

This code reproduces the frequency responses shown in Figs. 5.4 and 5.5.

Exercise 5.19: Using the Cascade Algorithm

In Exercise 4.31 we constructed a new mother wavelet $\hat{\psi}$ for piecewise linear functions by finding constants $\alpha, \beta, \gamma, \delta$ so that

$$\hat{\psi} = \psi - \alpha\phi_{0,0} - \beta\phi_{0,1} - \delta\phi_{0,2} - \gamma\phi_{0,N-1}.$$

Use the cascade algorithm to plot $\hat{\psi}$. Do this by using the wavelet kernel for the piecewise linear wavelet (do not use the code above, since we have not implemented kernels for this wavelet yet).

Exercise 5.20: Implementing the Dual Filter Bank Transforms

a) Show that $A_\lambda^T = B_\lambda$ and $B_\lambda^T = A_\lambda$, i.e. that the transpose of an elementary lifting matrix of even/odd type is an elementary lifting matrix of odd/even type.

b) Assume that we have a factorization of H in terms of elementary lifting matrices. Use a) to show that the dual forward transform is obtained by replacing each A_λ with $B_{-\lambda}$, B_λ with $A_{-\lambda}$, and k with $1/k$ in the multiplication step. These replacements are done by the function `set_wav_props` in the file `find_wav_props.m` in the library, on an object which is accessed by the general kernel.

c) Previously we expressed the forward and reverse transforms of the piecewise linear wavelets in terms of elementary lifting operations. Use b) to write down the dual forward and reverse transforms of these two wavelets in terms of lifting matrices.

Exercise 5.21: Transpose of the DWT and IDWT

Explain why

- The transpose of the DWT can be computed with an IDWT with the kernel of the dual IDWT
- The transpose of the dual DWT can be computed with an IDWT with the kernel of the IDWT
- The transpose of the IDWT can be computed with a DWT with the kernel of the dual DWT
- The transpose of the dual IDWT can be computed with a DWT with the kernel of the DWT

5.3 Symmetric Extensions

In Sect. 3.5 we saw how to adapt symmetric filters to symmetric extensions, creating transformations diagonalized by the DCT. In this section we will revisit this in the setting of wavelets. There will be two major differences from Sect. 3.5.

First of all, due to Theorem 26 our adaptation to symmetric extensions will only affect the wavelet coefficients near the boundaries 0 and T, since that theorem stated that a discontinuity affects the decay of wavelet coefficients only locally. This is contrary to Fourier series where a discontinuity may have an impact globally.

Secondly, we will change our sampling strategy, to adapt to how this is handled in wavelet literature. At the start of Sect. 2.4 we mentioned two strategies, depending on whether the boundaries are sampled or not. We will now consider the second strategy, where the boundaries are sampled. To make this more concrete, if f has period M (so that $\breve{f}$ has period $2M$), we let

- $\boldsymbol{x} = \{f(n2^{-m})\}_{n=0}^{2^m M}$ be the samples of f over one period (i.e. $f(0)$ is the first sample, $f(M)$ the last),
- $\breve{\boldsymbol{x}} = \{\breve{f}(n2^{-m})\}_{n=0}^{2^m 2M-1}$ be the samples of $\breve{f}$ over one period (i.e. $f(0)$ is the first sample, but $f(2M)$ is not a sample).

If we commit the wavelet crime, $\boldsymbol{x}$ would be used as input to the DWT. We will now use $\breve{\boldsymbol{x}}$ rather that $\boldsymbol{x}$ as input. Due to Theorem 26, if the wavelet has one vanishing moment, the DWT coefficients corresponding to $t \approx M$ should be small. This is the entire point of working with symmetric extensions. Clearly $\breve{x}$ is a symmetric extension of x in the following sense (with $N = 2^m M + 1$).

Definition 10. *Symmetric Extension of a Vector.*
By the *symmetric extension* of $x \in \mathbb{R}^N$, we mean $\breve{x} \in \mathbb{R}^{2N-2}$ defined by

$$\breve{x}_k = \begin{cases} x_k & 0 \le k < N \\ x_{2N-2-k} & N \le k < 2N - 3 \end{cases} \tag{5.6}$$

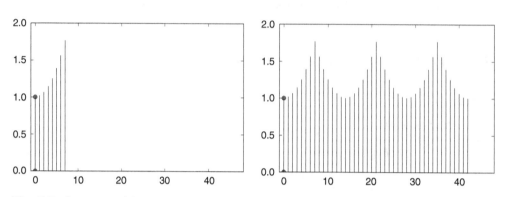

Fig. 5.7 A vector and its symmetric extension. The period of $\breve{x}$ is now $2N - 2$. For the vector in Fig. 2.5 it was $2N$

In Fig. 5.7 we have shown such an extension. This is different from the symmetric extension given by Definition 22 in that the boundary M is a sample in x. The lengths of x and $\breve{x}$ are N and $2N - 2$, so that we do not stay within vectors of lengths powers of two. This could be problematic if we depended on FFT algorithms, but it turns out that wavelet transforms can be implemented efficiently not using such algorithms. Let us now specialize to the case when all filters H_0, H_1, G_0, and G_1 are symmetric. In Theorem 14 we proved that symmetric filters preserved symmetric extensions as defined in Sect. 2.4. It is an exercise that symmetric filters preserve symmetric vectors in the sense of Definition 10 also.

Pre-filtering

As usual the vector $\breve{x}$ should be pre-filtered, in order to not commit the wavelet crime. As noted previously, the scaling function ϕ is symmetric (i.e. $\phi(t) = \phi(-t)$) when G_0 is symmetric, so that the filter used in the pre-filtering also is symmetric. As noted, we then know that the pre-filtered vector (which we also will denote by $\breve{x}$) will be a symmetric vector with the same symmetry.

H and G Preserve Symmetry

Again, the filters H_0 and H_1 will also preserve symmetry. To see that H will preserve symmetry as well, let us compare $(H\breve{x})_{N-1-k}$ and $(H\breve{x})_{N-1+k}$. The difference between $N - 1 - k$ and $N - 1 + k$ is even, and since $H_i\breve{x}$ is symmetric, for one i we have that

$$(H\breve{x})_{N-1-k} = (H_i\breve{x})_{N-1-k} = (H_i\breve{x})_{N-1+k} = (H\breve{x})_{N-1+k}.$$

Since this applies for all k, H also preserves symmetry. Since G is the inverse of H, it too must preserve symmetry.

Since a symmetric extension is uniquely defined from the values $x_0, \ldots, x_{N-1}$, we can view the restriction of H to symmetric extensions as a mapping from $\mathbb{R}^N$ to $\mathbb{R}^N$. In fact, for any mapping $S : \mathbb{R}^{2N-2} \to \mathbb{R}^{2N-2}$ which preserve symmetric extensions in the sense of Definition 10, we can define a unique mapping $S_r : \mathbb{R}^N \to \mathbb{R}^N$ by the requirement that $S_r \boldsymbol{x}$ should equal the first N components of $S\breve{\boldsymbol{x}}$ (parallel to the definition of S_r in Theorem 14). We can thus define mappings H_r, G_r, $(H_0)_r$, $(H_1)_r$, $(G_0)_r$, and $(G_1)_r$ from $\mathbb{R}^N$ to $\mathbb{R}^N$. Since the entries in $H_r \breve{\boldsymbol{x}}$ are assembled from the entries in $(H_i)_r \breve{\boldsymbol{x}}$, in the same way as the entries in $H\boldsymbol{x}$ are assembled from the entries in $H_i \boldsymbol{x}$, the following is clear.

Theorem 11. Symmetric Filters and Symmetric Extensions.

If the filters H_i and G_i are symmetric, then Theorem 2 holds with H, G, H_i, and G_i replaced by H_r, G_r, $(H_i)_r$, and $(G_i)_r$.

Lifting Factorizations

For our two piecewise linear wavelets, we could factor H and G in terms of elementary lifting matrices A_λ and B_λ. Clearly these are MRA-matrices with symmetric filters as well, so that they too preserve symmetric extensions, and we can define the restrictions $(A_\lambda)_r$, $(B_\lambda)_r$. Clearly, if $H = \prod_i A_{\lambda_{2i}} B_{\lambda_{2i+1}}$, then $H_r = \prod_i (A_{\lambda_{2i}})_r (B_{\lambda_{2i+1}})_r$. It is straightforward to find expressions for $(A_\lambda)_r$ and $(B_\lambda)_r$ (Exercise 5.25).

In conclusion, the library can be adapted to symmetric filters and their symmetric extensions by replacing A_λ and B_λ with $(A_\lambda)_r$ and $(B_\lambda)_r$. The purpose of the `bd_mode` parameter in the library is to control how the *boundaries* (i.e. the start and end of the input) are handled by the DWT and the IDWT, in particular whether symmetric or periodic extensions should be applied:

- If `bd_mode` equals `"symm"` (this is the default value), symmetric extension (i.e. $(A_\lambda)_r$ and $(B_\lambda)_r$) is applied.
- If `bd_mode` equals `"per"`, periodic extension (i.e. A_λ and B_λ) is applied.

There are also other ways one can handle the boundaries. The `bd_mode` parameter can be used for some of those as well, but we will not say more about them here, however, only remark that other supported boundary modes can be inspected in the implementations of `lifting_even_symm` and `lifting_odd_symm`. The `bd_mode` parameter is passed to these functions from the wavelet kernels.

Exercise 5.22: Symmetric Filters Preserve Symmetric Extensions

Prove that symmetric filters preserve symmetric vectors in the sense of Definition 10 also. Show also more generally that, if $\boldsymbol{x} \in \mathbb{R}^{2N}$ is symmetric around any "half-number" (i.e. $1/2$, 1, $3/2$, $2,\ldots$), then $S\boldsymbol{x}$ is symmetric around the same half-number. Use Theorem 14 as a guide.

Exercise 5.23: Implementing Symmetric Extensions

a) The function you implemented in Exercise 3.28 computed circular convolution, i.e. it assumed a periodic extension of the input. Reimplement that function so that it also takes a third parameter `bd_mode`. If this equals `"per"` a periodic extension should be assumed. If it equals `"symm"` a symmetric extension should be assumed, as described above. Again, the function `filter_impl` in the library can serve as a template for your code.

b) Implement functions

```
dwt_kernel_filters(x, bd_mode, wav_props)
idwt_kernel_filters(x, bd_mode, wav_props)
```

which return DWT and IDWT kernels that apply the function `filter_impl` from a) together with Theorem 2. The parameter `wav_props` should be an object which gives access to the filters by means of writing `wav_props.h0`, `wav_props.h1`, `wav_props.g0`, and `wav_props.g1`.

Using these functions one can define standard DWT and IDWT kernels in the following way, assuming that the `wav_props` has been computed:

```
dwt_kernel  = @(x, bd_mode) dwt_kernel_filters(x, bd_mode, wav_props);
idwt_kernel = @(x, bd_mode) idwt_kernel_filters(x, bd_mode, wav_props);
```

Exercise 5.24: The Boundary Mode in the Cascade Algorithm

In Exercise 5.19 we did not use symmetric extensions (the `bd_mode`-argument was set to `per`). Attempt to use symmetric extensions instead (set the `bd_mode`-argument to `"symm"`), and observe the new plots you obtain. Can you explain why these new plots do not show the correct functions, while the previous plots are correct?

Exercise 5.25: Restricted Matrices for Elementary Lifting

Show that

$$(A_\lambda)_r = \begin{pmatrix} 1 & 2\lambda & 0 & 0 & \cdots & 0 & 0 & 0 \\ 0 & 1 & 0 & 0 & \cdots & 0 & 0 & 0 \\ 0 & \lambda & 1 & \lambda & \cdots & 0 & 0 & 0 \\ \vdots & \vdots & \vdots & \vdots & \vdots & \vdots & \vdots & \vdots \\ 0 & 0 & 0 & 0 & \cdots & \lambda & 1 & \lambda \\ 0 & 0 & 0 & 0 & \cdots & 0 & 0 & 1 \end{pmatrix}$$

$$(B_\lambda)_r = \begin{pmatrix} 1 & 0 & 0 & 0 & \cdots & 0 & 0 & 0 \\ \lambda & 1 & \lambda & 0 & \cdots & 0 & 0 & 0 \\ 0 & 0 & 1 & 0 & \cdots & 0 & 0 & 0 \\ \vdots & \vdots & \vdots & \vdots & \vdots & \vdots & \vdots & \vdots \\ 0 & 0 & 0 & 0 & \cdots & 0 & 1 & 0 \\ 0 & 0 & 0 & 0 & \cdots & 0 & 2\lambda & 1 \end{pmatrix}.$$

Also, change the implementations of `lifting_even_symm` and `lifting_odd_symm` so that these expressions are used (rather than A_λ, B_λ) when the `bd_mode` parameter equals `"symm"`.

Exercise 5.26: Expression for S_r

Show that, with

$$S = \begin{pmatrix} S_1 & S_2 \\ S_3 & S_4 \end{pmatrix} \in \mathbb{R}^{2N-2} \times \mathbb{R}^{2N-2}$$

a symmetric filter, with $S_1 \in \mathbb{R}^N \times \mathbb{R}^N$, $S_2 \in \mathbb{R}^N \times \mathbb{R}^{N-2}$, we have that

$$S_r = S_1 + \left(0 \ (S_2)^{\leftrightarrow} \ 0\right).$$

Use the proof of Theorem 15 as a guide.

Exercise 5.27: Orthonormal Basis for the Symmetric Extensions

In this exercise we will establish an orthonormal basis for the symmetric extensions in the sense of Definition 10. This parallels Theorem 23.

a) Show that

$$\frac{1}{\sqrt{2N-2}} \cos\left(\frac{2\pi \cdot 0 \cdot k}{2N-2}\right),$$

$$\left\{\frac{1}{\sqrt{N-1}} \cos\left(\frac{2\pi nk}{2N-2}\right)\right\}_{n=1}^{N-2},$$

$$\frac{1}{\sqrt{2N-2}} \cos\left(\frac{2\pi(N-1)k}{2N-2}\right)$$

is an orthonormal basis for the symmetric extensions in $\mathbb{R}^{2N-2}$.

b) Assume that S is symmetric. Show that the vectors listed in a) are eigenvectors for S_r, when the vectors are viewed as vectors in $\mathbb{R}^N$, and that they are linearly independent. This shows that S_r is diagonalizable.

Exercise 5.28: Diagonalizing S_r

Let us explain how the matrix S_r can be diagonalized, similarly to how we previously diagonalized using the DCT. In Exercise 5.27 we showed that the vectors

$$\left\{\cos\left(\frac{2\pi nk}{2N-2}\right)\right\}_{n=0}^{N-1}$$

in $\mathbb{R}^N$ is a basis of eigenvectors for S_r when S is symmetric. S_r itself is not symmetric, however, so that this basis can not possibly be orthogonal (S is symmetric if and only if it is orthogonally digonalizable). However, when the vectors are viewed in $\mathbb{R}^{2N-2}$ we showed in Exercise 5.27a) that

$$\sum_{k=0}^{2N-3} \cos\left(\frac{2\pi n_1 k}{2N-2}\right) \cos\left(\frac{2\pi n_2 k}{2N-2}\right)$$

$$= (N-1) \times \begin{cases} 2 & \text{if } n_1 = n_2 \in \{0, N-1\} \\ 1 & \text{if } n_1 = n_2 \notin \{0, N-1\} \\ 0 & \text{if } n_1 \neq n_2 \end{cases}.$$

a) Show that

$$(N-1) \times \begin{cases} 1 & \text{if } n_1 = n_2 \in \{0, N-1\} \\ \frac{1}{2} & \text{if } n_1 = n_2 \notin \{0, N-1\} \\ 0 & \text{if } n_1 \neq n_2 \end{cases}$$

$$= \frac{1}{2} \cos\left(\frac{2\pi n_1 \cdot 0}{2N-2}\right) \cos\left(\frac{2\pi n_2 \cdot 0}{2N-2}\right)$$

$$+ \sum_{k=1}^{N-2} \cos\left(\frac{2\pi n_1 k}{2N-2}\right) \cos\left(\frac{2\pi n_2 k}{2N-2}\right)$$

$$+ \frac{1}{2} \cos\left(\frac{2\pi n_1(N-1)}{2N-2}\right) \cos\left(\frac{2\pi n_2(N-1)}{2N-2}\right).$$

Hint

Use that $\cos x = \cos(2\pi - x)$ to pair the summands k and $2N-2-k$.

b) Define the vectors $\{\boldsymbol{x}_n\}_{n=0}^{N-1}$ by

$$\left\{ \frac{1}{\sqrt{2}} \cos\left(\frac{2\pi n \cdot 0}{2N-2}\right), \left\{ \cos\left(\frac{2\pi n k}{2N-2}\right) \right\}_{k=1}^{N-2}, \frac{1}{\sqrt{2}} \cos\left(\frac{2\pi n(N-1)}{2N-2}\right) \right\},$$

and the vectors $\{\boldsymbol{d}_n^{(\mathrm{I})}\}_{n=0}^{N-1}$ by

$$\boldsymbol{d}_n^{(\mathrm{I})} = \begin{cases} \sqrt{\frac{1}{N-1}}\boldsymbol{x}_0 & n = 0 \\ \sqrt{\frac{2}{N-1}}\boldsymbol{x}_n & 1 \leq n < N-1 \\ \sqrt{\frac{1}{N-1}}\boldsymbol{x}_{N-1} & n = N-1. \end{cases}$$

Explain that the vectors $\boldsymbol{d}_n^{(\mathrm{I})}$ are orthonormal, and that the matrix

$$\sqrt{\frac{2}{N-1}} \begin{pmatrix} 1/\sqrt{2} & 0 & \cdots & 0 & 0 \\ 0 & 1 & \cdots & 0 & 0 \\ \vdots & \vdots & \vdots & \vdots & \vdots \\ 0 & 0 & \cdots & 1 & 0 \\ 0 & 0 & \cdots & 0 & 1/\sqrt{2} \end{pmatrix} \left(\cos\left(\frac{2\pi n k}{2N-2}\right) \right) \begin{pmatrix} 1/\sqrt{2} & 0 & \cdots & 0 & 0 \\ 0 & 1 & \cdots & 0 & 0 \\ \vdots & \vdots & \vdots & \vdots & \vdots \\ 0 & 0 & \cdots & 1 & 0 \\ 0 & 0 & \cdots & 0 & 1/\sqrt{2} \end{pmatrix}$$

has the $d_n^{(I)}$ as columns, and is therefore orthogonal. Orthogonality in $\mathbb{R}^N$ could thus be achieved by a scaling of the first and last components in the vectors. This orthogonal matrix is denoted $\mathrm{DCT}_N^{(I)}$. It is much used, just as the $\mathrm{DCT}_N^{(II)}$ of Sect. 2.4.

c) Prove that $\left(\cos\left(\frac{2\pi nk}{2N-2}\right)\right)^{-1}$ can be written as

$$\frac{2}{N-1}\begin{pmatrix} 1/2 & 0 & \cdots & 0 & 0 \\ 0 & 1 & \cdots & 0 & 0 \\ \vdots & \vdots & \vdots & \vdots & \vdots \\ 0 & 0 & \cdots & 1 & 0 \\ 0 & 0 & \cdots & 0 & 1/2 \end{pmatrix}\left(\cos\left(\frac{2\pi nk}{2N-2}\right)\right)\begin{pmatrix} 1/2 & 0 & \cdots & 0 & 0 \\ 0 & 1 & \cdots & 0 & 0 \\ \vdots & \vdots & \vdots & \vdots & \vdots \\ 0 & 0 & \cdots & 1 & 0 \\ 0 & 0 & \cdots & 0 & 1/2 \end{pmatrix}$$

Since S_r can be diagonalized as $\left(\cos\left(\frac{2\pi nk}{2N-2}\right)\right)D\left(\cos\left(\frac{2\pi nk}{2N-2}\right)\right)^{-1}$, we can substitute here with a simple expression for the inverse. Thus, even if the matrix $\left(\cos\left(\frac{2\pi nk}{2N-2}\right)\right)$ was not orthogonal, this represents no problem in the diagonalization/transition to the frequency domain.

5.4 A Generalization of the Filter Representation, and Its Use in Audio Coding

It turns out that the filter representation (H_0, H_1, G_0, G_1) can be generalized so that it is useful in audio coding. In this section we will explain how the MP3 standard uses such a generalization. Also the generalization will be called *filter bank transforms*, or simply *filter banks*, but there will be more filters than the (H_0, H_1, G_0, G_1):

Definition 12. *Forward Filter Bank Transform.*
    Let $H_0, H_1, \ldots, H_{M-1}$ be $N \times N$-filters. An M-channel *forward filter bank transform* H produces output $z \in \mathbb{R}^N$ from the input $x \in \mathbb{R}^N$ in the following way:

- $z_{iM} = (H_0 x)_{iM}$ for any i so that $0 \leq iM < N$.
- $z_{iM+1} = (H_1 x)_{iM+1}$ for any i so that $0 \leq iM + 1 < N$.
- $\ldots$
- $z_{iM+(M-1)} = (H_{M-1} x)_{iM+(M-1)}$ for any i so that $0 \leq iM + (M-1) < N$.

In other words, the output of a forward filter bank transform is computed by applying filters $H_0, H_1, \ldots, H_{M-1}$ to the input, and by downsampling and assembling these so that we obtain the same number of output samples as input samples. Also

- $H_0, H_1, \ldots, H_{M-1}$ are called the *analysis filter components*,
- the output from the filter H_i is called (the subband samples of) *channel i*,
- M is the number of channels.

    While the DWT splits the input c_m into two channels (c_{m-1} and w_{m-1}), an M-channel transform splits into M channels. In the matrix of a forward filter bank transform, rows and columns repeat cyclically with period M, similarly to MRA-matrices. In practice the filters are chosen so that they concentrate on specific frequency ranges, although the filters are usually not ideal band-pass filters. For wavelets, the two frequency ranges were simply denoted as low and high frequencies. Using a filter bank to split a signal into frequency components is also called *subband coding*. In Chap. 7 we will say more on how one can construct filter banks which can be used for subband coding.

Let us now turn to reverse filter bank transforms.

Definition 13. *Reverse Filter Bank Transforms.*

Let $G_0, G_1, \ldots, G_{M-1}$ be $N \times N$-filters. An M-channel *reverse filter bank transform* G produces $\boldsymbol{x} \in \mathbb{R}^N$ from $\boldsymbol{z} \in \mathbb{R}^N$ in the following way: First define $\boldsymbol{z}_k \in \mathbb{R}^N$ as the vector where $(\boldsymbol{z}_k)_{iM+k} = \boldsymbol{z}_{iM+k}$ for all i so that $0 \leq iM + k < N$, and $(\boldsymbol{z}_k)_s = 0$ for all other s, and then

$$\boldsymbol{x} = G_0 \boldsymbol{z}_0 + G_1 \boldsymbol{z}_1 + \ldots + G_{M-1} \boldsymbol{z}_{M-1}. \tag{5.7}$$

$G_0, G_1, \ldots, G_{M-1}$ are also called the *synthesis filter components*.

Again, this generalizes the IDWT, and also here the rows/columns repeat cyclically with period M. The filters G_i are in general different from the H_i, but also in this setting there are important special cases where they are equal, and where the frequency responses are simply shifts of one another, as we saw for the Haar wavelet.

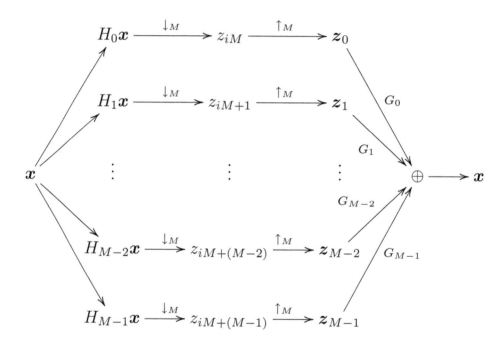

Fig. 5.8 Illustration of forward and reverse filter bank transforms

It is clear that Definitions 12 and 13 give the diagram in Fig. 5.8 for computing forward and reverse transforms. This is very similar to Fig. 5.3. Here $\downarrow_M$ means that we extract every M'th element in the vector (i.e. down-sampling), and $\uparrow_M$ means that we add $M-1$ zeros between the elements (i.e. up-sampling). These are straightforward generalizations of $\downarrow_2$ and $\uparrow_2$.

A reverse filter bank transform may not invert a forward filter bank transform exactly, but if it does we also in this setting use the term *perfect reconstruction*. As before, since H introduces undesired frequencies (i.e. aliasing), these have to be canceled by

G (alias cancellation), in order for perfect reconstruction. Dual filter bank transforms (Definition 4) have a natural generalization to the M-channel setting as well: The dual forward filter bank transform has filters $(G_0)^T, \ldots, (G_{N-1})^T$, the dual reverse filter bank transform has filters $(H_0)^T, \ldots, (H_{N-1})^T$. It is clear that these correspond to the matrices G^T and H^T, respectively.

Now, let us turn to how the MP3 standard is connected to M-channel filter bank transforms. The standard document states procedures for how one should encode and decode video, but with little mentioning of how everything is connected to filter bank transforms. The next two subsections will make this connection, and establish that the filter bank transforms in question are in fact cosine modulated, as they were termed in Exercise 3.39. We will make this connection by "reverse-engineering" the steps stated in the standard, so that they clearly can be seen to take the form of a cosine modulated filter bank. This reverse-engineering involves some technical computations, although not too difficult. We remark that we say more on how to deduce these filter banks in Sect. 7.3, where we state a form for these filter banks which is much more intuitive, and useful for analysis.

5.4.1 Forward Filter Bank Transform in the MP3 Standard

The MP3 standard states that one should encode audio by first performing the following steps iteratively (see p. 67 of the standard document. The procedure is slightly modified with mathematical terminology adapted to this book):

1. Input 32 new audio samples, and build an input sample vector $X \in \mathbb{R}^{512}$, where the 32 new samples are placed first, and all other samples are delayed with 32 elements. In particular the 32 last samples are taken out.
2. Multiply X component-wise with a vector C (this vector is defined through a table in the standard), to obtain a vector $Z \in \mathbb{R}^{512}$. The standard calls this *windowing*.[1]
3. Compute the vector $Y \in \mathbb{R}^{64}$ where $Y_i = \sum_{j=0}^{7} Z_{i+64j}$. The standard calls this a *partial calculation*.
4. Multiply Y with the 32×64- matrix with entries $\cos((2i+1)(k-16)\pi/64)$. The output $S \in \mathbb{R}^{32}$ is called the vector of output samples, or output subband samples. The standard calls this *matrixing*.[2]

The standard does not motivate these steps very well, nor any relation to filter banks. Note also that

- the standard does not explained how the values in the vector C have been obtained,
- the order of the input samples has been reversed from what we are used to, and
- the standard says that the signal should be padded with zeros at the boundaries, in order to apply the algorithm at the start and end of the signal. No other attempts are made to address the behavior at the boundaries.

---

[1] The name windowing is a bit strange. Here this does not correspond to applying a window to the sound samples as we explained in Exercise 3.9. We will see that it rather corresponds to applying a filter coefficient to a sound sample.

[2] It seems strange to use the name matrixing, for something which obviously is matrix multiplication. The reason must be that the procedure has been established outside a linear algebra framework.

Let us prove that the steps above really correspond to a forward filter bank transform, and let us find the filters. The procedure computes 32 outputs in each iteration (the vector S). Therefore, from the standard we would guess that we have $M = 32$ channels, and we want to find 32 filters $H_0, H_1, \ldots, H_{31}$. Clearly, the procedure defines a linear transformation. The input are the audio samples, which we will denote by a vector $\boldsymbol{x}$. At iteration s of the procedure the input audio samples are $x_{32s-512}, x_{32s-511}, \ldots, x_{32s-1}$, which are reversed into a vector $X \in \mathbb{R}^{512}$ by computing $X_i = x_{32s-i-1}$, $0 \le i < 512$. The output to the transformation at iteration s are the $S_0, \ldots, S_{31}$. We assemble these into a vector $\boldsymbol{z}$, so that the output at iteration s are $z_{32(s-1)} = S_0$, $z_{32(s-1)+1} = S_1, \ldots, z_{32(s-1)+31} = S_{31}$.

We will have use for the following cosine-property, which is easily verified:

$$\cos\left(2\pi(n + 1/2)(k + 2Nr)/(2M)\right) = (-1)^r \cos\left(2\pi(n + 1/2)k/(2M)\right) \qquad (5.8)$$

With the terminology above and using Property (5.8) the transformation can be written as

$$z_{32(s-1)+n}$$

$$= \sum_{k=0}^{63} \cos(2\pi(n + 1/2)(k - 16)/(2M))Y_k$$

$$= \sum_{k=0}^{63} \cos(2\pi(n + 1/2)(k - 16)/(2M)) \sum_{j=0}^{7} Z_{k+64j}$$

$$= \sum_{k=0}^{63} \sum_{j=0}^{7} (-1)^j \cos(2\pi(n + 1/2)(k + 64j - 16)/(2M))Z_{k+64j}$$

$$= \sum_{k=0}^{63} \sum_{j=0}^{7} \cos(2\pi(n + 1/2)(k + 64j - 16)/(2M))(-1)^j C_{k+64j}X_{k+64j}$$

$$= \sum_{k=0}^{63} \sum_{j=0}^{7} \cos(2\pi(n + 1/2)(k + 64j - 16)/(2M))(-1)^j C_{k+64j}x_{32s-(k+64j)-1}.$$

Now, if we define $\{h_r\}_{r=0}^{511}$ by $h_{k+64j} = (-1)^j C_{k+64j}$, $0 \le j < 8, 0 \le k < 64$, and $h^{(n)}$ as the filter with coefficients $\{\cos(2\pi(n + 1/2)(k - 16)/64)h_k\}_{k=0}^{511}$, the above can be simplified as

$$z_{32(s-1)+n} = \sum_{k=0}^{511} \cos(2\pi(n + 1/2)(k - 16)/64)h_k x_{32s-k-1} = \sum_{k=0}^{511} (h^{(n)})_k x_{32s-k-1}$$

$$= (h^{(n)}\boldsymbol{x})_{32s-1} = (E_{n-31}h^{(n)}\boldsymbol{x})_{32(s-1)+n},$$

where we again used the notation E_d for time delay. Comparing with Definition 12 we can state the following result.

Theorem 14. Forward Filter Bank Transform for the MP3 Standard.

Define $\{h_r\}_{r=0}^{511}$ by $h_{k+64j} = (-1)^j C_{k+64j}$, $0 \le j < 8, 0 \le k < 64$, and $h^{(n)}$ as the filter with coefficients $\{\cos(2\pi(n + 1/2)(k - 16)/(2M))h_k\}_{k=0}^{511}$. The encoding procedure stated in the MP3 standard corresponds to applying a forward filter bank transform with filters $H_n = E_{n-31}h^{(n)}$.

The filter bank from Exercise 3.39 was said to be cosine modulated, and the theorem gives precisely such a bank (with k replaced by $k - 16$, i.e. a shift in phase). $\{h_r\}_{r=0}^{511}$ is called the *prototype* of the bank. It may seem strange that there is a different delay $(n-31)$ for each filter. But it is straightforward to see that this delay aligns 32 succeeding rows (with 512 filter coefficients each) above one another in H, and this means that there will be no difference in the time delay for the 32 values in the same iteration. In fact, here our notation differs from standard filter bank terminology, where one not would write down the delay $n-31$ here, as 32 succeeding samples are considered to correspond to the same instance in time. We, on the other hand, consider all rows/column to correspond to a uniform partition of time.

The cosine multiplication simply shifted the frequency response, so that, if H_0 zooms in on one frequency range, the other filters will zoom in on the other frequency ranges. It is now straightforward to reproduce the plots from Example 3.33. The library has a function `mp3_ctable()`, which returns the table C. The following code finds the prototype filter by multiplying the table C with ± 1 as we have described.

```
C = mp3_ctable();
signs = [ repmat(1, 64, 1); repmat(-1, 64, 1)];
signs = repmat(signs, 4, 1);
h=signs.*C;
```

The frequency response of h from the left part of Fig. 3.7 can now be found as follows. We have reorganized the DFT values in order to get values in the range $[-\pi, \pi]$.

```
N = 1024;
omega = linspace(-pi, pi, N);
figure();
fft_vals = abs(fft( [h; zeros(N-length(h),1)] ));
plot(omega, [fft_vals((N/2+1):end); fft_vals(1:N/2)], 'k-');
axis([-0.7 0.7 0 3])
```

The plot in the right part can be found by cosine modulating as follows

```
h_n = h.*cos(2*pi*(n+1/2)*((0:511)'-16)/64);
```

Operation Count and Memory Requirements

The steps in the procedure are clearly motivated by reducing computations. For instance by multiplying with the h_n first, one eliminates the need to do this for every output sample. Also, adding the values in the partial calculation (step 3) spares the need to multiply each value with a cosine. Let us make a simple arithmetic count for the stated steps. For every 32 output samples, we have that

- Step 2 computes 512 multiplications,
- Step 3 computes 64 sums of 8 elements each, i.e. a total of $7 \times 64 = 448$ additions (note that $q = 512/64 = 8$).

The standard says nothing about how the matrix multiplication in the final step can be computed efficiently. A direct approach would yield $32 \times 64 = 2048$ multiplications. multiplications, leaving a total number of multiplications at 2560. In a direct implementation of the forward filter bank transform, the computation of 32 samples would need $32 \times 512 = 16,384$ multiplications, so that the procedure sketched in the standard gives a big reduction.

One can also reduce the number of multiplications in the final step further, since it clearly can use a DCT-type implementation. We already have an efficient implementation for multiplication with a 32×32 type-III cosine matrix (this is simply the IDCT), and this implementation can be chosen to reduce the number of multiplications to $\frac{N}{2} \log_2 N = 80$ (with $M = 32$), so that the total number of multiplications is $512 + 80 = 592$. Clearly then, step 2 is the computationally most intensive part.

It is also possible to reduce the number of arithmetic operations further by reorganizing step 1 in the procedure. In Exercise 7.16 you will be guided through the details in this.

Again we have not taken the computation of the cosines into account, since these can be pre-computed, as with the DCT. Note also that all steps in the procedure are well suited for parallel processing.

Besides the number of arithmetic operations, it is also instructive to consider the memory requirements of the encoding procedure. We do not count the cosine values and the coefficients of the prototype filter in this respect, as these are fixed for all input. Besides the input samples (stored in X), step 2 needs an additional vector $Z \in \mathbb{R}^{512}$, step 3 needs a vector $Y \in \mathbb{R}^{64}$, and step 4 needs an output vector $S \in \mathbb{R}^{32}$. It is actually possible to eliminate the need for all additional memory, so that the output can be computed directly into the memory of the input, i.e. in-place, just as for the FFT and the wavelet transforms we have considered:

- The additional memory from step 3 can be eliminated, by storing the result in the first 64 components of Z.
- The additional memory from step 4 can be eliminated by storing the output samples at the indices corresponding to the audio samples which will be taken out in the next iteration.

The hardest thing is to see that the additional memory from step 2 can be eliminated. After a minor change to the filter bank transform, this will be addressed in Sect. 7.3: in Exercise 7.16 it is explained that this really is a generalization of in-place computation of elementary lifting.

5.4.2 Reverse Filter Bank Transform in the MP3 Standard

For decoding audio, the MP3 standard states the following steps, without saying if they really reverse the steps of the forward transform. We will return to this in Sect. 7.3:

1. Input 32 new subband samples as the vector S.
2. Change vector $V \in \mathbb{R}^{512}$, so that all elements are delayed with 64 elements. In particular the 64 last elements are taken out.
3. Set the first 64 elements of V as $NS \in \mathbb{R}^{64}$, where N is the 64×32- matrix where $N_{ik} = \cos((16 + i)(2k + 1)\pi/64)$. This is also called matrixing.
4. Build the vector $U \in \mathbb{R}^{512}$ from V from the formulas $U_{64i+j} = V_{128i+j}$, $U_{64i+32+j} = V_{128i+96+j}$ for $0 \le i \le 7$ and $0 \le j \le 31$, i.e. U is the vector where V is first split into segments of length 132, and U is constructed by assembling the first and last 32 elements of each of these segments.
5. Multiply U component-wise with a vector D (this vector is defined in the standard), to obtain a vector $W \in \mathbb{R}^{512}$. This is also called windowing.
6. Compute the 32 next sound samples as $\sum_{i=0}^{15} W_{32i+j}$.

To interpret this as a reverse filter bank transform, rewrite first steps 4 to 6 as

$$x_{32(s-1)+j} = \sum_{i=0}^{15} W_{32i+j} = \sum_{i=0}^{15} D_{32i+j} U_{32i+j}$$

$$= \sum_{i=0}^{7} D_{64i+j} U_{64i+j} + \sum_{i=0}^{7} D_{64i+32+j} U_{64i+32+j}$$

$$= \sum_{i=0}^{7} D_{64i+j} V_{128i+j} + \sum_{i=0}^{7} D_{64i+32+j} V_{128i+96+j}. \qquad (5.9)$$

The elements in V are obtained by "matrixing" different segments of the vector $\mathbf{z}$. More precisely, at iteration s we have that

$$\begin{pmatrix} V_{64r} \\ V_{64r+1} \\ \vdots \\ V_{64r+63} \end{pmatrix} = N \begin{pmatrix} z_{32(s-r-1)} \\ z_{32(s-r-1)+1} \\ \vdots \\ z_{32(s-r-1)+31} \end{pmatrix}$$

for $0 \le r < 8$, so that

$$V_{64r+j} = \sum_{k=0}^{31} \cos(2\pi(16+j)(k+1/2)/(2M)) z_{32(s-r-1)+k}$$

for $0 \le j \le 63$. Since also

$$V_{128i+j} = V_{64(2i)+j} \qquad\qquad V_{128i+96+j} = V_{64(2i+1)+j+32},$$

we can rewrite Eq. (5.9) as

$$\sum_{i=0}^{7} D_{64i+j} \sum_{k=0}^{31} \cos(2\pi(16+j)(k+1/2)/(2M)) z_{32(s-2i-1)+k}$$

$$+ \sum_{i=0}^{7} D_{64i+32+j} \sum_{k=0}^{31} \cos(2\pi(16+j+32)(k+1/2)/(2M)) z_{32(s-2i-2))+k}.$$

Again using Relation (5.8), this can be written as

$$\sum_{k=0}^{31} \sum_{i=0}^{7} (-1)^i D_{64i+j} \cos(2\pi(16+64i+j)(k+1/2)/(2M)) z_{32(s-2i-1)+k}$$

$$+ \sum_{k=0}^{31} \sum_{i=0}^{7} (-1)^i D_{64i+32+j} \cos(2\pi(16+64i+j+32)(k+1/2)/(2M)) z_{32(s-2i-2)+k}.$$

Now, if we define $\{g_r\}_{r=0}^{511}$ by $g_{64i+s} = (-1)^i D_{64i+s}$, $0 \leq i < 8, 0 \leq s < 64$, and $g^{(k)}$ as the filter with coefficients $\{\cos(2\pi(r+16)(k+1/2)/(2M))g_r\}_{r=0}^{511}$, the above can be simplified as

$$\sum_{k=0}^{31}\sum_{i=0}^{7}(g^{(k)})_{64i+j}z_{32(s-2i-1)+k} + \sum_{k=0}^{31}\sum_{i=0}^{7}(g^{(k)})_{64i+j+32}z_{32(s-2i-2)+k}$$

$$= \sum_{k=0}^{31}\left(\sum_{i=0}^{7}(g^{(k)})_{32(2i)+j}z_{32(s-2i-1)+k} + \sum_{i=0}^{7}(g^{(k)})_{32(2i+1)+j}z_{32(s-2i-2)+k}\right)$$

$$= \sum_{k=0}^{31}\sum_{r=0}^{15}(g^{(k)})_{32r+j}z_{32(s-r-1)+k},$$

where we observed that $2i$ and $2i+1$ together run through the values from 0 to 15 when i runs from 0 to 7. Since z has the same values as z_k on the indices $32(s-r-1)+k$, this can be written as

$$= \sum_{k=0}^{31}\sum_{r=0}^{15}(g^{(k)})_{32r+j}(z_k)_{32(s-r-1)+k}$$

$$= \sum_{k=0}^{31}(g^{(k)}z_k)_{32(s-1)+j+k} = \sum_{k=0}^{31}((E_{-k}g^{(k)})z_k)_{32(s-1)+j}.$$

By substituting a general s and j we see that $x = \sum_{k=0}^{31}(E_{-k}g^{(k)})z_k$. We have thus proved the following.

Theorem 15. Reverse Filter Bank Transform for the MP3 Standard.
  Define $\{g_r\}_{r=0}^{511}$ by $g_{64i+s} = (-1)^i D_{64i+s}$, $0 \leq i < 8, 0 \leq s < 64$, and $g^{(k)}$ as the filter with coefficients $\{\cos(2\pi(r+16)(k+1/2)/(2M))g_r\}_{r=0}^{511}$. The decoding procedure stated in the MP3 standard corresponds to applying a reverse filter bank transform with filters $G_k = E_{-k}g^{(k)}$.

  Thus, the reverse filter bank transform is also cosine-modulated, and with prototype $\{g_r\}_{r=0}^{511}$. Since different tables C and D are used, the prototype filters in the forward and reverse transforms may be different, and a comparison of the tables leads us to believe that they indeed are so. At closer inspection, however, one sees connection: If you multiply the values in the C-table with $M = 32$ (i.e. the number of filters in the bank), you get values very close to those in the table D. This indicates that the prototype filters are equal, up to a scalar. It seems strange that the MP3 standard uses a couple of pages for each table, when they are practically equal. This connection between the two tables will be explained in Sect. 7.3, together with some other relationships.

  While the decoding steps seem a bit more complex than the encoding steps, they are clearly also motivated by the desire to reduce the number of arithmetic operations.

Exercise 5.29: Plotting Frequency Responses

In addition to the function `mp3_ctable()`, the library has a function `mp3_dtable()` which returns the table D used by the reverse transform. Verify the connection we have stated between the tables C and D, i.e. that $D_i = 32C_i$ for all i.

Exercise 5.30: Implementing Forward and Reverse Filter Bank Transforms

It is not too difficult to implement the encoding and decoding steps as explained in the MP3 standard. Write a function which implements the encoding steps, and one which implements the decoding steps. In your code, assume for simplicity that the input and output vectors to your methods all have lengths which are multiples of 32, and use the functions `mp3_ctable`, `mp3_dtable`.

In the library the functions `mp3_forward_fbt` and `mp3_reverse_fbt` implement these encoding- and decoding steps. Use these as a template for your code.

Exercise 5.31: Perfect Reconstruction for Forward and Reverse M-Channel Filter Bank Transforms

In this exercise H and G are forward and reverse M-channel filter bank transforms, respectively.

a) Generalize the proof of Theorem 3 to show that

$$H\phi_r = \frac{1}{M} \left(\sum_{t=0}^{M-1} \sum_{s=0}^{M-1} \lambda_{H_s,r} e^{-2\pi its/M} \right) \phi_{r+tN/M}$$

$$G\phi_r = \frac{1}{M} \left(\sum_{t=0}^{M-1} \sum_{s=0}^{M-1} \lambda_{G_s,r+tN/M} e^{-2\pi ist/M} \right) \phi_{r+tN/M}$$

where the ϕ_r are the Fourier basis vectors. This shows that the output from ϕ_r is a sum of the $\{\phi_{r+sN/M}\}_{s=0}^{M-1}$, so that at most $M-1$ frequencies can contribute with aliasing.

b) Find a formula for $GH\phi_r$, and conclude that, in terms of continuous frequency responses, we have alias cancellation if and only if

$$\sum_{s=0}^{M-1} \lambda_{H_s}(\omega + 2\pi q/M)\lambda_{G_s}(\omega)e^{2\pi iqs/M} = 0$$

for all $0 \le q < M/2$, and perfect reconstruction if and only if in addition

$$\lambda_{H_0}(\omega)\lambda_{G_0}(\omega) + \cdots + \lambda_{H_{M-1}}(\omega)\lambda_{G_{M-1}}(\omega) = M.$$

These expressions clearly generalize the conditions obtained for $M = 2$ in Exercise 5.9.

Exercise 5.32: Time Domain View of Perfect Reconstruction

Assume that H_i and G_i are the filters of forward and reverse M-channel filter bank transforms. Show that GH can be computed with the formula

$$z_n = \sum_l \left(\sum_{k=0}^{M-1} G_k^{(n-k)} * H_k^{(k-l)} \right)_0 x_l,$$

where $G_k^{(n-k)}$ is polyphase component $n-k$ of G_k, i.e. $(G_k^{(n-k)})_r = (G_k)_{rM+n-k}$. This can be though of as a time domain version of the previous exercise, in that alias cancellation and perfect reconstruction can be spotted in terms of time domain polyphase components. We also obtain an expression for the matrix GH as

$$(GH)_{n,l} = \left(\sum_{k=0}^{M-1} (G_k^{(n-k)}) * (H_k^{(k-l)}) \right)_0.$$

Exercise 5.33: Connection Between Cosine-Modulated Forward and Reverse Transforms

Show that, if H is a cosine-modulated forward transform with prototype $\{h_r\}_{r=0}^{511}$, then $G = E_{481}H^T$ is a cosine-modulated reverse transform with prototype $\{g_r\}_{r=0}^{511}$, with $g_r = h_{512-r}$ (i.e. the prototype filter is reversed). Deduce also from this that $H = G^T E_{481}$.

In Exercise 7.13 we construct an example showing that the forward and reverse transforms of this section do not invert each other exactly. In Sect. 7.3, we also will see how one can construct similar filter banks with perfect reconstruction. It may seem strange that the MP3 standard does not do this. A partial explanation may be that, while perfect reconstruction is important, other design criteria may be even more important.

5.5 Summary

We started this chapter by showing that the DWT and the IDWT can be realized in terms of filters, and therefore also called them filter bank transforms. Filter bank transforms provide an entirely different view: instead of constructing function spaces with certain properties and deducing filters from these, we can use other known techniques to produce filters with certain properties (such as alias cancellation, perfect reconstruction, and low-pass and high-pass characteristics). Some such filters may be meaningful in a wavelet context, but they may also be very useful in other contexts. We will have more on this in Chap. 7. Filter bank theory is an entire field in itself, developed independently from wavelets, and a variety of different filter banks have been developed [76, 61].

We stated requirements for filter bank matrices to invert each other: The frequency responses of the low-pass filters must satisfy a certain equation. The high-pass filters can then be obtained from the low-pass filters similarly to how we converted between low-pass- and high-pass filters in Chap. 3.

We also generalized filter bank transforms from 2 to $M > 2$ channels, and such transforms had an interpretation as splitting the input into frequency bands. We saw that the MP3 standard applied a 32-channel cosine modulated filter bank transform. This had a very efficient implementations, but is largely out-dated for a couple of reasons:

- It is too simple in the sense that the frequency bands are uniform. The human auditory system is more sensitive to certain frequencies, and more recent standards address this by using non-uniform frequency partitions.
- It does not give perfect reconstruction (although it provides alias cancellation and no phase distortion).

The MP3 standard was chosen here because it is well-known and simple to present, using the theory we have gone through.

The MP3 standard does not mention how the coefficients of the prototype filter can be found by adding a sign which alternates for every 64th element. Nor does it mention the simple connection we noted between the tables C and D, or how the values in these tables were found. Lookabaugh and Perkins [36] and Pan [55] explain some of these things. The origin of the filter bank of the MP3 standard can be traced back to [63].

The motivation behind filter bank transforms is that their output is more suitable for further processing, such as compression with lossless- and lossy methods such as Huffman- and arithmetic coding, or playback in an audio system. They also have efficient implementations.

One difference in the presentations of wavelet transforms and filter bank transforms so far is that wavelet transforms have been applied in stages, while filter bank transforms have not. Filter bank transforms in stages are also of theoretical interest, however, giving rise to what is called *tree-structured filter banks* [76]. Another difference lies in the use of the term perfect reconstruction. In wavelet theory perfect reconstruction follows by construction, since the DWT and the IDWT correspond to changes of coordinates. In signal processing one has a wider perspective, since one can design many more useful systems with fast implementations when one replaces the requirement of perfect reconstruction with alias cancellation and near perfect reconstruction. Classical QMF filter banks are examples. The original definition of classical QMF filter banks are from [11], and differ only in a sign from our definition. 2-channel filter bank transforms with perfect reconstruction have been around since [2], and the alternative QMF filter banks since [67]. While wavelets certainly give new perspectives to filter bank transforms, it is a good question if they provide anything new regarding the design of filters. Ramstad et al. [61] points to a negative answer here, since many filters that appeared in a wavelet setting already were known from filter bank transforms, as is apparent from the citations above.

All filters we encounter for wavelets and filter banks in this book are FIR. This is just done to limit the exposition. Much useful theory has been developed using IIR-filters also.

Applying the DCT to the blocks of the input can also be viewed as a filter bank transform. Just as the Haar wavelet, this gives a block-diagonal filter bank transform, with all diagonal blocks being equal to the DCT. This is contrary to the filter bank transform of the MP3 standard, which is not block-diagonal. Considering the success of the DCT, it is a legitimate question if more general filter bank transforms actually give improvements over *block transforms* such as the DCT. In [61] a partially affirmative answer to this is given, because of the flexibility one has in designing filter bank transforms.

What You Should Have Learned in This Chapter

- How one can find the filters of a wavelet transform by considering its matrix and its inverse.
- Forward and reverse filter bank transforms.
- How one can implement the DWT and the IDWT with the help of the filters.
- Plot of the frequency responses for the filters of the wavelets we have considered, and their interpretation as low-pass and high-pass filters.

Chapter 6
Constructing Interesting Wavelets

Previously we have associated several MRA's with filter bank transforms. Since filter bank transforms also can be constructed outside the setting of wavelets, it would be interesting to see when we can make the association the opposite way: Which filter bank transforms appear in the context of a multiresolution analysis? An answer to this question certainly could transfer much theory between wavelets and filters. Also, it may be easier to construct good filter bank transforms than good wavelet bases.

In this chapter we will give a partial answer to this question, in that we will find conditions on the filters so that they arise in the setting of an MRA. We will also state conditions on the filters in order for the corresponding mother wavelet to have vanishing moments, or the scaling function to be differentiable. In particular we will see how we can find such filters with the fewest coefficients. Initially we have much flexibility in choosing the filters, but requirements on the number of vanishing moments and on the number of filter coefficients will narrow down the possibilities a lot. This chapter is rather technical regarding these things, but it should be possible to obtain a good understanding of the main result, even if you skip substantial parts of the technical details.

To establish the results in this section we need some more theory on Fourier transforms. To be more precise, we need to express a non-periodic function on the entire real line in terms of frequencies. It is then not enough to consider only a discrete set of frequencies. Rather a function should be expressed in terms of frequencies as an integral such as $f(t) = \int_{-\infty}^{\infty} \hat{f}(\nu)e^{i\nu t}d\nu$, where we integrate over all frequencies ν, with $\hat{f}(\nu)$ the contribution at frequency ν. This will be the starting point for what is called the continuous time Fourier transform.

Several of the wavelets we consider in this chapter enjoy a widespread use in applications. Two of them are used in the JPEG2000 standard, one for lossless compression, and one for lossy compression. These wavelets have symmetric filters. We will also look at orthonormal wavelets with different number of vanishing moments, where the filters are not symmetric.

All functions in this chapter are defined on the entire real line, as they always are in a multiresolution analysis.

© Springer Nature Switzerland AG 2019

Ø. Ryan, *Linear Algebra, Signal Processing, and Wavelets - A Unified Approach*,
Springer Undergraduate Texts in Mathematics and Technology,
https://doi.org/10.1007/978-3-030-01812-2_6

6.1 From Filters to Scaling Functions and Mother Wavelets

The first question we have to ask regards which properties the filters H_0, H_1, G_0, and G_1 must fulfill, in order to give an association with scaling functions and mother wavelets. We should demand that ϕ and ψ satisfy the two-scale equations

$$\phi(t) = \sum_{n=0}^{2N-1} (G_0)_{n,0}\phi_{1,n}(t) \qquad\qquad \psi(t) = \sum_{n=0}^{2N-1} (G_1)_{n,1}\phi_{1,n}(t). \qquad (6.1)$$

To construct such functions we will have use for the *Continuous-time Fourier Transform*, which we briefly mentioned in Chap. 1:

Theorem 1. Continuous-Time Fourier Transform.
Let f be a function defined on $(-\infty, \infty)$. The Continuous-time Fourier Transform (or CTFT) of f is the function defined on $(-\infty, \infty)$ by

$$\hat{f}(\nu) = \frac{1}{\sqrt{2\pi}}\int_{-\infty}^{\infty} f(t)e^{-i\nu t}dt.$$

ν here denotes frequency, and can be any value on the real line. We will not use ω for frequency, since it is reserved for angular frequency. We will relate the CTFT to frequency responses of digital filters, which is why the frequency responses below take frequency ν as input, rather than angular frequency ω.

Just as the DFT could be interpreted as the frequency response of a digital filter, the CTFT can be interpreted as the frequency response of an analog filter (see Theorem 27). In the setting of analog filters we restricted to f with compact support, but note that the CTFT integral may exist even when the support of f is unbounded, as long as f decays quickly enough as $t \to \pm\infty$. If not, the CTFT is not defined. Just as for the convergence of Fourier series, finding conditions for when the CTFT integral converges is a difficult issue. We will restrict to functions in $L^2(\mathbb{R})$, for which we have the following result, which we state without proof.

Theorem 2. The Continuous-Time Fourier Transform Is Unitary on $L^2(\mathbb{R})$.
The CTFT is a unitary transform, i.e. for all f, g in $L^2(\mathbb{R})$,

$$\langle f, g \rangle = \int_{-\infty}^{\infty} f(t)\overline{g(t)}dt = \int_{-\infty}^{\infty} \hat{f}(\nu)\overline{\hat{g}(\nu)}d\nu = \langle \hat{f}, \hat{g} \rangle.$$

As a consequence, $f \in L^2(\mathbb{R})$ if and only if $\hat{f} \in L^2(\mathbb{R})$. When $f \in L^2(\mathbb{R})$ we have that

$$f(t) = \frac{1}{\sqrt{2\pi}}\int_{-\infty}^{\infty} \hat{f}(\nu)e^{i\nu t}d\nu. \qquad (6.2)$$

Equation (6.2) can be interpreted as writing f as a sum of frequencies, with $\hat{f}(\nu)$ the contribution at frequency ν. The CTFT satisfies several properties similar to those of Fourier series (Theorem 17) and the DFT (Theorem 7). We will have use for some of these in the following:

Theorem 3. Properties of the CTFT.
The following hold

1. $\hat{f}(\nu) = \overline{\hat{f}(-\nu)}$.
2. *If $f(t) = f(-t)$ (i.e. f is symmetric), then $\hat{f}(\nu)$ is real for all ν.*

3. If $f(t) = -f(-t)$ (i.e. f is antisymmetric), then $\hat{f}(\nu)$ is purely imaginary for all ν.
4. If $g(t) = f(t - d)$, then $\hat{g}(\nu) = e^{-id\nu}\hat{f}(\nu)$.
5. If $g(t) = e^{idt}f(t)$, then $\hat{g}(\nu) = \hat{f}(\nu - d)$.

It is a simple exercise to prove these properties. Assuming that ϕ satisfies (6.1), and that $\hat{\phi}$ exists, we can use some of these properties to obtain

$$
\begin{aligned}
\hat{\phi}(\nu) &= \frac{1}{\sqrt{2\pi}} \int_{-\infty}^{\infty} \phi(t)e^{-i\nu t}dt = \frac{1}{\sqrt{2\pi}} \int_{-\infty}^{\infty} \left(\sum_n (G_0)_{n,0}\sqrt{2}\phi(2t - n) \right) e^{-i\nu t}dt \\
&= \frac{1}{2\sqrt{\pi}} \sum_n \int_{-\infty}^{\infty} (G_0)_{n,0}\phi(t)e^{-i\nu(t+n)/2}dt \\
&= \frac{1}{\sqrt{2}} \left(\sum_n (G_0)_{n,0}e^{-i(\nu/2)n} \right) \frac{1}{\sqrt{2\pi}} \int_{-\infty}^{\infty} \phi(t)e^{-i(\nu/2)t}dt \\
&= \frac{\lambda_{G_0}(\nu/2)}{\sqrt{2}}\hat{\phi}(\nu/2).
\end{aligned}
\tag{6.3}
$$

Continuing this expression recursively and defining

$$
g_{N,0}(\nu) = \prod_{s=1}^{N} \frac{\lambda_{G_0}(\nu/2^s)}{\sqrt{2}}\chi_{[0,2\pi]}(2^{-N}\nu),
\tag{6.4}
$$

we see that $\hat{\phi}(\nu) = g_{N,0}(\nu)\hat{\phi}(\nu/2^N)$ for $\nu \in [0, 2\pi 2^N]$. We can now prove the following.

Lemma 4. $g_{N,0}(\nu)$ Converges.
   Assume that $\sum_n (G_0)_n = \sqrt{2}$ (i.e. $\lambda_{G_0}(0) = \sqrt{2}$), and that G_0 is a FIR-filter. Then $g_{N,0}(\nu)$ converges pointwise as $N \to \infty$ to an infinitely differentiable function.

Proof. We need to verify that the infinite product $\prod_{s=1}^{\infty} \frac{\lambda_{G_0}(\nu/2^s)}{\sqrt{2}}$ converges. The function $r(\nu) = \frac{\lambda_{G_0}(\nu)}{\sqrt{2}}$ is differentiable in ν, so that $|r'(\nu)| \le c$ for some number c. Since also $r(0) = 1$ we must have that $|r(\nu)| \le 1 + c|\nu|$, so that $|r(\nu)| \le e^{c|\nu|}$. This means that the infinite product $\prod_{s=1}^{\infty} \frac{\lambda_{G_0}(\nu/2^s)}{\sqrt{2}}$ is bounded by $e^{\sum_{n=1} c|\nu|/2^n} = e^{c|\nu|}$. It follows from the Weierstrass M-test[1] that the infinite product converges uniformly. It is also well known that a uniform limit of differentiable functions is differentiable. The result now follows from the fact that each finite product is infinitely differentiable. □

We clearly have that $g_{N+1,0}(\nu) = \frac{\lambda_{G_0}(\nu/2)}{\sqrt{2}}g_{N,0}(\nu/2)$. Taking limits on both sides of this and defining

$$
g_0(\nu) = \lim_{N \to \infty} g_{N,0}(\nu) = \lim_{N \to \infty} \prod_{s=1}^{N} \frac{\lambda_{G_0}(\nu/2^s)}{\sqrt{2}}\chi_{[0,2\pi]}(2^{-N}\nu),
$$

---

[1] The Weierstrass M-test is usually formulated for sums in the literature. To get from products to sums one simply takes logarithms.

it is clear that $g_0(\nu) = \frac{\lambda_{G_0}(\nu/2)}{\sqrt{2}}g_0(\nu/2)$. Thus, if we can find a ϕ so that $g_0 = \hat{\phi}$, we have a scaling function which satisfies the first equation in (6.1) (note that $g_0(0) = \hat{\phi}(0) = 1$). The problem is, however, that there may not exist such a ϕ. If we assume that $g_0 \in L^2(\mathbb{R})$, however, we can take an inverse CTFT (6.2) in order find a $\phi \in L^2(\mathbb{R})$ so that $\hat{\phi} = g_0$. There are several ways to secure that $g_0 \in L^2(\mathbb{R})$. One way is by requiring that $g_0(\nu) \leq C(1 + |\nu|)^{-1/2-\epsilon}$ for some $\epsilon > 0$. To see why, note that $(g_0(\nu))^2 \leq C^2/(1 + |\nu|)^{-1-2\epsilon}$, so that

$$\int_0^\infty (g_0(\nu))^2 d\nu \leq \left[-\frac{1}{2\epsilon}(1+\nu)^{-2\epsilon} \right]_0^\infty = \frac{1}{2\epsilon},$$

and similarly for the integral from $-\infty$ to 0.

Similarly to (6.3) we get

$$\hat{\psi}(\nu) = \frac{1}{\sqrt{2}}\left(\sum_n (G_1)_{n,1}e^{-i(\nu/2)n} \right)\hat{\phi}(\nu/2)$$

$$= \frac{1}{\sqrt{2}}\left(\sum_n (G_1)_{n-1,0}e^{-i(\nu/2)n} \right)\hat{\phi}(\nu/2)$$

$$= \frac{1}{\sqrt{2}}\left(\sum_n (G_1)_{n,0}e^{-i(\nu/2)(n+1)} \right)\hat{\phi}(\nu/2)$$

$$= e^{-i\nu/2}\frac{\lambda_{G_1}(\nu/2)}{\sqrt{2}}\hat{\phi}(\nu/2).$$

It follows that, as long as $g_0(\nu) \leq C(1 + |\nu|)^{-1/2-\epsilon}$, there exists a $\psi \in L^2(\mathbb{R})$ which satisfies the second equation in (6.1) also. Now, defining

$$g_{N,1}(\nu) = e^{-i\nu/2}\frac{\lambda_{G_1}(\nu/2)}{\sqrt{2}}\prod_{s=2}^{N-1}\frac{\lambda_{G_0}(\nu/2^s)}{\sqrt{2}}\chi_{[0,2\pi]}(2^{-N}\nu),$$

we have that $g_{N+1,1}(\nu) = e^{-i\nu/2}\frac{\lambda_{G_1}(\nu/2)}{\sqrt{2}}g_{N,0}(\nu/2)$. Taking limits as $N \to \infty$ on both sides, and denoting the limit by g_1, we obtain that $g_1(\nu) = e^{-i\nu/2}\frac{\lambda_{G_1}(\nu/2)}{\sqrt{2}}g_0(\nu/2)$, and we would like to define ψ by $\hat{\psi} = g_1$.

g_0 and g_1 are uniquely defined from the filters G_0 and G_1. Unfortunately, it is easy to find filters so that g_0 and g_1 are not square integrable, so that we should be careful in how to choose the filters. In our main theorem, which we will shortly state, we will find conditions on the filter G_0 which secure that $g_i(\nu) \leq C(1 + |\nu|)^{-1/2-\epsilon}$.

In Chap. 5 we defined the dual filter bank transform, which used the filters $(H_0)^T$ and $(H_1)^T$ instead of G_0 and G_1. For the dual transform we would like to find functions $\tilde{\phi}, \tilde{\psi}$ so that

$$\tilde{\phi}(t) = \sum_{n=0}^{2N-1}((H_0)^T)_{n,0}\tilde{\phi}_{1,n}(t) \qquad \tilde{\psi}(t) = \sum_{n=0}^{2N-1}((H_1)^T)_{n,1}\tilde{\phi}_{1,n}(t). \qquad (6.5)$$

If we repeat the analysis above for the dual transform, one obtains functions $\tilde{g}_{N,0}$, $\tilde{g}_{N,1}$ satisfying

$$\tilde{g}_{N+1,0}(\nu) = \frac{\lambda_{(H_0)^T}(\nu/2)}{\sqrt{2}} \tilde{g}_{N,0}(\nu/2) \quad \tilde{g}_{N+1,1}(\nu) = e^{-i\nu/2} \frac{\lambda_{(H_1)^T}(\nu/2)}{\sqrt{2}} \tilde{g}_{N,0}(\nu/2),$$

infinitely differentiable functions $\tilde{g}_0$, $\tilde{g}_1$ satisfying

$$\tilde{g}_0(\nu) = \frac{\lambda_{(H_0)^T}(\nu/2)}{\sqrt{2}} \tilde{g}_0(\nu/2) \qquad \tilde{g}_1(\nu) = e^{-i\nu/2} \frac{\lambda_{(H_1)^T}(\nu/2)}{\sqrt{2}} \tilde{g}_0(\nu/2),$$

with $\tilde{\phi}$ and $\tilde{\psi}$ (satisfying $\hat{\tilde{\phi}} = \tilde{g}_0$, $\hat{\tilde{\psi}} = \tilde{g}_1$) both in $L^2(\mathbb{R})$ as long as $\tilde{g}_0(\nu) \leq C(1 + |\nu|)^{-1/2-\epsilon}$ for some $\epsilon > 0$.

Our goal is thus to find filters so that the derived functions g_i and $\tilde{g}_i$ are bounded by $C(1 + |\nu|)^{-1/2-\epsilon}$. We would also like the constructed functions ϕ, ψ, $\tilde{\phi}$, $\tilde{\psi}$ to give rise to wavelet bases which span all of $L^2(\mathbb{R})$ (i.e. the first requirement of a multiresolution analysis). These bases may not be orthonormal, however, such as the bases for the piecewise linear wavelet. Nevertheless, we would like to put these bases in a framework similar to that of orthogonality. To establish the concepts for this, let first $\mathcal{H}$ be an inner product space. Recall that a *Cauchy sequence* in $\mathcal{H}$ is a sequence $\boldsymbol{x}_n$ from $\mathcal{H}$ so that $\|\boldsymbol{x}_n - \boldsymbol{x}_m\| \to 0$ as $m, n \to \infty$. If any Cauchy sequence in $\mathcal{H}$ converges in $\mathcal{H}$ (i.e. we can find $\boldsymbol{x} \in \mathcal{H}$ so that $\|\boldsymbol{x}_n - \boldsymbol{x}\| \to 0$ as $n \to \infty$), $\mathcal{H}$ is called a *Hilbert space*. We will need the following definitions.

Definition 5. *Frame.*

Let $\mathcal{H}$ be a Hilbert space. A set of vectors $\{\boldsymbol{u}_n\}_n$ is called a *frame* of $\mathcal{H}$ if there exist constants $A > 0$ and $B > 0$ so that, for any $\boldsymbol{f} \in \mathcal{H}$,

$$A\|\boldsymbol{f}\|^2 \leq \sum_n |\langle \boldsymbol{f}, \boldsymbol{u}_n \rangle|^2 \leq B\|\boldsymbol{f}\|^2.$$

If $A = B$, the frame is said to be *tight*. A and B are called the *lower and upper frame bounds*.

Note that for a frame any $\boldsymbol{f} \in \mathcal{H}$ is uniquely characterized by the inner products $\langle \boldsymbol{f}, \boldsymbol{u}_n \rangle$. Indeed, if both $\boldsymbol{f}_1, \boldsymbol{f}_2 \in \mathcal{H}$ have the same inner products, then $\boldsymbol{f}_1 - \boldsymbol{f}_2 \in \mathcal{H}$ have inner products 0, which implies that $\boldsymbol{f}_1 = \boldsymbol{f}_2$ from the left inequality.

Many of the spaces we consider are finite-dimensional, and it is interesting to interpret the frame condition for the case $\mathcal{H} = \mathbb{R}^m$. We will go through much of this in the exercises, and the central ingredient used there is called the *singular value decomposition*, see Appendix A. Assuming that there are n vectors in the frame, and letting U be the $m \times n$-matrix with columns $\boldsymbol{u}_i$, the frame condition can be rephrased as

$$A\|\boldsymbol{f}\|^2 \leq \|U^T \boldsymbol{f}\|^2 \leq B\|\boldsymbol{f}\|^2.$$

It will follow from this in the exercises that $\{\boldsymbol{u}_n\}_n$ is a frame if and only if UU^T is non-singular, and that this happens if and only if the first m singular values of U^T are nonzero. UU^T is also called the *frame operator*. A consequence of this will be that the frame vectors span $\mathbb{R}^m$.

For every frame it is possible to find a *dual frame* $\{\tilde{\boldsymbol{u}}_n\}_n$ which satisfies

$$\frac{1}{B}\|\boldsymbol{f}\|^2 \leq \sum_n |\langle \boldsymbol{f}, \tilde{\boldsymbol{u}}_n \rangle|^2 \leq \frac{1}{A}\|\boldsymbol{f}\|^2,$$

and

$$\boldsymbol{f} = \sum_n \langle \boldsymbol{f}, \boldsymbol{u}_n \rangle \tilde{\boldsymbol{u}}_n = \sum_n \langle \boldsymbol{f}, \tilde{\boldsymbol{u}}_n \rangle \boldsymbol{u}_n \qquad (6.6)$$

for any $\boldsymbol{f} \in \mathcal{H}$. From the first condition it follows that the dual frame is tight if the frame is tight. From the second it follows that the frame and the dual frame have the same number of vectors. To interpret the dual frame conditions for the case $\mathcal{H} = \mathbb{R}^m$, let $\tilde{U}$ be the $m \times n$-matrix with columns $\tilde{\boldsymbol{u}}_i$. Equation (6.6) can then be rephrased as $U(\tilde{U})^T = \tilde{U}U^T = I_m$ (from which it is seen that one equality implies the other), and the lower and upper frame bounds can be rephrased as

$$\frac{1}{B}\|\boldsymbol{f}\|^2 \le \|\tilde{U}^T \boldsymbol{f}\|^2 \le \frac{1}{A}\|\boldsymbol{f}\|^2.$$

We will continue to work with these rephrased versions in the exercises, to see how a dual frame can be constructed. It will turn out that there may not be a unique dual frame, but we will review a concrete construction which always produces a dual frame.

A frame is called a *Riesz basis* if all its vectors also are linearly independent. For the case $\mathcal{H} = \mathbb{R}^m$, this means that there must be m vectors in the frame, so that the dual frame is uniquely defined from the equation $U(\tilde{U})^T = I_m$. In particular the dual frame is also a Riesz basis, a result which can be shown to hold more generally. The dual frame is then called the *dual Riesz basis*. Since $U(\tilde{U})^T = I_m$ when $\mathcal{H} = \mathbb{R}^m$ we also have $(\tilde{U})^T U = I_m$, i.e. $\langle \boldsymbol{u}_n, \tilde{\boldsymbol{u}}_m \rangle = 1$ when $n = m$, and 0 otherwise. Due to linear independence and Eq. (6.6), this holds more generally also. We say that the vectors $\{\boldsymbol{u}_n\}_n$ and $\{\tilde{\boldsymbol{u}}_m\}_m$ are *biorthogonal*. For $\mathcal{H} = \mathbb{R}^m$ this is captured by the matrix identity $(\tilde{U})^T U = I_m$. Clearly any orthonormal basis

- is biorthogonal,
- is a Riesz basis,
- has a dual Riesz basis which is equal to itself.

The point of using Riesz bases is that, even though a basis is not orthonormal, it may still be a Riesz basis, and have an interesting dual Riesz basis. The important connection between frames and wavelets is that, in a very general sense and regardless of whether the wavelet is orthonormal, wavelet bases are Riesz bases, and the dual Riesz basis is also a wavelet basis. The following is the main result regarding this, and is also the main result of the chapter. It is a restriction of propositions 4.8 and 4.9 in [9] to the setting we need:

Theorem 6. Creating Wavelet Bases from Filters.
Assume that G_0, G_1, H_0, H_1 are FIR filters so that $GH = I$, $\lambda_{G_0}(0) = \lambda_{H_0}(0) = \sqrt{2}$, and that the frequency responses λ_{G_0} and λ_{H_0} can be written as

$$\frac{\lambda_{H_0}(\nu)}{\sqrt{2}} = \left(\frac{1+e^{-i\nu}}{2}\right)^L \mathcal{F}(\nu) \qquad \frac{\lambda_{G_0}(\nu)}{\sqrt{2}} = \left(\frac{1+e^{-i\nu}}{2}\right)^{\tilde{L}} \tilde{\mathcal{F}}(\nu), \qquad (6.7)$$

where $\mathcal{F}$ and $\tilde{\mathcal{F}}$ are trigonometric polynomials of finite degree. Assume also that, for some $k, \tilde{k} > 0$,

$$B_k = \max_\nu \left|\mathcal{F}(\nu) \cdots \mathcal{F}(2^{k-1}\nu)\right|^{1/k} < 2^{L-1/2} \qquad (6.8)$$

$$\tilde{B}_k = \max_\nu \left|\tilde{\mathcal{F}}(\nu) \cdots \tilde{\mathcal{F}}(2^{\tilde{k}-1}\nu)\right|^{1/\tilde{k}} < 2^{\tilde{L}-1/2} \qquad (6.9)$$

Then there exist functions ϕ, ψ, $\tilde{\phi}$, $\tilde{\psi}$, all in $L^2(\mathbb{R})$, so that (6.1) and (6.5) hold. Moreover,

- *ψ has L vanishing moments and $\tilde{\psi}$ has $\tilde{L}$ vanishing moments.*
- *the bases ϕ_0 and $\tilde{\phi}_0$ are biorthogonal.*
- *$\{\psi_m\}_{m=0}^{\infty}$ is a Riesz basis for $L^2(\mathbb{R})$, and $\{\tilde{\psi}_m\}_{m=0}^{\infty}$ is its dual Riesz basis, i.e.*

$$f = \sum_{m,n} \langle f, \tilde{\psi}_{m,n} \rangle \psi_{m,n} = \sum_{m,n} \langle f, \psi_{m,n} \rangle \tilde{\psi}_{m,n} \qquad (6.10)$$

for any $f \in L^2(\mathbb{R})$. If also

$$B_k < 2^{L-1-m} \qquad\qquad \tilde{B}_k < 2^{\tilde{L}-1-\tilde{m}}, \qquad (6.11)$$

then ϕ, ψ are $\tilde{m}$ times differentiable, and $\tilde{\phi}$, $\tilde{\psi}$ are m times differentiable.

The proof for Theorem 6 is long and technical, and split in many stages. The entire proof can be found in [9]. We will not go through all of it, only address some simple parts in the following. After that we will see how we can find FIR filters G_0, H_0 so that Eqs. (6.7)–(6.9) are satisfied. First some remarks are in order.

1. From the theorem one obtains an MRA with scaling function ϕ, and a dual MRA, with scaling function $\tilde{\phi}$, as well as corresponding mother wavelets ψ and $\tilde{\psi}$. Resolution spaces and their dual counterparts satisfying

$$V_0 \subset V_1 \subset V_2 \subset \cdots \subset V_m \subset \cdots \qquad\qquad \tilde{V}_0 \subset \tilde{V}_1 \subset \tilde{V}_2 \subset \cdots \subset \tilde{V}_m \subset \cdots$$

can then be defined as before, as well as detail spaces W_m and dual counterparts $\tilde{W}_m$ satisfying

$$V_m = V_{m-1} \oplus W_{m-1} \qquad\qquad \tilde{V}_m = \tilde{V}_{m-1} \oplus \tilde{W}_{m-1}.$$

In general V_m is different from $\tilde{V}_m$, except when $\phi = \tilde{\phi}$, when all bases are orthonormal.

2. Equation (6.10) comes from eliminating the $\phi_{m,n}$ by letting $m \to \infty$, i.e. taking an infinite-level DWT. Applying only m levels for $f \in V_m$ gives

$$f(t) = \sum_n \langle f(t), \tilde{\phi}_{m,n} \rangle \phi_{m,n} = \sum_n \langle f(t), \tilde{\phi}_{0,n} \rangle \phi_{0,n} + \sum_{m' < m, n} \langle f(t), \tilde{\psi}_{m',n} \rangle \psi_{m',n},$$

It follows that

- the input to a DWT are the inner products $c_{m,n} = \langle f, \tilde{\phi}_{m,n} \rangle$,
- the output the inner products $c_{0,n} = \langle f, \tilde{\phi}_{0,n} \rangle$ and $w_{m',n} = \langle f, \tilde{\psi}_{m',n} \rangle$.

It is easy to see that the entries in the DWT matrix also are the inner products $\langle \phi_{1,k}, \tilde{\phi}_{0,l} \rangle$ and $\langle \phi_{1,k}, \tilde{\psi}_{0,l} \rangle$.

3. When $\phi = \tilde{\phi}$ (i.e. orthonormal MRA's), best approximations from V_m are easily computed. When $\phi \neq \tilde{\phi}$ an approximation to f from V_m can be computed as

$$\sum_n \langle f(t), \tilde{\phi}_{m,n} \rangle \phi_{m,n},$$

but this is in general different from the best approximation.

4. While the results here apply for functions in $L^2(\mathbb{R})$, they can be restricted to functions in $L^2[0, N]$ following what we did in Sect. 4.3.2. The result on the vanishing

moments does not extend to the restricted bases, however. One can, however, alter some of the basis functions in order to achieve this [8]. One can think of this as a better extension strategy than periodic and symmetric extension. The latter suffer from non-differentiable extensions at the boundary, so that the corresponding wavelet coefficients may be large, even though the wavelet has many vanishing moments.

5. The result gives some insight into how the filters should be chosen in order to obtain vanishing moments and regularity for ϕ and ψ. But these may easily give undesirable properties for $\tilde{\phi}$ and $\tilde{\psi}$. Also, the result says nothing about how the properties of the filters relate to orthogonality of the wavelet basis functions.

Now, let us go into parts of the proof of Theorem 6. We will not address why the constructed wavelet functions are linearly independent, and why they are dual Riesz bases. These details are quite technical, and can be found in [9].

Vanishing Moments

The CTFT of the mother wavelet ψ is

$$\hat{\psi}(\nu) = \frac{1}{\sqrt{2\pi}} \int_{-\infty}^{\infty} \psi(t) e^{-i\nu t} dt.$$

It is known that one can differentiate this by differentiating the integrand, as long as this differentiation leads to a new integrand bounded by an integrable function. Here this bound is easy to establish since ψ has compact support, so that we can write

$$\hat{\psi}^{(k)}(\nu) = \frac{1}{\sqrt{2\pi}} \int_{-\infty}^{\infty} \psi(t)(-it)^k e^{-i\nu t} dt.$$

Evaluating this at zero one obtains

$$\hat{\psi}^{(k)}(0) = \frac{(-i)^k}{\sqrt{2\pi}} \int_{-\infty}^{\infty} t^k \psi(t) dt.$$

Thus, ψ has k vanishing moments if and only if $\hat{\psi}$ has a zero of multiplicity k at 0. Now, since

$$\hat{\psi}(\nu) = e^{-i\nu/2} \frac{\lambda_{G_1}(\nu/2)}{\sqrt{2}} \hat{\phi}(\nu/2).$$

and since $g_0(0) = \hat{\phi}(0) = 1$, it follows that ψ has k vanishing moments if and only if $\lambda_{G_1}(\nu)$ has a zero of multiplicity k at 0. Due to Exercise 5.10, this is the case if and only if $\lambda_{H_0}(\nu)$ has a zero of multiplicity k at π. This means that, when $\lambda_{H_0}(\nu)$ has the form stated in (6.7), ψ must have L vanishing moments. Repeating this for the dual transform we also obtain that $\tilde{\psi}$ has $\tilde{L}$ vanishing moments when $\lambda_{G_0}(\nu)$ has the form stated.

The result also proves why G_0, H_0 should be low-pass filters with flat frequency responses near high frequencies ($\nu = \pi$). The flatter they are, the better ψ and $\tilde{\psi}$ are for approximation of functions. This is analogous to the low-pass filters we constructed from rows in Pascal's triangle previously, as the row index there equaled the multiplicity of a zero at π. The frequency response for the Haar wavelet had just a simple zero at π, so that it could not represent differentiable functions efficiently.

From Theorem 6 the following generalization to Theorem 26 also follows.

Theorem 7. Vanishing Moments, Regularity, and the Decay of Wavelet Coefficients, General Version.

The following hold when $\tilde{\psi}$ has k vanishing moments.

- *If f is k times continuously differentiable, then there exists a constant D so that $|w_{m,n}| \leq D2^{-m/2-mk}$ for all m.*
- *If f is a polynomial of degree less than or equal to $k-1$ then $w_{m,n} = 0$ for all m.*

A Closer Look on the Conditions (6.8) and (6.9)

Interestingly, the theorem splits the conditions on the filters in two parts: One part which secures the number of vanishing moments, i.e. $\left(\frac{1+e^{-i\nu}}{2}\right)^L$, and another which secures the regularity of ϕ and ψ, i.e. $\mathcal{F}(\nu)$. The second part has to be sufficiently small in order for ϕ, ψ to give rise to Riesz bases ((6.8) and (6.9)), or to be regular (6.11). By examining the term $\mathcal{F}(\nu)$ carefully for the wavelets in this book, one can verify that all our wavelets actually give rise to MRA's, and how regular the corresponding basis functions are.

Conditions (6.8) and (6.9) seem rather complex, and in [9] they are the most general conditions stated. In particular these conditions imply that $\hat{\phi}(\nu) \leq C(1+|\nu|)^{-1/2-\epsilon}$ for an $\epsilon > 0$, from which it follows that $\hat{\phi} \in L^2(\mathbb{R})$, as we have seen. To see why, first write

$$\prod_{s=1}^{\infty} \left(\frac{1+e^{-i\nu/2^s}}{2}\right) = \prod_{s=1}^{\infty} \left(e^{-i\nu/2^{s+1}} \cos\left(\nu/2^{s+1}\right)\right)$$

$$= e^{-i\nu/2} \prod_{s=1}^{\infty} \frac{\sin\left(\nu/2^s\right)}{2\sin\left(\nu/2^{s+1}\right)} = e^{-i\nu/2} \frac{\sin(\nu/2)}{\nu/2}.$$

Inserting this in (6.7) we get

$$\hat{\phi}(\nu) = \prod_{s=1}^{\infty} \frac{\lambda_{G_0}(\nu/2^s)}{\sqrt{2}} = e^{-i\tilde{L}\nu/2} \left(\frac{\sin(\nu/2)}{\nu/2}\right)^{\tilde{L}} \prod_{s=1}^{\infty} \tilde{\mathcal{F}}(\nu/2^s).$$

so that

$$|\hat{\phi}(\nu)| = \left|\frac{\sin(\nu/2)}{\nu/2}\right|^{\tilde{L}} \left|\prod_{s=1}^{\infty} \tilde{\mathcal{F}}(\nu/2^s)\right|.$$

Let us bound each of the terms here. Since $|\sin(t)| \leq 1$, it is straightforward to find a constant C so that the first term is bounded by $\left|\frac{\sin(\nu/2)}{\nu/2}\right|^{\tilde{L}} \leq C(1+|\nu|)^{-\tilde{L}}$. For the second term, note first that we must have $\mathcal{F}(0) = \tilde{\mathcal{F}}(0) = 1$, in order to obtain 1 on both sides in (6.7). From the proof of Lemma 4 it therefore follows that the infinite product $\prod_{s=1}^{\infty} \tilde{\mathcal{F}}(\nu/2^s)$ converges uniformly to a differentiable function. In particular, this product can be uniformly bounded for $|\nu| < 1$.

Assume instead that $|\nu| > 1$. Then there exists a unique integer l_0 so that $2^{kl_0} \leq |\nu| < 2^{k(l_0+1)}$ (this implies that $kl_0 \leq \log_2 |\nu|$). Split the infinite product into

$$\prod_{s=1}^{\infty} \tilde{\mathcal{F}}(\nu/2^s) = \prod_{s=1}^{k(l_0+1)} \tilde{\mathcal{F}}(\nu/2^s) \prod_{s=k(l_0+1)+1}^{\infty} \tilde{\mathcal{F}}(\nu/2^s). \tag{6.12}$$

For the first term in (6.12) we obtain

$$\prod_{s=1}^{k(l_0+1)} \tilde{\mathcal{F}}(\nu/2^s) = \prod_{l=0}^{l_0} \left(\tilde{\mathcal{F}}(\nu/2^{lk+1})\tilde{\mathcal{F}}(\nu/2^{lk+2}) \cdots \tilde{\mathcal{F}}(\nu/2^{lk+k}) \right)$$

$$\leq (\tilde{B}_k)^{k(l_0+1)} \leq \tilde{B}_k^{k+\log_2|\nu|} = \tilde{B}_k^k |\nu|^{\log_2 \tilde{B}_k}.$$

One can also reformulate this bound as

$$\prod_{s=1}^{k(l_0+1)} \tilde{\mathcal{F}}(\nu/2^s) \leq C(1+|\nu|)^{\log_2 \tilde{B}_k}$$

for some number C. For the second term in (6.12) we obtain

$$\prod_{s=k(l_0+1)+1}^{\infty} \tilde{\mathcal{F}}(\nu/2^s) = \prod_{s=1}^{\infty} \tilde{\mathcal{F}}\left(\frac{\nu}{2^{k(l_0+1)}} 2^{-s} \right).$$

Here $\left| \frac{\nu}{2^{k(l_0+1)}} \right| < 1$ by choice of l_0, so that this is an infinite product where $\tilde{\mathcal{F}}$ is evaluated for $|\nu| < 1$, and above we obtained a uniform bound for this. Thus, (6.12) has a bound on the form $C(1+|\nu|)^{\log_2 \tilde{B}_k}$. Combining the bounds for the first and second terms we obtain

$$C(1+|\nu|)^{-\tilde{L}}(1+|\nu|)^{\log_2 \tilde{B}_k} = C(1+|\nu|)^{-\tilde{L}+\log_2 \tilde{B}_k}.$$

Writing this as $C(1+|\nu|)^{-1/2-(\tilde{L}-\log_2 \tilde{B}_k-1/2)}$, we see hat $\hat{\phi} \in L^2(\mathbb{R})$ when $\epsilon = \tilde{L} - \log_2 \tilde{B}_k - 1/2 > 0$, i.e. when $\log_2 \tilde{B}_k < \tilde{L} - 1/2$, i.e. when $\tilde{B}_k < 2^{\tilde{L}-1/2}$, which is seen to be condition (6.8).

Regarding the requirement (6.11), increasing $\tilde{m}$ leads to a decrease in $\log_2 \tilde{B}_k$ with the same amount, so that the exponent α in the bound $|\hat{\phi}| \leq C(1+|\nu|)^{-\alpha}$ increases with the same amount. When one has such a bound, the regularity of ϕ increases with α. To see this, differentiate the integrand in the expression for the inverse CTFT to obtain

$$\phi^{(k)}(t) = \frac{1}{\sqrt{2\pi}} \int_{-\infty}^{\infty} (i\nu)^k \hat{\phi}(\nu)e^{i\nu t} d\nu.$$

Again, this is legal as long as the integrand remains bounded by an integrable function. Here the bound takes the form $\left| (i\nu)^k \hat{\phi}(\nu)e^{i\nu t} \right| \leq C|\nu|^k(1+|\nu|)^{-\alpha}$, which clearly is integrable as long as k is small enough, and depending on α.

Sketch of Proof of Biorthogonality

Recall that we defined functions $g_{N,i}$, $\tilde{g}_{N,i}$ satisfying

$$g_{N+1,0}(\nu) = \frac{\lambda_{G_0}(\nu/2)}{\sqrt{2}} g_{N,0}(\nu/2) \qquad g_{N+1,1}(\nu) = e^{-i\nu/2} \frac{\lambda_{G_1}(\nu/2)}{\sqrt{2}} g_{N,0}(\nu/2)$$

$$\tilde{g}_{N+1,0}(\nu) = \frac{\lambda_{(H_0)^T}(\nu/2)}{\sqrt{2}} \tilde{g}_{N,0}(\nu/2) \qquad \tilde{g}_{N+1,1}(\nu) = e^{-i\nu/2} \frac{\lambda_{(H_1)^T}(\nu/2)}{\sqrt{2}} \tilde{g}_{N,0}(\nu/2),$$

and that the idea was to take limits as $N \to \infty$ to obtain functions $g_0 = \hat{\phi}$, $g_1 = \hat{\psi}$, $\tilde{g}_0 = \hat{\tilde{\phi}}$, $\tilde{g}_1 = \hat{\tilde{\psi}}$. The $g_{N,i}$, $\tilde{g}_{N,i}$ are compactly supported, and equal to trigonometric polynomials on their support, so that $g_{N,i}, \tilde{g}_{N,i} \in L^2(\mathbb{R})$. It follows that there exist functions $u_{N,i}, \tilde{u}_{N,i} \in L^2(\mathbb{R})$ so that $g_{N,i} = \hat{u}_{N,i}$, $\tilde{g}_{N,i} = \hat{\tilde{u}}_{N,i}$. Since the above relationship equals that of Eq. (6.3) with $\hat{\phi}$, $\hat{\psi}$ replaced with $g_{N,i}$, we must have that

$$u_{N+1,0}(t) = \sum_n (G_0)_{n,0}\sqrt{2}u_{N,0}(2t-n)$$

$$u_{N+1,1}(t) = \sum_n (G_1)_{n,1}\sqrt{2}u_{N,0}(2t-n)$$

$$\tilde{u}_{N+1,0}(t) = \sum_n ((H_0)^T)_{n,0}\sqrt{2}\tilde{u}_{N,0}(2t-n)$$

$$\tilde{u}_{N+1,1}(t) = \sum_n ((H_1)^T)_{n,1}\sqrt{2}\tilde{u}_{N,0}(2t-n).$$

This means that

$$\begin{pmatrix} u_{N+1,0}(t-0) \\ u_{N+1,1}(t-0) \\ u_{N+1,0}(t-1) \\ u_{N+1,1}(t-1) \\ \vdots \end{pmatrix} = G^T \begin{pmatrix} \sqrt{2}u_{N,0}(2t-0) \\ \sqrt{2}u_{N,0}(2t-1) \\ \sqrt{2}u_{N,0}(2t-2) \\ \sqrt{2}u_{N,0}(2t-3) \\ \vdots \end{pmatrix}$$

$$\begin{pmatrix} \tilde{u}_{N+1,0}(t-0) \\ \tilde{u}_{N+1,1}(t-0) \\ \tilde{u}_{N+1,0}(t-1) \\ \tilde{u}_{N+1,1}(t-1) \\ \vdots \end{pmatrix} = H \begin{pmatrix} \sqrt{2}\tilde{u}_{N,0}(2t-0) \\ \sqrt{2}\tilde{u}_{N,0}(2t-1) \\ \sqrt{2}\tilde{u}_{N,0}(2t-2) \\ \sqrt{2}\tilde{u}_{N,0}(2t-3) \\ \vdots \end{pmatrix}. \tag{6.13}$$

If we define

$$\mathcal{B}_N = \{u_{N+1,0}(t-0), u_{N+1,1}(t-0), u_{N+1,0}(t-1), u_{N+1,1}(t-1), \ldots\}$$
$$\tilde{\mathcal{B}}_N = \{\tilde{u}_{N+1,0}(t-0), \tilde{u}_{N+1,1}(t-0), \tilde{u}_{N+1,0}(t-1), \tilde{u}_{N+1,1}(t-1), \ldots\}$$
$$\mathcal{C}_N = \{\sqrt{2}u_{N,0}(2t-0), \sqrt{2}u_{N,0}(2t-1), \sqrt{2}u_{N,0}(2t-2), \sqrt{2}u_{N,0}(2t-3), \ldots\}$$
$$\tilde{\mathcal{C}}_N = \{\sqrt{2}\tilde{u}_{N,0}(2t-0), \sqrt{2}\tilde{u}_{N,0}(2t-1), \sqrt{2}\tilde{u}_{N,0}(2t-2), \sqrt{2}\tilde{u}_{N,0}(2t-3), \ldots\},$$

this means that

$$\left(\langle \mathcal{B}_N, \tilde{\mathcal{B}}_N \rangle\right) = G^T \left(\langle \mathcal{C}_N, \tilde{\mathcal{C}}_N \rangle\right) H^T, \tag{6.14}$$

where $\left(\langle \mathcal{B}_N, \tilde{\mathcal{B}}_N \rangle\right)$ is the Gramm matrix with entries consisting of all inner products between vectors in $\mathcal{B}_N$ and $\tilde{\mathcal{B}}_N$ (see Appendix A). Now, using that $g_{0,0}(\nu) = \tilde{g}_{0,0}(\nu) = \chi_{[0,2\pi]}(\nu)$, and that $g_{0,0} = \hat{u}_{0,0}$, $\tilde{g}_{0,0} = \hat{\tilde{u}}_{0,0}$, we get

$$\int_{\infty}^{-\infty} u_{0,0}(t-k)\overline{\tilde{u}_{0,0}(t-l)}dt = \int_{\infty}^{-\infty} g_{0,0}(\nu)e^{2\pi ik\nu}\overline{\tilde{g}_{0,0}(\nu)}e^{2\pi il\nu}d\nu$$

$$= \int_0^{2\pi} e^{2\pi i(k-l)\nu}d\nu = \delta_{k,l}.$$

After a change of variables $u = 2t$ it follows that $(\langle \mathcal{C}_0, \tilde{\mathcal{C}}_0 \rangle) = I$. Inserting this is Eq. (6.14) and using that $GH = I$ it follows that $(\langle \mathcal{B}_0, \tilde{\mathcal{B}}_0 \rangle) = I$ also. In particular it follows that $(\langle \mathcal{C}_1, \tilde{\mathcal{C}}_1 \rangle) = I$ also (change of variables $u = 2t$ again), and so on. By induction it follows that, for any given N, $\mathcal{B}_N$ and $\tilde{\mathcal{B}}_N$ are biorthogonal, and $\mathcal{C}_N$ and $\tilde{\mathcal{C}}_N$ are biorthogonal. The biorthogonality of $\mathcal{B}_N$ and $\tilde{\mathcal{B}}_N$ would imply that of (ϕ_0, ψ_0) and $(\tilde{\phi}_0, \tilde{\psi}_0)$, while the biorthogonality of $\mathcal{C}_N$ and $\tilde{\mathcal{C}}_N$ would imply that of ϕ_1 and $\tilde{\phi}_1$, assuming that $u_{N,0} \to \phi$, $u_{N,1} \to \psi$, $\tilde{u}_{N,0} \to \tilde{\phi}$, $\tilde{u}_{N,1} \to \tilde{\psi}$. By repeating the decomposition (6.13), the existence of these limits would also imply the biorthogonality of

$$(\phi_0, \psi_0, \psi_1, \psi_2, \dots \psi_{m-1}) \text{ and } (\tilde{\phi}_0, \tilde{\psi}_0, \tilde{\psi}_1, \tilde{\psi}_2, \dots \tilde{\psi}_{m-1}).$$

Exercise 6.1: Properties of the CTFT

Prove the properties of the Continuous time Fourier transform listed in Theorem 3.

Exercise 6.2: The CTFT of a Pulse

a) Show that the CTFT of $\chi_{[-a,a]}(t)$ is $\frac{1}{\sqrt{2\pi}} \frac{\sin(\nu a)}{\nu/2}$ for any a.

b) Use a) to show that the CTFT of $\frac{\sin(\pi t/T_s)}{\pi t/T_s}$ is $\frac{T_s}{\sqrt{2\pi}} \chi_{[-\pi/T_s, \pi/T_s]}(t)$.

Exercise 6.3: Mixed Continuous-Discrete Expressions

Let $f(t) = \sum_k x_k h(t - nT_s)$. Show that $\hat{f}(\nu) = \hat{x}(\nu T_s)\hat{h}(\nu)$. This can be considered a mixed continuous-discrete expression for that convolution in time to corresponds to multiplication in frequency.

Exercise 6.4: The Frame Inequalities and Singular Values

Show that

$$A\|\boldsymbol{f}\|^2 \le \|U^T \boldsymbol{f}\|^2 \le B\|\boldsymbol{f}\|^2.$$

for some constants $A, B > 0$ if and only if UU^T is non-singular. Show also in this case that the optimal values for A and B are $A = \sigma_m^2$ and $B = \sigma_1^2$, where σ_i are the singular values of U^T.

Hint

Consult Appendix A about the singular value decomposition.

Exercise 6.5: Finding a Dual Frame

Assume that UU^T is non-singular, so that $\{u_n\}_n$ is a frame. From the previous exercise it follows from this that $\sigma_m > 0$. From Appendix A it also follows that the rank of U is m, i.e. U has full column rank. Show that $\tilde{U} = (U^T)^\dagger$ satisfies the requirements of a dual frame, where $(U^T)^\dagger$ is the generalized inverse of U^T, defined in the appendix. The appendix then says that $\tilde{U} = (U^T)^\dagger = (UU^T)^{-1}U$. It follows that $(UU^T)^{-1}$ maps the frame vectors u_k to the dual frame vectors $\tilde{u}_k$, and that the frame operator UU^T maps the dual frame vectors to the frame vectors.

Exercise 6.6: Tight Frames in Finite-Dimensional Spaces

Let $\{u_i\}_{i=1}^n$ be a set of vectors in $\mathbb{R}^m$. Show that $\{u_i\}_{i=1}^n$ is a tight frame with frame bound A is equivalent to any of the two following statements.

a) $UU^T = AI_m$.

b) $x = \frac{1}{A}\sum_{i=1}^n \langle x, u_i\rangle u_i$ for all $x \in \mathbb{R}^m$.

Exercise 6.7: Riesz Bases for Finite-Dimensional Spaces

Let $\{u_i\}_{i=1}^m$ be a Riesz basis for $\mathbb{R}^m$.

a) Show that the dual frame of $\{u_i\}_{i=1}^m$ are the rows of U^{-1}.

b) Show that $\{u_i\}_{i=1}^m$ is a tight frame if and only if the vectors are orthogonal, and that the dual frame is a scaled version of the frame itself.

6.2 Characterization of Wavelets w.r.t. Number of Vanishing Moments

In this section we will attempt to characterize wavelets with a given number of vanishing moments. In particular we will characterize the simplest such, i.e. where there are fewest filter coefficients. There are two particular cases we will look at. First we will consider the case when all filters are symmetric, then the case of orthonormal wavelets. It turns out that these two cases are mutually disjoint, but that there is a common result which can be used to characterize both. Recall that the number of vanishing moments equaled the multiplicities of the zeros of $\lambda_{H_0}, \lambda_{G_0}$ at π.

6.2.1 Symmetric Filters

The main result when the filters are symmetric looks as follows.

Theorem 8. Conditions for Vanishing Moments When the Filters Are Symmetric.
Assume that H_0, H_1, G_0, G_1 are symmetric FIR filters so that $GH = I$, and

- λ_{H_0} has a zero of multiplicity N_1 at π,
- λ_{G_0} has a zero of multiplicity N_2 at π.

Then N_1 and N_2 are even, and there exists a polynomial Q which satisfies

$$u^{(N_1+N_2)/2}Q(1-u) + (1-u)^{(N_1+N_2)/2}Q(u) = 2, \qquad (6.15)$$

and so that $\lambda_{H_0}(\omega), \lambda_{G_0}(\omega)$ can be written on the form

$$\lambda_{H_0}(\omega) = \left(\frac{1}{2}(1+\cos\omega)\right)^{N_1/2} Q_1\left(\frac{1}{2}(1-\cos\omega)\right) \qquad (6.16)$$

$$\lambda_{G_0}(\omega) = \left(\frac{1}{2}(1+\cos\omega)\right)^{N_2/2} Q_2\left(\frac{1}{2}(1-\cos\omega)\right), \qquad (6.17)$$

where $Q = Q_1 Q_2$.

Note that we did not include the conditions $\lambda_{G_0}(0) = \lambda_{H_0}(0) = \sqrt{2}$ here. These are only necessary for Theorem 6, when one needs the infinite products for $\hat{\phi}$, $\hat{\psi}$ therein to converge.

Proof. Since the filters are symmetric, $\lambda_{H_0}(\omega) = \lambda_{H_0}(-\omega)$ and $\lambda_{G_0}(\omega) = \lambda_{G_0}(-\omega)$. Since $e^{in\omega} + e^{-in\omega} = 2\cos(n\omega)$, and since $\cos(n\omega)$ is the real part of $(\cos\omega + i\sin\omega)^n$, which is a polynomial in $\cos^k\omega\sin^l\omega$ with l even, and since $\sin^2\omega = 1 - \cos^2\omega$, λ_{H_0} and λ_{G_0} can both be written on the form $P(\cos\omega)$, with P a real polynomial.

A zero at π in λ_{H_0}, λ_{G_0} corresponds to a factor of the form $1 + e^{-i\omega}$, so that we can write

$$\lambda_{H_0}(\omega) = \left(\frac{1+e^{-i\omega}}{2}\right)^{N_1} f(e^{i\omega}) = e^{-iN_1\omega/2}\cos^{N_1}(\omega/2)f(e^{i\omega}),$$

where f is a polynomial. In order for this to be real, we must have that $f(e^{i\omega}) = e^{iN_1\omega/2}g(e^{i\omega})$ where g clearly is a real polynomial with coefficients symmetric around 0, so that $g(e^{i\omega})$ too can be written as a real polynomial in $\cos\omega$. This means that $\lambda_{H_0}(\omega) = \cos^{N_1}(\omega/2)P_1(\cos\omega)$, and similarly for $\lambda_{G_0}(\omega)$. Clearly these can be polynomials in $e^{i\omega}$ only if N_1 and N_2 are even, so that we can write

$$\lambda_{H_0}(\omega) = \cos^{N_1}(\omega/2)P_1(\cos\omega) = (\cos^2(\omega/2))^{N_1/2}P_1(1 - 2\sin^2(\omega/2))$$
$$= (\cos^2(\omega/2))^{N_1/2}Q_1(\sin^2(\omega/2)),$$

where we have used that $\cos\omega = 1 - 2\sin^2(\omega/2)$, and defined Q_1 by the relation $Q_1(x) = P_1(1-2x)$. Similarly we can write $\lambda_{G_0}(\omega) = (\cos^2(\omega/2))^{N_2/2}Q_2(\sin^2(\omega/2))$ for another polynomial Q_2. Using the identities

$$\cos^2\frac{\omega}{2} = \frac{1}{2}(1+\cos\omega) \qquad\qquad \sin^2\frac{\omega}{2} = \frac{1}{2}(1-\cos\omega),$$

we see that λ_{H_0} and λ_{G_0} satisfy Eqs. (6.16) and (6.17). With $Q = Q_1 Q_2$, the perfect reconstruction condition from Exercise 5.10 can now be rewritten as

$$2 = \lambda_{G_0}(\omega)\lambda_{H_0}(\omega) + \lambda_{G_0}(\omega + \pi)\lambda_{H_0}(\omega + \pi)$$
$$= \left(\cos^2(\omega/2)\right)^{(N_1+N_2)/2} Q(\sin^2(\omega/2))$$
$$+ \left(\cos^2((\omega + \pi)/2)\right)^{(N_1+N_2)/2} Q(\sin^2((\omega + \pi)/2))$$
$$= (\cos^2(\omega/2))^{(N_1+N_2)/2} Q(\sin^2(\omega/2)) + (\sin^2(\omega/2))^{(N_1+N_2)/2} Q(\cos^2(\omega/2))$$
$$= (\cos^2(\omega/2))^{(N_1+N_2)/2} Q(1 - \cos^2(\omega/2))$$
$$+ (1 - \cos^2(\omega/2))^{(N_1+N_2)/2} Q(\cos^2(\omega/2))$$

Setting $u = \cos^2(\omega/2)$ we see that Q must fulfill the equation

$$u^{(N_1+N_2)/2} Q(1 - u) + (1 - u)^{(N_1+N_2)/2} Q(u) = 2,$$

which is Eq. (6.15). This completes the proof. $\square$

This result says nothing about how we can find the filters with fewest coefficients. The polynomial Q decides the length of the filters H_0, G_0, however, so that we could try to find the polynomial Q of smallest degree. In this direction, note first that the polynomials $u^{N_1+N_2}$ and $(1 - u)^{N_1+N_2}$ have no zeros in common. Bezout's theorem, proved in Sect. 6.2.3, states that the equation

$$u^N q_1(u) + (1 - u)^N q_2(u) = 1 \tag{6.18}$$

has unique solutions q_1, q_2 with $\deg(q_1), \deg(q_2) < (N_1 + N_2)/2$. To find these solutions, substituting $1 - u$ for u gives the following equations:

$$u^N q_1(u) + (1 - u)^N q_2(u) = 1$$
$$u^N q_2(1 - u) + (1 - u)^N q_1(1 - u) = 1,$$

and uniqueness in Bezout's theorem gives that $q_1(u) = q_2(1 - u)$, and $q_2(u) = q_1(1 - u)$. Equation (6.18) can thus be stated as

$$u^N q_2(1 - u) + (1 - u)^N q_2(u) = 1,$$

and comparing with Eq. (6.15) (set $N = (N_1 + N_2)/2$) we see that $Q(u) = 2q_2(u)$. $u^N q_1(u) + (1 - u)^N q_2(u) = 1$ now gives

$$q_2(u) = (1 - u)^{-N}(1 - u^N q_1(u)) = (1 - u)^{-N}(1 - u^N q_2(1 - u))$$
$$= \left(\sum_{k=0}^{N-1} \binom{N + k - 1}{k} u^k + O(u^N)\right)(1 - u^N q_2(1 - u))$$
$$= \sum_{k=0}^{N-1} \binom{N + k - 1}{k} u^k + O(u^N),$$

where we have used the first N terms in the Taylor series expansion of $(1 - u)^{-N}$ around 0. Since q_2 is a polynomial of degree $N - 1$, we must have that

$$Q(u) = 2q_2(u) = 2\sum_{k=0}^{N-1} \binom{N + k - 1}{k} u^k. \tag{6.19}$$

Define $Q^{(N)}(u) = 2\sum_{k=0}^{N-1} \binom{N+k-1}{k} u^k$. The first $Q^{(N)}$ are

$$Q^{(1)}(u) = 2 \qquad\qquad\qquad Q^{(2)}(u) = 2 + 4u$$
$$Q^{(3)}(u) = 2 + 6u + 12u^2 \qquad Q^{(4)}(u) = 2 + 8u + 20u^2 + 40u^3,$$

for which we compute

$$Q^{(1)}\left(\frac{1}{2}(1 - \cos\omega)\right) = 2$$

$$Q^{(2)}\left(\frac{1}{2}(1 - \cos\omega)\right) = -e^{-i\omega} + 4 - e^{i\omega}$$

$$Q^{(3)}\left(\frac{1}{2}(1 - \cos\omega)\right) = \frac{3}{4}e^{-2i\omega} - \frac{9}{2}e^{-i\omega} + \frac{19}{2} - \frac{9}{2}e^{i\omega} + \frac{3}{4}e^{2i\omega}$$

$$Q^{(4)}\left(\frac{1}{2}(1 - \cos\omega)\right) = -\frac{5}{8}e^{-3i\omega} + 5e^{-2i\omega} - \frac{131}{8}e^{-i\omega} + 26 - \frac{131}{8}e^{i\omega} + 5e^{2i\omega}$$
$$- \frac{5}{8}e^{3i\omega},$$

Thus, in order to construct wavelets where $\lambda_{H_0}, \lambda_{G_0}$ have as many zeros at π as possible, and where there are as few filter coefficients as possible, we need to compute the polynomials above, factorize them into polynomials Q_1 and Q_2, and distribute these among λ_{H_0} and λ_{G_0}. Since we need real factorizations, we must also pair complex roots. If we do this we obtain the factorizations

$$Q^{(1)}\left(\frac{1}{2}(1 - \cos\omega)\right) = 2$$

$$Q^{(2)}\left(\frac{1}{2}(1 - \cos\omega)\right) = \frac{1}{3.7321}(e^{i\omega} - 3.7321)(e^{-i\omega} - 3.7321)$$

$$Q^{(3)}\left(\frac{1}{2}(1 - \cos\omega)\right) = \frac{3}{4}\frac{1}{9.4438}(e^{2i\omega} - 5.4255e^{i\omega} + 9.4438)$$
$$\times (e^{-2i\omega} - 5.4255e^{-i\omega} + 9.4438)$$

$$Q^{(4)}\left(\frac{1}{2}(1 - \cos\omega)\right) = \frac{5}{8}\frac{1}{3.0407}\frac{1}{7.1495}$$
$$\times (e^{i\omega} - 3.0407)(e^{2i\omega} - 4.0623e^{i\omega} + 7.1495)$$
$$\times (e^{-i\omega} - 3.0407)(e^{-2i\omega} - 4.0623e^{-i\omega} + 7.1495), \qquad (6.20)$$

The factors here can be distributed as factors in the frequency responses of $\lambda_{H_0}(\omega)$, and $\lambda_{G_0}(\omega)$. One possibility is to let one of the frequency responses absorb all the factors, so that one filter gets more filter coefficients. Another possibility is to split the factors as evenly as possible across the two. In the following examples, both factor distribution strategies will be considered. Note that it is straightforward to use your computer to factor Q into a product of polynomials Q_1 and Q_2. First the **roots** function can be used to find the roots in the polynomials. Then the **conv** function can be used to multiply together factors corresponding to different roots, to obtain the coefficients in the polynomials Q_1 and Q_2.

6.2.2 Orthonormal Wavelets

Now we turn to the case of orthonormal wavelets, i.e. where $G_0 = (H_0)^T$, $G_1 = (H_1)^T$. For simplicity we will assume $d = 0, \alpha = -1$ in the alias cancellation conditions of Exercise 5.10 (this corresponded to requiring $\lambda_{H_1}(\omega) = -\overline{\lambda_{H_0}(\omega + \pi)}$ in the definition of alternative QMF filter banks). We will also assume for simplicity that G_0 is causal (other solutions can be derived from this). We saw that the Haar wavelet was such an orthonormal wavelet. We have the following result:

Theorem 9. Conditions for Vanishing Moments in the Orthogonal Case.
  Assume that H_0, H_1, G_0, and G_1 are the filters of an alternative QMF filter bank, i.e. $H_0 = (G_0)^T$, $H_1 = (G_1)^T$), $\lambda_{H_1}(\omega) = -\overline{\lambda_{H_0}(\omega + \pi)}$. Assume also that $\lambda_{G_0}(\omega)$ has a zero of multiplicity N at π, and that G_0 is causal. Then there exists a polynomial Q which satisfies

$$u^N Q(1 - u) + (1 - u)^N Q(u) = 2, \tag{6.21}$$

so that if f is another polynomial which satisfies $f(e^{i\omega})f(e^{-i\omega}) = Q\left(\frac{1}{2}(1 - \cos\omega)\right)$, $\lambda_{G_0}(\omega)$ can be written on the form

$$\lambda_{G_0}(\omega) = \left(\frac{1 + e^{-i\omega}}{2}\right)^N f(e^{-i\omega}). \tag{6.22}$$

We avoided stating $\lambda_{H_0}(\omega)$ in this result, since the relation $H_0 = (G_0)^T$ gives that $\lambda_{H_0}(\omega) = \overline{\lambda_{G_0}(\omega)}$. In particular, $\lambda_{H_0}(\omega)$ also has a zero of multiplicity N at π. That G_0 is causal is included to simplify the expression further.

Proof. The proof is very similar to that of Theorem 8. N vanishing moments and that G_0 is causal means that we can write

$$\lambda_{G_0}(\omega) = \left(\frac{1 + e^{-i\omega}}{2}\right)^N f(e^{-i\omega}) = (\cos(\omega/2))^N e^{-iN\omega/2} f(e^{-i\omega}),$$

where f is a real polynomial. Also

$$\lambda_{H_0}(\omega) = \overline{\lambda_{G_0}(\omega)} = (\cos(\omega/2))^N e^{iN\omega/2} f(e^{i\omega}).$$

The perfect reconstruction condition from Exercise 5.10 now says that

$$2 = \lambda_{G_0}(\omega)\lambda_{H_0}(\omega) + \lambda_{G_0}(\omega + \pi)\lambda_{H_0}(\omega + \pi)$$
$$= (\cos^2(\omega/2))^N f(e^{i\omega})f(e^{-i\omega}) + (\sin^2(\omega/2))^N f(e^{i(\omega+\pi)})f(e^{-i(\omega+\pi)}).$$

The function $f(e^{i\omega})f(e^{-i\omega})$ is symmetric around 0, so that it can be written on the form $P(\cos\omega)$ with P a polynomial, so that

$$2 = (\cos^2(\omega/2))^N P(\cos\omega) + (\sin^2(\omega/2))^N P(\cos(\omega + \pi))$$
$$= (\cos^2(\omega/2))^N P(1 - 2\sin^2(\omega/2)) + (\sin^2(\omega/2))^N P(1 - 2\cos^2(\omega/2)).$$

If we as in the proof of Theorem 8 define Q by $Q(x) = P(1 - 2x)$, we can write this as

$$(\cos^2(\omega/2))^N Q(\sin^2(\omega/2)) + (\sin^2(\omega/2))^N Q(\cos^2(\omega/2)) = 2,$$

which again gives Eq. (6.15) for finding Q. We thus need to compute the polynomial $Q\left(\frac{1}{2}(1-\cos\omega)\right)$ as before, and consider the different factorizations of this on the form $f(e^{i\omega})f(e^{-i\omega})$. Since this polynomial is symmetric, a is a root if and only $1/a$ is, and if and only if $\bar{a}$ is. If the real roots are

$$b_1,\ldots,b_m,1/b_1,\ldots,1/b_m,$$

and the complex roots are

$$a_1,\ldots,a_n,\overline{a_1},\ldots,\overline{a_n} \text{ and } 1/a_1,\ldots,1/a_n,\overline{1/a_1},\ldots,\overline{1/a_n},$$

we can write

$$
\begin{aligned}
Q&\left(\frac{1}{2}(1-\cos\omega)\right)\\
&= K(e^{-i\omega}-b_1)\ldots(e^{-i\omega}-b_m)\\
&\times (e^{-i\omega}-a_1)(e^{-i\omega}-\overline{a_1})(e^{-i\omega}-a_2)(e^{-i\omega}-\overline{a_2})\cdots(e^{-i\omega}-a_n)(e^{-i\omega}-\overline{a_n})\\
&\times (e^{i\omega}-b_1)\ldots(e^{i\omega}-b_m)\\
&\times (e^{i\omega}-a_1)(e^{i\omega}-\overline{a_1})(e^{i\omega}-a_2)(e^{i\omega}-\overline{a_2})\cdots(e^{i\omega}-a_n)(e^{i\omega}-\overline{a_n}),
\end{aligned}
$$

where K is a constant, and define f by

$$
\begin{aligned}
f(e^{i\omega}) =&\sqrt{K}(e^{i\omega}-b_1)\ldots(e^{i\omega}-b_m)\\
&\times (e^{i\omega}-a_1)(e^{i\omega}-\overline{a_1})(e^{i\omega}-a_2)(e^{i\omega}-\overline{a_2})\cdots(e^{i\omega}-a_n)(e^{i\omega}-\overline{a_n})
\end{aligned}
$$

in order to obtain a factorization $Q\left(\frac{1}{2}(1-\cos\omega)\right) = f(e^{i\omega})f(e^{-i\omega})$. This concludes the proof. $\square$

In the previous proof we note that the polynomial f is not unique—we could pair the roots in many different ways. The new algorithm is as follows:

- Write $Q\left(\frac{1}{2}(1-\cos\omega)\right)$ as a polynomial in $e^{i\omega}$, and find the roots.
- Split the roots into the two classes

$$\{b_1,\ldots,b_m,a_1,\ldots,a_n,\overline{a_1},\ldots,\overline{a_n}\}$$

and

$$\{1/b_1,\ldots,1/b_m,1/a_1,\ldots,1/a_n,\overline{1/a_1},\ldots,\overline{1/a_n}\},$$

and form the polynomial f as above.

- Compute $\lambda_{G_0}(\omega) = \left(\frac{1+e^{-i\omega}}{2}\right)^N f(e^{-i\omega})$.

Clearly the filters obtained with this strategy are not symmetric since f is not symmetric. In Sect. 6.5 we will take a closer look at wavelets constructed in this way.

6.2.3 The Proof of Bezout's Theorem

In the previous subsections we used a theorem called *Bezout's theorem*. This can be formulated and proved as follows.

Theorem 10. Bezout's Theorem.

If p_1 and p_2 are two polynomials, of degrees n_1 and n_2 respectively, with no common zeros, then there exist unique polynomials q_1, q_2, of degree less than n_2, n_1, respectively, so that

$$p_1(x)q_1(x) + p_2(x)q_2(x) = 1. \tag{6.23}$$

Proof. We first establish the existence of q_1, q_2 satisfying Eq. (6.23). Denote by $\deg(P)$ the degree of the polynomial P. Renumber the polynomials if necessary, so that $n_1 \geq n_2$. By polynomial division, we can now write

$$p_1(x) = a_2(x)p_2(x) + b_2(x),$$

where $\deg(a_2) = \deg(p_1) - \deg(p_2)$, $\deg(b_2) < \deg(p_2)$. Similarly, we can write

$$p_2(x) = a_3(x)b_2(x) + b_3(x),$$

where $\deg(a_3) = \deg(p_2) - \deg(b_2)$, $\deg(b_3) < \deg(b_2)$. We can repeat this procedure, so that we obtain a sequence of polynomials $a_n(x), b_n(x)$ so that

$$b_{n-1}(x) = a_{n+1}(x)b_n(x) + b_{n+1}(x), \tag{6.24}$$

where $\deg a_{n+1} = \deg(b_{n-1}) - \deg(b_n)$, $\deg(b_{n+1} < \deg(b_n)$. Since $\deg(b_n)$ is strictly decreasing, we must have that $b_{N+1} = 0$ and $b_N \neq 0$ for some N, i.e. $b_{N-1}(x) = a_{N+1}(x)b_N(x)$. Since $b_{N-2} = a_N b_{N-1} + b_N$, it follows that b_{N-2} can be divided by b_N, and by induction that all b_n can be divided by b_N, in particular p_1 and p_2 can be divided by b_N. Since p_1 and p_2 have no common zeros, b_N must be a nonzero constant.

Using Eq. (6.24), we can write recursively

$$
\begin{aligned}
b_N &= b_{N-2} - a_N b_{N-1} \\
&= b_{N-2} - a_N(b_{N-3} - a_{N-1}b_{N-2}) \\
&= (1 + a_N a_{N-1})b_{N-2} - a_N b_{N-3}.
\end{aligned}
$$

By induction we can write

$$b_N = a_{N,k}^{(1)} b_{N-k} + a_{N,k}^{(2)} b_{N-k-1}.$$

We see that the leading order term for $a_{N,k}^{(1)}$ is $a_N \cdots a_{N-k+1}$, which has degree

$$
\begin{aligned}
(\deg(b_{N-2}) - \deg(b_{N-1})) + \cdots + (\deg(b_{N-k-1}) - \deg(b_{N-k}) \\
= \deg(b_{N-k-1}) - \deg(b_{N-1}),
\end{aligned}
$$

while the leading order term for $a_{N,k}^{(2)}$ is $a_N \cdots a_{N-k+2}$, which similarly has order $\deg(b_{N-k}) - \deg(b_{N-1})$. For $k = N - 1$ we find

$$b_N = a_{N,N-1}^{(1)} b_1 + a_{N,N-1}^{(2)} b_0 = a_{N,N-1}^{(1)} p_2 + a_{N,N-1}^{(2)} p_1, \tag{6.25}$$

with $\deg(a_{N,N-1}^{(1)}) = \deg(p_1) - \deg(b_{N-1}) < \deg(p_1)$ (since by construction $\deg(b_{N-1}) > 0$), and $\deg(a_{N,N-1}^{(2)}) = \deg(p_2) - \deg(b_{N-1}) < \deg(p_2)$. From Eq. (6.25) it follows that $q_1 = a_{N,N-1}^{(2)}/b_N$ and $q_2 = a_{N,N-1}^{(1)}/b_N$ satisfy Eq. (6.23), and that they satisfy the required degree constraints.

Now we turn to the uniqueness of the solutions q_1 and q_2. Assume that r_1 and r_2 are two other solutions to Eq. (6.23). Then

$$p_1(q_1 - r_1) + p_2(q_2 - r_2) = 0.$$

Since p_1 and p_2 have no zeros in common this means that every zero of p_2 is a zero of $q_1 - r_1$, with at least the same multiplicity. If $q_1 \neq r_1$, this means that $\deg(q_1 - r_1) \geq \deg(p_2)$, which is impossible since $\deg(q_1) < \deg(p_2)$, $\deg(r_1) < \deg(p_2)$. Hence $q_1 = r_1$. Similarly we have that $q_2 = r_2$, and we have established uniqueness. $\square$

Exercise 6.8: Computing Filters

Compute the filters H_0, G_0 in Theorem 8 when $N = N_1 = N_2 = 4$, and $Q_1 = Q^{(4)}$, $Q_2 = 1$. Compute also filters H_1, G_1 so that we have perfect reconstruction (note that these are not unique).

6.3 A Design Strategy Suitable for Lossless Compression

If we choose $Q_1 = Q$, $Q_2 = 1$ in Theorem 8, there is no need to find factors in Q, and the frequency responses of the filters can be written

$$\lambda_{H_0}(\omega) = \left(\frac{1}{2}(1+\cos\omega)\right)^{N_1/2} Q^{(N)}\left(\frac{1}{2}(1-\cos\omega)\right)$$

$$\lambda_{G_0}(\omega) = \left(\frac{1}{2}(1+\cos\omega)\right)^{N_2/2}, \tag{6.26}$$

where $N = (N_1+N_2)/2$. Since $Q^{(N)}$ has degree $N-1$, λ_{H_0} has degree $N_1+N_1+N_2-2 = 2N_1 + N_2 - 2$, and λ_{G_0} has degree N_2. These are both even numbers, so that the filters have odd length. The corresponding wavelets are indexed by the filter lengths and are called *Spline wavelets*. Let us explain how these wavelets are connected to splines. We have that

$$\lambda_{G_0}(\omega) = \frac{1}{2^{N_2/2}}(1+\cos\omega)^{N_2/2} = \cos(\omega/2)^{N_2}.$$

We compute

$$\prod_{s=1}^{k}\lambda_{G_0}(\nu/2^s) = \prod_{s=1}^{k}\cos(\nu/2^{s+1})^{N_2} = \prod_{s=1}^{k}\left(\frac{\sin(\nu/2^s)}{2\sin(\nu/2^{s+1})}\right)^{N_2}$$

$$= \left(\frac{\sin(\nu/2)}{2^k\sin(\nu/2^{k+1})}\right)^{N_2} \to \left(\frac{\sin(\nu/2)}{\nu/2}\right)^{N_2},$$

where we have used the identity $\cos\omega = \frac{\sin(2\omega)}{2\sin\omega}$, and took the limit as $k \to \infty$. On the other hand, the CTFT of $\chi_{[-1/2,1/2)}(t)$ is

$$\int_{-1/2}^{1/2}e^{-i\nu t}dt = \left[\frac{1}{-i\nu}e^{-i\nu t}\right]_{-1/2}^{1/2} = \frac{1}{-i\nu}(e^{-\nu/2} - e^{i\nu/2}) = \frac{\sin(\nu/2)}{\nu/2}.$$

Now, the *convolution of two functions* on $\mathbb{R}$ is defined as

$$(f * g)(t) = \int_{-\infty}^{\infty} f(x)g(t-x)dx,$$

and it is known that, when the integrals exist, the CTFT of $f * g$ is $\hat{f}\hat{g}$. For the current case this means that the CTFT of $*_{k=1}^{N_2}\chi_{[-1/2,1/2)}(t)$ is $\prod_{s=1}^{\infty}\lambda_{G_0}(\nu/2^s)$. If G_0 is scaled so that $\lambda_{G_0} = \sqrt{2}$, we can obtain an actual scaling function so that $\hat{\phi}(\nu) = \prod_{s=1}^{\infty}\frac{\lambda_{G_0}(\nu/2^s)}{\sqrt{2}}$, and this would imply that $\phi(t) = *_{k=1}^{N_2}\chi_{[-1/2,1/2)}(t)$ by the uniqueness of the CTFT. This function is called the *B-spline of order* N_2

In Exercise 6.10 you will be asked to show that this scaling function gives rise to the multiresolution analysis of functions which are piecewise polynomials and differentiable at the borders, also called *splines*. To be more precise, the resolution spaces are as follows.

Definition 11. *Resolution Spaces of Piecewise Polynomials.*

We define V_m as the subspace of functions which are $r-1$ times continuously differentiable and equal to a polynomial of degree r on any interval of the form $[n2^{-m}, (n+1)2^{-m}]$.

Note that the piecewise linear wavelet can be considered as the first spline wavelet. This is further considered in the following example.

6.3.1 The Spline 5/3 Wavelet

For the case of $N_1 = N_2 = 2$ when the first design strategy is used, Eqs. (6.16) and (6.17) take the form

$$\lambda_{G_0}(\omega) = \frac{1}{2}(1 + \cos\omega) = \frac{1}{4}e^{i\omega} + \frac{1}{2} + \frac{1}{4}e^{-i\omega}$$

$$\lambda_{H_0}(\omega) = \frac{1}{2}(1 + \cos\omega)Q^{(1)}\left(\frac{1}{2}(1 - \cos\omega)\right)$$

$$= \frac{1}{4}(2 + e^{i\omega} + e^{-i\omega})(4 - e^{i\omega} - e^{-i\omega})$$

$$= -\frac{1}{4}e^{2i\omega} + \frac{1}{2}e^{i\omega} + \frac{3}{2} + \frac{1}{2}e^{-i\omega} - \frac{1}{4}e^{-2i\omega}.$$

The filters G_0, H_0 are thus

$$G_0 = \left\{\frac{1}{4}, \frac{1}{2}, \frac{1}{4}\right\} \qquad\qquad H_0 = \left\{-\frac{1}{4}, \frac{1}{2}, \frac{3}{2}, \frac{1}{2}, -\frac{1}{4}\right\}$$

The length of the filters are 3 and 5 in this case, so that this wavelet is called the *Spline 5/3 wavelet.*

Regarding the filters G_1 and H_1, Exercise 5.10 says that we simply can swap the filters G_0 and H_0, and add an alternating sign. We thus can set

$$G_1 = \left\{-\frac{1}{4}, -\frac{1}{2}, \frac{3}{2}, -\frac{1}{2}, -\frac{1}{4}\right\} \qquad\qquad H_1 = \left\{-\frac{1}{4}, \frac{1}{2}, -\frac{1}{4}\right\}.$$

We have now found all the filters. Up to a constant, the filters are seen to be the same as those of the alternative piecewise linear wavelet, see Example 5.3. This means that we get the same scaling function and mother wavelet.

The coefficients for the Spline wavelets are always dyadic fractions. They are therefore suitable for lossless compression, as all arithmetic operations can be carried out using binary numbers only. The particular Spline wavelet from Sect. 6.3.1 is used for lossless compression in the JPEG2000 standard.

Exercise 6.9: Plotting Frequency Responses of Spline Wavelets

The library represents a Spline wavelet with x and y vanishing moments (i.e. N_1 and N_2) with the name "splinex.y".

a) Listen to the low-resolution approximations and detail components in sound for the "spline4.4" wavelet.

b) Plot all scaling functions and mother wavelets (using the cascade algorithm), and frequency responses for the "spline4.4" wavelet.

Exercise 6.10: Wavelets Based on Higher Degree Polynomials

Show that $B_r(t) = *_{k=1}^r \chi_{[-1/2,1/2)}(t)$, the B-spline of order r, is $r-2$ times differentiable, and equals a polynomial of degree $r-1$ on subintervals of the form $[n, n+1]$. Explain why these functions can be used as bases for the resolution spaces from Definition 11.

6.4 A Design Strategy Suitable for Lossy Compression

Let us now instead split factors as evenly as possible among Q_1 and Q_2 in Theorem 8. We now need to factorize Q into a product of real polynomials. This can be done by finding all roots, and pairing the complex conjugate roots into real second degree polynomials (if Q is real, its roots come in conjugate pairs), and then distribute these as evenly as possible among Q_1 and Q_2. These filters are called the CDF-wavelets, after Cohen, Daubechies, and Feauveau, who discovered them.

Example 6.11: The CDF 9/7 Wavelet

We now choose $N_1 = N_2 = 4$. In Eq. (6.20) we pair inverse terms to obtain

$$Q^{(3)}\left(\frac{1}{2}(1-\cos\omega)\right)$$
$$= \frac{5}{8}\frac{1}{3.0407}\frac{1}{7.1495}(e^{i\omega}-3.0407)(e^{-i\omega}-3.0407)$$

$$\times (e^{2i\omega} - 4.0623e^{i\omega} + 7.1495)(e^{-2i\omega} - 4.0623e^{-i\omega} + 7.1495)$$

$$= \frac{5}{8}\frac{1}{3.0407}\frac{1}{7.1495}(-3.0407e^{i\omega} + 10.2456 - 3.0407e^{-i\omega})$$

$$\times (7.1495e^{2i\omega} - 33.1053e^{i\omega} + 68.6168 - 33.1053e^{-i\omega} + 7.1495e^{-2i\omega}).$$

We can write this as $Q_1 Q_2$ with $Q_1(0) = Q_2(0)$ when

$$Q_1(\omega) = -1.0326e^{i\omega} + 3.4795 - 1.0326e^{-i\omega}$$
$$Q_2(\omega) = 0.6053e^{2i\omega} - 2.8026e^{i\omega} + 5.8089 - 2.8026e^{-i\omega} + 0.6053e^{-2i\omega},$$

from which we obtain

$$\lambda_{G_0}(\omega) = \left(\frac{1}{2}(1 + \cos\omega)\right)^2 Q_1(\omega)$$
$$= -0.0645e^{3i\omega} - 0.0407e^{2i\omega} + 0.4181e^{i\omega} + 0.7885$$
$$+ 0.4181e^{-i\omega} - 0.0407e^{-2i\omega} - 0.0645e^{-3i\omega}$$
$$\lambda_{H_0}(\omega) = \left(\frac{1}{2}(1 + \cos\omega)\right)^2 40Q_2(\omega)$$
$$= 0.0378e^{4i\omega} - 0.0238e^{3i\omega} - 0.1106e^{2i\omega} + 0.3774e^{i\omega} + 0.8527$$
$$+ 0.3774e^{-i\omega} - 0.1106e^{-2i\omega} - 0.0238e^{-3i\omega} + 0.0378e^{-4i\omega}.$$

The filters G_0, H_0 are thus

$$G_0 = \{0.0645, 0.0407, -0.4181, \underline{-0.7885}, -0.4181, 0.0407, 0.0645\}$$
$$H_0 = \{-0.0378, 0.0238, 0.1106, -0.3774, \underline{-0.8527}, -0.3774, 0.1106, 0.0238, -0.0378\}.$$

The corresponding frequency responses are plotted in Fig. 6.1. It is seen that both filters are low-pass also here, and that they are closer to an ideal band-pass filter. The frequency responses seem very flat near π.

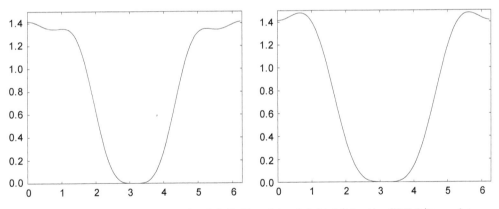

Fig. 6.1 The frequency responses $\lambda_{H_0}(\omega)$ (left) and $\lambda_{G_0}(\omega)$ (right) for the CDF 9/7 wavelet

We also get

$$G_1 = \{-0.0378, -0.0238, 0.1106, 0.3774, \underline{-0.8527}, 0.3774, 0.1106, -0.0238, -0.0378\}$$
$$H_1 = \{-0.0645, 0.0407, 0.4181, \underline{-0.7885}, 0.4181, 0.0407, -0.0645\}.$$

The length of the filters are 9 and 7 in this case, which is why this wavelet is called the *CDF 9/7 wavelet*.

In Example 5.3 we saw that we had analytical expressions for the scaling functions and the mother wavelet, but that we could not obtain this for the dual functions. For the CDF 9/7 wavelet it turns out that none of the four functions have analytical expressions. Let us therefore use the cascade algorithm to plot them. Note first that since G_0 has 7 filter coefficients, and G_1 has 9 filter coefficients, it follows from Exercise 5.16 that $\text{Supp}(\phi) = [-3, 3]$, $\text{Supp}(\psi) = [-3, 4]$, $\text{Supp}(\tilde{\phi}) = [-4, 4]$, and $\text{Supp}(\tilde{\psi}) = [-3, 4]$. The scaling functions and mother wavelets over these supports are shown in Fig. 6.2. Again they have irregular shapes, but now at least the functions and dual functions resemble each other more. The functions would be equal to their dual counterparts if the wavelet was orthogonal, so this suggests that this wavelet is close to being orthogonal. This may also explain partially why the CDF 9/7 wavelet works well for compression purposes: It is in fact used for lossy compression with the JPEG2000 standard.

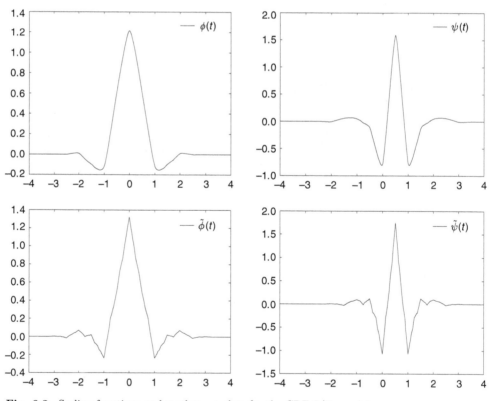

Fig. 6.2 Scaling functions and mother wavelets for the CDF 9/7 wavelet

In the above example there was a unique way of factoring Q into a product of real polynomials. For higher degree polynomials there is no unique way to distribute the factors, and we will not go into what strategy can be used for this. In general, we must go through the following steps:

- Compute the polynomial Q, and find its roots.
- Pair complex conjugate roots into real second degree polynomials, and form polynomials Q_1, Q_2.
- Compute the coefficients in Eqs. (6.16) and (6.17).

6.5 Orthonormal Wavelets

The filters we constructed for orthonormal wavelets were not symmetric. In Theorem 9 we characterized orthonormal wavelets where G_0 was causal. All our filters have an even number, say $2L$, of filter coefficients. It turns out that we can also translate the filters so that

- H_0 has a minimum possible overweight of coefficients with negative indices,
- H_1 has a minimum possible overweight of coefficients with positive indices,

i.e. that the filters can be written with the following compact notation:

$$H_0 = \{t_{-L}, \ldots, t_{-1}, \underline{t_0}, t_1, \ldots, t_{L-1}\} \quad H_1 = \{s_{-L+1}, \ldots, s_{-1}, \underline{s_0}, s_1, \ldots, s_L\}. \quad (6.27)$$

To see why, Exercise 5.10 says that we first can shift the coefficients of H_0 so that they have this form, but we then need to shift G_0 in the opposite direction. H_1 and G_1 can then be defined by setting $\alpha = 1$ and $d = 0$. We will follow this convention in what follows.

The polynomials $Q^{(0)}$, $Q^{(1)}$, and $Q^{(2)}$ require no further action to obtain the factorization $f(e^{i\omega})f(e^{-i\omega}) = Q\left(\frac{1}{2}(1 - \cos\omega)\right)$. The polynomial $Q^{(3)}$ in Eq. (6.20) can be factored further as

$$Q^{(3)}\left(\frac{1}{2}(1 - \cos\omega)\right)$$
$$= \frac{5}{8}\frac{1}{3.0407}\frac{1}{7.1495}(e^{-3i\omega} - 7.1029e^{-2i\omega} + 19.5014^{-i\omega} - 21.7391)$$
$$\times (e^{3i\omega} - 7.1029e^{2i\omega} + 19.5014^{i\omega} - 21.7391),$$

which gives that

$$f(e^{i\omega}) = \sqrt{\frac{5}{8}\frac{1}{3.0407}\frac{1}{7.1495}}(e^{3i\omega} - 7.1029e^{2i\omega} + 19.5014^{i\omega} - 21.7391).$$

This factorization is not unique, however. This gives the frequency response $\lambda_{G_0}(\omega) = \left(\frac{1+e^{-i\omega}}{2}\right)^N f(e^{-i\omega})$ as

$$\frac{1}{2}(e^{-i\omega} + 1)\sqrt{2}$$

$$\frac{1}{4}(e^{-i\omega} + 1)^2\sqrt{\frac{1}{3.7321}}(e^{-i\omega} - 3.7321)$$

$$\frac{1}{8}(e^{-i\omega}+1)^3\sqrt{\frac{3}{4}\frac{1}{9.4438}}(e^{-2i\omega}-5.4255e^{-i\omega}+9.4438)$$

$$\frac{1}{16}(e^{-i\omega}+1)^4\sqrt{\frac{5}{8}\frac{1}{3.0407}\frac{1}{7.1495}}(e^{-3i\omega}-7.1029e^{-2i\omega}+19.5014^{-i\omega}-21.7391),$$

which gives the filters

$$G_0=(H_0)^T=(\sqrt{2}/2,\sqrt{2}/2)$$
$$G_0=(H_0)^T=(-0.4830,\underline{-0.8365},-0.2241,0.1294)$$
$$G_0=(H_0)^T=(0.3327,0.8069,\underline{0.4599},-0.1350,-0.0854,0.0352)$$
$$G_0=(H_0)^T=(-0.2304,-0.7148,-0.6309,\underline{0.0280},0.1870,-0.0308,-0.0329,0.0106)$$

so that we get 2, 4, 6 and 8 filter coefficients in $G_0=(H_0)^T$. We see that the filters for $N=1$ are those of the Haar wavelet. The filters for $N=2,3$, and 4, are new. The filter $G_1=(H_1)^T$ can be obtained from the relation $\lambda_{G_1}(\omega)=-\overline{\lambda_{G_0}(\omega+\pi)}$, i.e. by reversing the elements and adding an alternating sign, plus an extra minus sign, so that

$$G_1=(H_1)^T=(\sqrt{2}/2,\underline{-\sqrt{2}/2})$$
$$G_1=(H_1)^T=(0.1294,0.2241,\underline{-0.8365},0.4830)$$
$$G_1=(H_1)^T=(0.0352,0.0854,-0.1350,\underline{-0.4599},0.8069,-0.3327)$$
$$G_1=(H_1)^T=(0.0106,0.0329,-0.0308,-0.1870,\underline{0.0280},0.6309,-0.7148,0.2304).$$

The corresponding frequency responses are shown in Fig. 6.3 for $N=1$ to $N=6$. It is seen that they get increasingly flatter as N increases. The frequency responses are now complex-valued, so their magnitudes are plotted.

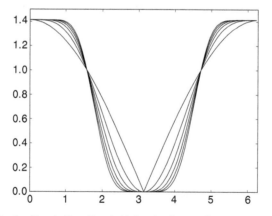

Fig. 6.3 The magnitudes $|\lambda_{G_0}(\omega)|=|\lambda_{H_0}(\omega)|$ for the first orthonormal wavelets

Clearly these filters are low-pass, and the high-pass responses are simply shifts of the low-pass responses. The frequency response also gets flatter near the high and low frequencies as N increases, and gets closer to an ideal low-pass/high-pass filter. One can verify that this is the case also when N is increased further.

The way we have defined the filters, one can show in the same way as in Exercise 5.16 that, when all filters have $2N$ coefficients, $\phi=\tilde{\phi}$ has support $[-N+1,N]$, $\psi=\tilde{\psi}$ has support $[-N + 1/2, N - 1/2]$ (i.e. the support of ψ is symmetric about the origin). In particular we have that

- for $N = 2$: $\mathrm{Supp}(\phi) = \mathrm{Supp}(\psi) = [-1, 2]$,
- for $N = 3$: $\mathrm{Supp}(\phi) = \mathrm{Supp}(\psi) = [-2, 3]$,
- for $N = 4$: $\mathrm{Supp}(\phi) = \mathrm{Supp}(\psi) = [-3, 4]$.

The scaling functions and mother wavelets are shown in Fig. 6.4. All functions have been plotted over $[-4, 4]$, so that all these support sizes can be verified. Also here we have used the cascade algorithm to approximate the functions.

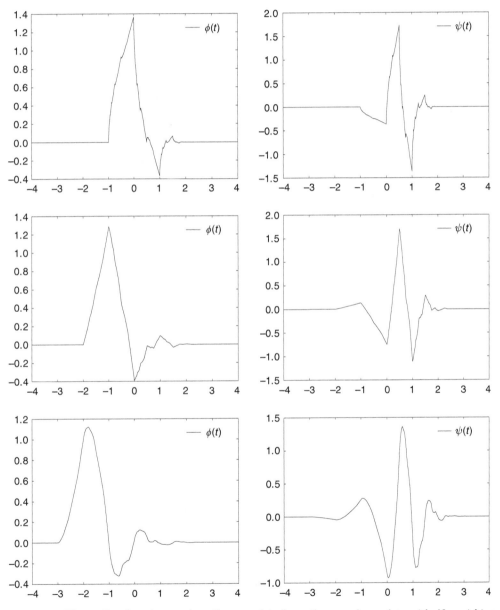

Fig. 6.4 The scaling functions and mother wavelets for orthonormal wavelets with N vanishing moments, for different values of N

6.6 Summary

We started this chapter by showing how scaling functions and mother wavelets could be constructed from filters. The starting point was the two-scale equations, expressing how the basis functions in ϕ_0 and ψ_0 should be expressed in terms of the basis functions in ϕ_1, using the filter coefficients. Taking continuous time Fourier transforms, the two-scale equations were transformed to a form where $\hat{\phi}$ and $\hat{\psi}$ were the unknowns, and this form turned out to be more suitable for analysis. The end result was that, starting with filters H_0, H_1, G_0, and G_1 satisfying certain conditions, and so that $GH = I$, we obtained functions ϕ, ψ, $\tilde{\phi}$, and $\tilde{\psi}$, which are valid scaling functions and mother wavelets and correspond to the wavelet transform and its dual, and which satisfied a biorthogonality relation.

We also took a closer look at some wavelets which are used in image compression, plotting their frequency responses, scaling functions, and mother wavelets. We could not find exact expressions for these functions, and depended therefore on the cascade algorithm to plot them. The importance of these functions are that they are particularly suited for approximation of regular functions, providing a compact representation of such functions which is localized in time. We also considered orthonormal wavelets with compact support. These were first constructed in [12]. We made the assumption that the scaling function and mother wavelet of these both have support $[-N + 1, N]$, which is a standard assumption in the literature. The general construction of (non-orthonormal) wavelets with compact support and vanishing moments can be found in [9]. To limit the scope of this book, we have skipped many things proved in these papers, in particular that the wavelet bases we construct actually are Riesz bases for $L^2(\mathbb{R})$.

We have mentioned previously that the common way to prove the sampling theorem in the literature is by using an argument involving both the CTFT and the DTFT. Having defined the CTFT, we can now elaborate on this. First rewrite the reconstruction formula (2.10) of the sampling theorem (Theorem 13) as

$$\sum_{k=-\infty}^{\infty} x_k \frac{\sin(\pi(t - kT_s)/T_s)}{\pi(t - kT_s)/T_s}, \tag{6.28}$$

where $x_k = f(kT_s)$. If we can show that the CTFT of this equals $\hat{f}$, the result would follow from the uniqueness of the CTFT. With $h(t) = \sin(\pi t/T_s)/(\pi t/T_s)$, using Exercise 6.3 the CTFT of (6.28) is

$$\hat{x}(\nu T_s)\hat{h}(\nu), \tag{6.29}$$

where $\hat{x}$ is a DTFT, while $\hat{f}$ and $\hat{h}$ are CTFT's. It is also known that the CTFT of f and the DTFT of its samples x are related by

$$\hat{x}(\nu T_s) = \frac{\sqrt{2\pi}}{T_s} \sum_k \hat{f}(\nu - 2\pi k/T_s),$$

and from Exercise 6.2 it follows that $\hat{h}(\nu) = \frac{T_s}{\sqrt{2\pi}}\chi_{[-\pi/T_s, \pi/T_s]}$. Inserting these equalities in (6.29) gives

$$\sum_k \hat{f}(\nu - 2\pi k/T_s)\chi_{[-\pi/T_s, \pi/T_s]}.$$

The assumption of the sampling theorem is that $\hat{f}(\nu) = 0$ for $|\nu| > \pi/T_s$. This means that only $k = 0$ contributes in the sum, so that the sum equals $\hat{f}(\nu)$. This completes the proof of the sampling theorem.

The first "non-trivial" family of orthonormal wavelets were discovered by Y. Meyer in 1985 [46], and were rather surprising at the time. These wavelets have later been coined *Meyer wavelets*, and are different from the orthogonal ones in this chapter in that they are compactly supported in frequency, rather than in time (i.e. $\hat{\phi}$ is compactly supported, while ϕ is not). The orthogonal wavelets from this chapter were discovered later by Daubechies [12], who together with A. Cohen and J. -C. Feauveau also laid the foundation for biorthogonal wavelets as we have described [9]. Y. Meyer has also collected some of his contributions to wavelets in two books [48, 50], while I. Daubechies has collected many of her contributions in [13]. Many other advanced texts on wavelets also exist, such as [4].

Chapter 7
The Polyphase Representation of Filter Bank Transforms

In Chap. 5 we expressed wavelet transforms, and more generally filter bank transforms, in terms of filters. Through this one obtains intuition on a how a wavelet transform splits the input using low-pass and high-pass filters, and how the filters in the MP3 standard split the input into frequency bands.

We have also seen that the transforms for the two piecewise linear wavelets could be factored in terms of what we called elementary lifting matrices. In this chapter we will see that such lifting factorizations apply more generally, that they reduce the number of arithmetic operations, and provide for in-place implementations. These are the main reasons why we apply lifting in our wavelet kernels.

An essential ingredient in obtaining a lifting factorization will be the concept of the *polyphase representation* of a matrix, which we now will define. This will not only be used to obtain lifting factorizations, but also as a general tool for filter bank transforms: In Sect. 7.3 we will see that the polyphase representation of M-channel filter bank transforms sheds some light on how one can construct orthogonal filter bank transforms. Note that, in Exercise 5.31 we found conditions which secure perfect reconstruction for M-channel filter bank transforms, but these were in terms of the frequency response. The considerations in Sect. 7.3 will mostly be in the time-domain, however. This makes it easier to address requirements such as the number of filter coefficients, for example, in order to limit the complexity.

Definition 1. *Polyphase Representation.*

Assume that S is a matrix. By the *polyphase components* of S we mean the matrices $S^{(i,j)}$ defined by $S^{(i,j)}_{r_1,r_2} = S_{i+r_1 M, j+r_2 M}$, i.e. the matrices obtained by taking every M'th component of S. The block matrix with entries $S^{(i,j)}$ is also called the *polyphase representation* of S.

As an example, the 6×6 MRA-matrix

$$S = \begin{pmatrix} 2\,3\,0\,0\,0\,1 \\ 4\,5\,6\,0\,0\,0 \\ 0\,1\,2\,3\,0\,0 \\ 0\,0\,4\,5\,6\,0 \\ 0\,0\,0\,1\,2\,3 \\ 6\,0\,0\,0\,4\,5 \end{pmatrix}. \tag{7.1}$$

© Springer Nature Switzerland AG 2019
Ø. Ryan, *Linear Algebra, Signal Processing, and Wavelets - A Unified Approach*,
Springer Undergraduate Texts in Mathematics and Technology,
https://doi.org/10.1007/978-3-030-01812-2_7

has polyphase representation

$$
\begin{pmatrix} S^{(0,0)} & S^{(0,1)} \\ S^{(1,0)} & S^{(1,1)} \end{pmatrix} = \left(\begin{array}{ccc|ccc} 2 & 0 & 0 & 3 & 0 & 1 \\ 0 & 2 & 0 & 1 & 3 & 0 \\ 0 & 0 & 2 & 0 & 1 & 3 \\ \hline 4 & 6 & 0 & 5 & 0 & 0 \\ 0 & 4 & 6 & 0 & 5 & 0 \\ 6 & 0 & 4 & 0 & 0 & 5 \end{array} \right).
$$

Since polyphase representations are block matrices, they can be multiplied and transposed as such (see Appendix A): For matrices A and B, with $A^{(i,j)}$, $B^{(i,j)}$ the corresponding polyphase components, we have that

- A^T has polyphase components $(A^{(j,i)})^T$,
- $C = AB$ has polyphase components $C^{(i,j)} = \sum_k A^{(i,k)} B^{(k,j)}$.

We will also refer to the polyphase components of a vector $\boldsymbol{x}$ as this was defined in Sect. 2.3, i.e. the p'th polyphase component of $\boldsymbol{x}$, denoted $\boldsymbol{x}^{(p)}$, is the vector with components x_{p+rM}. A matrix is in fact similar to its polyphase representation, through a permutation matrix which lists the polyphase components of the input in order. Due to this similarity we will use the same notation for the polyphase- and the filter representation: which one we use will usually be clear from the context.

Polyphase components can be defined for any matrix, but we will always restrict to filter bank transforms. The following is clear:

Theorem 2. The Polyphase Components Are Filters.

The polyphase components $S^{(i,j)}$ of a filter bank transform are filters, i.e. the polyphase representation is an $M \times M$-block matrix where all blocks are filters.

Also, the transpose, and composition of filter bank transforms are again filter bank transforms. The polyphase components $H^{(i,j)}$, $G^{(i,j)}$ of a filter bank transform are in general different from the filters $H_0, H_1, \ldots$ and $G_0, G_1, \ldots$, from the filter representation.

We will start this chapter by using the polyphase representation to show that some conditions for perfect reconstruction deduced in the previous chapter are in fact necessary. We then move to the lifting factorization, show that most MRA matrices have such a factorization, and find them for the wavelets constructed in Chap. 6 (for the CDF 9/7 wavelet, and for orthonormal wavelets). Finally we consider general M-channel filter bank transforms, and show how the polyphase representation can be used to dismantle the cosine-modulated filter bank transform from the MP3 standard.

7.1 The Polyphase Representation and Perfect Reconstruction

In Exercise 5.10 some conditions for perfect reconstruction were proved to be sufficient. We will now prove that, for FIR filters, they are also necessary. The result will be stated in polyphase form, and the proof of the equivalence with the statement from Exercise 5.10 is left as Exercise 7.1.

Theorem 3. Criteria for Perfect Reconstruction.

Let H_0, H_1, G_0, G_1 be FIR filters which give perfect reconstruction (i.e. $GH = I$), and let $H^{(i,j)}$ and $G^{(i,j)}$ be the polyphase components of H and G, respectively. Then

$$\begin{pmatrix} G^{(0,0)} & G^{(0,1)} \\ G^{(1,0)} & G^{(1,1)} \end{pmatrix} = \alpha E_{-2d} \begin{pmatrix} H^{(1,1)} & -H^{(0,1)} \\ -H^{(1,0)} & H^{(0,0)} \end{pmatrix}, \tag{7.2}$$

where the scalar α and the delay d are uniquely given by $G^{(0,0)}G^{(1,1)} - G^{(0,1)}G^{(1,0)} = \alpha E_{-d}$.

Proof. $GH = I$ can be stated in polyphase form as

$$\begin{pmatrix} G^{(0,0)} & G^{(0,1)} \\ G^{(1,0)} & G^{(1,1)} \end{pmatrix} \begin{pmatrix} H^{(0,0)} & H^{(0,1)} \\ H^{(1,0)} & H^{(1,1)} \end{pmatrix} = \begin{pmatrix} I & 0 \\ 0 & I \end{pmatrix}.$$

If we here multiply with $\begin{pmatrix} G^{(1,1)} & -G^{(0,1)} \\ -G^{(1,0)} & G^{(0,0)} \end{pmatrix}$ on both sides to the left, or with

$\begin{pmatrix} H^{(1,1)} & -H^{(0,1)} \\ -H^{(1,0)} & H^{(0,0)} \end{pmatrix}$ on both sides to the right, we get these two matrices on the right hand side, and on the left hand side we obtain the matrices

$$\begin{pmatrix} (G^{(0,0)}G^{(1,1)} - G^{(1,0)}G^{(0,1)})H^{(0,0)} & (G^{(0,0)}G^{(1,1)} - G^{(1,0)}G^{(0,1)})H^{(0,1)} \\ (G^{(0,0)}G^{(1,1)} - G^{(1,0)}G^{(0,1)})H^{(1,0)} & (G^{(0,0)}G^{(1,1)} - G^{(1,0)}G^{(0,1)})H^{(1,1)} \end{pmatrix}$$

$$\begin{pmatrix} (H^{(0,0)}H^{(1,1)} - H^{(1,0)}H^{(0,1)})G^{(0,0)} & (H^{(0,0)}H^{(1,1)} - H^{(1,0)}H^{(0,1)})G^{(0,1)} \\ (H^{(0,0)}H^{(1,1)} - H^{(1,0)}H^{(0,1)})G^{(1,0)} & (H^{(0,0)}H^{(1,1)} - H^{(1,0)}H^{(0,1)})G^{(1,1)} \end{pmatrix}$$

Now since $G^{(0,0)}G^{(1,1)} - G^{(1,0)}G^{(0,1)}$ and $H^{(0,0)}H^{(1,1)} - H^{(1,0)}H^{(0,1)}$ also are circulant Toeplitz matrices, the expressions above give that

$$l(H^{(0,0)}) \le l(G^{(1,1)}) \le l(H^{(0,0)})$$
$$l(H^{(0,1)}) \le l(G^{(0,1)}) \le l(H^{(0,1)})$$
$$l(H^{(1,0)}) \le l(G^{(1,0)}) \le l(H^{(1,0)}),$$

where $l(S)$ is the length of a filter as defined at the beginning of Chap. 3. This means that we must have equality here, and with both

$$G^{(0,0)}G^{(1,1)} - G^{(1,0)}G^{(0,1)} \text{ and } H^{(0,0)}H^{(1,1)} - H^{(1,0)}H^{(0,1)}$$

having only one nonzero diagonal. In particular we can define the diagonal matrix $D = G^{(0,0)}G^{(1,1)} - G^{(0,1)}G^{(1,0)}$ which is on the form αE_{-d}, for uniquely given α, d, We also obtain

$$\begin{pmatrix} G^{(1,1)} & -G^{(0,1)} \\ -G^{(1,0)} & G^{(0,0)} \end{pmatrix} = \begin{pmatrix} \alpha E_{-d} H^{(0,0)} & \alpha E_{-d} H^{(0,1)} \\ \alpha E_{-d} H^{(1,0)} & \alpha E_{-d} H^{(1,1)} \end{pmatrix},$$

which also can be written as

$$\begin{pmatrix} G^{(0,0)} & G^{(0,1)} \\ G^{(1,0)} & G^{(1,1)} \end{pmatrix} = \begin{pmatrix} \alpha E_{-d} H^{(1,1)} & -\alpha E_{-d} H^{(0,1)} \\ -\alpha E_{-d} H^{(1,0)} & \alpha E_{-d} H^{(0,0)} \end{pmatrix}$$

$$= \alpha E_{-2d} \begin{pmatrix} H^{(1,1)} & -H^{(0,1)} \\ -H^{(1,0)} & H^{(0,0)} \end{pmatrix}.$$

This completes the proof. $\square$

In Sect. 7.3 we will have use for the following version of this result, where perfect reconstruction also allows for a delay, and where the values α and d may be known due to a given form for H and G.

Theorem 4. *Assume that $H^{(i,j)}$ and $G^{(i,j)}$ are the polyphase components of forward and reverse filter bank transforms H and G, and assume that there exist α and d so that*

$$\begin{pmatrix} G^{(0,0)} & G^{(0,1)} \\ G^{(1,0)} & G^{(1,1)} \end{pmatrix} = \alpha E_{-2d} \begin{pmatrix} H^{(1,1)} & -H^{(0,1)} \\ -H^{(1,0)} & H^{(0,0)} \end{pmatrix}.$$

Then we always have alias cancellation. Moreover, if

$$H^{(0,0)} H^{(1,1)} - H^{(0,1)} H^{(1,0)} = \alpha^{-1} E_{d'}$$

for some other d', then we have that $GH = E_{2(d'-d)}$ (i.e. perfect reconstruction with delay $2(d' - d)$).

Proof. We compute

$$GH = \alpha E_{-2d} \begin{pmatrix} S & 0 \\ 0 & S \end{pmatrix},$$

with $S = H^{(0,0)} H^{(1,1)} - H^{(0,1)} H^{(1,0)}$. It follows from Exercise 7.3 that we have alias cancellation. If also $S = \alpha^{-1} E_{d'}$ for some other d', we get

$$\begin{pmatrix} S & 0 \\ 0 & S \end{pmatrix} = \alpha^{-1} E_{2d'},$$

so that $GH = \alpha E_{-2d} \alpha^{-1} E_{2d'} = E_{2(d'-d)}$. $\square$

The case when $H^{(0,0)} H^{(1,1)} - H^{(0,1)} H^{(1,0)}$ only is approximately $\alpha^{-1} E_{d'}$ is also of practical interest. Such approximate cases may be easier to construct, and will provide "near perfect reconstruction"; With $S = \{\ldots, t_{-1}, \underline{t_0}, t_1, \ldots\}$ we have that

$$\begin{pmatrix} S & 0 \\ 0 & S \end{pmatrix} = \{\ldots, t_{-1}, 0, \underline{t_0}, 0, t_1, \ldots\} \approx \alpha^{-1} E_{2d'},$$

and the frequency response of this is $\sum_k t_k e^{-i2k\omega}$. If the magnitude of this is close to constant, we have near perfect reconstruction.

Exercise 7.1: The Frequency Responses of the Polyphase Components

Let H and G be 2-channel forward and reverse filter bank transforms, with filters H_0, H_1, G_0, G_1, and polyphase components $H^{(i,j)}$, $G^{(i,j)}$.

a) Show that

$$\lambda_{H_0}(\omega) = \lambda_{H^{(0,0)}}(2\omega) + e^{i\omega} \lambda_{H^{(0,1)}}(2\omega)$$
$$\lambda_{H_1}(\omega) = \lambda_{H^{(1,1)}}(2\omega) + e^{-i\omega} \lambda_{H^{(1,0)}}(2\omega)$$

$$\lambda_{G_0}(\omega) = \lambda_{G^{(0,0)}}(2\omega) + e^{-i\omega}\lambda_{G^{(1,0)}}(2\omega)$$
$$\lambda_{G_1}(\omega) = \lambda_{G^{(1,1)}}(2\omega) + e^{i\omega}\lambda_{G^{(0,1)}}(2\omega).$$

b) In the proof of the last part of Exercise 5.10, we deferred the last part, namely that

$$\lambda_{H_1}(\omega) = \alpha^{-1}e^{-2id\omega}\lambda_{G_0}(\omega + \pi)$$
$$\lambda_{G_1}(\omega) = \alpha e^{2id\omega}\lambda_{H_0}(\omega + \pi).$$

follow from

$$\begin{pmatrix} G^{(0,0)} & G^{(0,1)} \\ G^{(1,0)} & G^{(1,1)} \end{pmatrix} = \begin{pmatrix} \alpha E_{-d}H^{(1,1)} & -\alpha E_{-d}H^{(0,1)} \\ -\alpha E_{-d}H^{(1,0)} & \alpha E_{-d}H^{(0,0)} \end{pmatrix}.$$

Prove this based on the result from a).

Exercise 7.2: Polyphase View of Time Delay

a) Write down the polyphase representation for an arbitrary delay E_d, and for any given M.

b) Assume that d is a multiple of M. Show that E_d commutes with any M-channel filter bank transform. Will this be the case if d is not a multiple of M?

Exercise 7.3: Polyphase View of Alias Cancellation

How we can spot alias cancellation in terms of the polyphase representation will be useful to us in the next section when we explain why the filter bank transform of the MP3 standard provides such cancellation. We will there end up with a polyphase representation on the form given in a) of this exercise.

a) Assume that the polyphase representation of S is a $k \times k$-block diagonal matrix where the diagonal elements are $M \times M$-filters which are all equal. Show that S is a $(kM) \times (kM)$-filter. Thus, if GH has this form for forward and reverse filter bank transforms H and G, we have alias cancellation.

b) If S is a general $(kM) \times (kM)$-filter, give a general description of its polyphase representation.

Exercise 7.4: Polyphase Components for Symmetric Filters

Assume that the filters H_0, H_1 of a forward filter bank transform are symmetric, and let $H^{(i,j)}$ be the corresponding polyphase components. Show that

- $H^{(0,0)}$ and $H^{(1,1)}$ are symmetric filters,
- the filter coefficients of $H^{(1,0)}$ have symmetry about $-1/2$,
- the filter coefficients of $H^{(0,1)}$ have symmetry about $1/2$.

Also show a similar statement for reverse filter bank transforms.

Exercise 7.5: Classical QMF Filter Banks

Recall from Exercise 5.12 that we defined a classical QMF filter bank as one where $M = 2$, $G_0 = H_0$, $G_1 = H_1$, and $\lambda_{H_1}(\omega) = \lambda_{H_0}(\omega + \pi)$. Show that the forward and reverse filter bank transforms of a classical QMF filter bank have polyphase representations of the form

$$H = G = \begin{pmatrix} A & -E_1 B \\ B & A \end{pmatrix}.$$

Exercise 7.6: Alternative QMF Filter Banks

Recall from Exercise 5.13 that we defined an alternative QMF filter bank as one where $M = 2$, $G_0 = (H_0)^T$, $G_1 = (H_1)^T$, and $\lambda_{H_1}(\omega) = \overline{\lambda_{H_0}(\omega + \pi)}$.

a) Show that the forward and reverse filter bank transforms of an alternative QMF filter bank have polyphase representations of the form

$$H = \begin{pmatrix} A^T & B^T \\ -B & A \end{pmatrix} \qquad G = \begin{pmatrix} A & -B^T \\ B & A^T \end{pmatrix} = \begin{pmatrix} A^T & B^T \\ -B & A \end{pmatrix}^T.$$

b) Show that A and B give rise to an alternative QMF filter bank with perfect reconstruction as above if and only if $A^T A + B^T B = I$.

c) Consider alternative QMF filter banks where we use an additional sign, so that $\lambda_{H_1}(\omega) = -\overline{\lambda_{H_0}(\omega + \pi)}$ (the Haar wavelet was an example of such a filter bank, see Exercise 5.15). Show that such forward and reverse filter bank transforms have polyphase representations of the form

$$H = \begin{pmatrix} A^T & B^T \\ B & -A \end{pmatrix} \qquad G = \begin{pmatrix} A & B^T \\ B & -A^T \end{pmatrix} = \begin{pmatrix} A^T & B^T \\ B & -A \end{pmatrix}^T.$$

This sign change does not substantially change the properties of alternative QMF filter banks, so that we will also call these new filter banks alternative QMF filter banks.

7.2 The Polyphase Representation and the Lifting Factorization

Let us now turn to how the polyphase representation sheds light on the lifting factorization. We define *lifting matrices of even and odd types* by having polyphase representations

$$\begin{pmatrix} I & S \\ 0 & I \end{pmatrix} \text{ and } \begin{pmatrix} I & 0 \\ S & I \end{pmatrix},$$

respectively. It is straightforward to check that this is a generalization of the previous definitions of elementary lifting matrices of even and odd type (see Eqs. (4.36) and (4.28), since the polyphase representations of these are, using compact filter notation

$$\begin{pmatrix} I & \lambda_i\{\underline{1},1\} \\ 0 & I \end{pmatrix} \text{ and } \begin{pmatrix} I & 0 \\ \lambda_i\{1,\underline{1}\} & I \end{pmatrix},$$

respectively. Just as for elementary lifting matrices, general lifting matrices are easily inverted as well, as one can easily check that

$$\begin{pmatrix} I & S \\ 0 & I \end{pmatrix}^{-1} = \begin{pmatrix} I & -S \\ 0 & I \end{pmatrix}, \text{ and } \begin{pmatrix} I & 0 \\ S & I \end{pmatrix}^{-1} = \begin{pmatrix} I & 0 \\ -S & I \end{pmatrix}$$

Just as multiplication with elementary lifting matrices can be computed in-place (see Sect. 4.3), the same is true for general lifting matrices also. Also

$$\begin{pmatrix} I & S \\ 0 & I \end{pmatrix}^T = \begin{pmatrix} I & 0 \\ S^T & I \end{pmatrix}, \text{ and } \begin{pmatrix} I & 0 \\ S & I \end{pmatrix}^T = \begin{pmatrix} I & S^T \\ 0 & I \end{pmatrix},$$

and

$$\begin{pmatrix} I & S_1 \\ 0 & I \end{pmatrix}\begin{pmatrix} I & S_2 \\ 0 & I \end{pmatrix} = \begin{pmatrix} I & S_1 + S_2 \\ 0 & I \end{pmatrix}$$

$$\begin{pmatrix} I & 0 \\ S_1 & I \end{pmatrix}\begin{pmatrix} I & 0 \\ S_2 & I \end{pmatrix} = \begin{pmatrix} I & 0 \\ S_1 + S_2 & I \end{pmatrix},$$

Due to this one can assume that odd and even types of lifting matrices appear in alternating order in a lifting factorization, since matrices of the same type can be grouped together. The following result states that any non-singular MRA matrix can be factored into a product of lifting matrices.

Theorem 5. Factorization in Terms of Lifting Matrices.
Any non-singular MRA matrix $S = \begin{pmatrix} S^{(0,0)} & S^{(0,1)} \\ S^{(1,0)} & S^{(1,1)} \end{pmatrix}$ can be written on the form

$$S = \Lambda_1 \cdots \Lambda_n \begin{pmatrix} \alpha_0 E_p & 0 \\ 0 & \alpha_1 E_q \end{pmatrix}, \tag{7.3}$$

where Λ_i are lifting matrices, p, q are integers, α_0, α_1 are nonzero scalars, and E_p, E_q are time delays. The inverse is given by

$$S^{-1} = \begin{pmatrix} \alpha_0^{-1} E_{-p} & 0 \\ 0 & \alpha_1^{-1} E_{-q} \end{pmatrix}(\Lambda_n)^{-1} \cdots (\Lambda_1)^{-1}. \tag{7.4}$$

Factorizations on the form (7.3) will be called *lifting factorizations*. From the proof below it will be see that a lifting factorization is far from unique.

Proof. Assume that S is non-singular. We will incrementally find lifting matrices Λ_i, with filter S_i in the lower left/upper right corner, so that $\Lambda_i S$ has filters of lower length in the first column. Assume first that $l(S^{(0,0)}) \geq l(S^{(1,0)})$. If Λ_i is of even type, then the first column in $\Lambda_i S$ is

$$\begin{pmatrix} I & S_i \\ 0 & I \end{pmatrix}\begin{pmatrix} S^{(0,0)} \\ S^{(1,0)} \end{pmatrix} = \begin{pmatrix} S^{(0,0)} + S_i S^{(1,0)} \\ S^{(1,0)} \end{pmatrix}. \tag{7.5}$$

S_i can now be chosen so that $l(S^{(0,0)} + S_i S^{(1,0)}) < l(S^{(1,0)})$. To see how, recall that we in Sect. 3.2 stated that multiplying filters corresponds to multiplying polynomials. S_i can thus be found from polynomial division with remainder, by first finding polynomials Q and P with $l(P) < l(S^{(1,0)})$ so that $S^{(0,0)} = QS^{(1,0)} + P$, and so that the length of P is less than $l(S^{(1,0)})$ (finally set $S_i = -Q$). The same can be said if Λ_i is of odd type, in which case the first and second components are simply swapped. This procedure can be continued until we arrive at a product

$$\Lambda_n \cdots \Lambda_1 S$$

where either the first or the second component in the first column is 0. If the first component in the first column is 0, the identity

$$\begin{pmatrix} I & 0 \\ -I & I \end{pmatrix} \begin{pmatrix} I & I \\ 0 & I \end{pmatrix} \begin{pmatrix} 0 & X \\ Y & Z \end{pmatrix} = \begin{pmatrix} Y & X+Z \\ 0 & -X \end{pmatrix}$$

explains that we can bring the matrix to a form where the second element in the first column is zero instead, with the help of the additional lifting matrices

$$\Lambda_{n+1} = \begin{pmatrix} I & I \\ 0 & I \end{pmatrix} \text{ and } \Lambda_{n+2} = \begin{pmatrix} I & 0 \\ -I & I \end{pmatrix},$$

so that we always can assume that the second element in the first column is 0, i.e.

$$\Lambda_n \cdots \Lambda_1 S = \begin{pmatrix} P & Q \\ 0 & R \end{pmatrix},$$

for some matrices P, Q, R. In the proof of Theorem 3 we stated that, for S to be non-singular, we must have that $S^{(0,0)}S^{(1,1)} - S^{(0,1)}S^{(1,0)} = -\alpha^{-1}E_d$ for some nonzero scalar α and integer d. Since

$$\begin{pmatrix} P & Q \\ 0 & R \end{pmatrix}$$

is also non-singular, we must thus have that PR must be on the form αE_n. When the filters have a finite number of filter coefficients, the only possibility for this to happen is when $P = \alpha_0 E_p$ and $R = \alpha_1 E_q$ for some p, q, α_0, α_1. Using this, and also isolating S on one side, we obtain that

$$S = (\Lambda_1)^{-1} \cdots (\Lambda_n)^{-1} \begin{pmatrix} \alpha_0 E_p & Q \\ 0 & \alpha_1 E_q \end{pmatrix}, \tag{7.6}$$

Noting that

$$\begin{pmatrix} \alpha_0 E_p & Q \\ 0 & \alpha_1 E_q \end{pmatrix} = \begin{pmatrix} 1 & \frac{1}{\alpha_1} E_{-q} Q \\ 0 & 1 \end{pmatrix} \begin{pmatrix} \alpha_0 E_p & 0 \\ 0 & \alpha_1 E_q \end{pmatrix},$$

we can rewrite Eq. (7.6) as

$$S = (\Lambda_1)^{-1} \cdots (\Lambda_n)^{-1} \begin{pmatrix} 1 & \frac{1}{\alpha_1} E_{-q} Q \\ 0 & 1 \end{pmatrix} \begin{pmatrix} \alpha_0 E_p & 0 \\ 0 & \alpha_1 E_q \end{pmatrix},$$

which is the lifting factorization we stated. The last matrix in the lifting factorization is not really an lifting matrix, but it too can easily be inverted, so that we arrive at Eq. (7.4). This completes the proof.   □

Assume that we have applied Theorem 5 in order to get a factorization of the MRA matrix H on the form

$$\Lambda_n \cdots \Lambda_2 \Lambda_1 H = \begin{pmatrix} \alpha & \mathbf{0} \\ \mathbf{0} & \beta \end{pmatrix}. \tag{7.7}$$

We now obtain the following factorizations.

$$H = (\Lambda_1)^{-1}(\Lambda_2)^{-1} \cdots (\Lambda_n)^{-1} \begin{pmatrix} \alpha & \mathbf{0} \\ \mathbf{0} & \beta \end{pmatrix} \tag{7.8}$$

$$G = \begin{pmatrix} 1/\alpha & \mathbf{0} \\ \mathbf{0} & 1/\beta \end{pmatrix} \Lambda_n \cdots \Lambda_2 \Lambda_1 \tag{7.9}$$

$$H^T = \begin{pmatrix} \alpha & \mathbf{0} \\ \mathbf{0} & \beta \end{pmatrix} ((\Lambda_n)^{-1})^T ((\Lambda_{n-1})^{-1})^T \cdots ((\Lambda_1)^{-1})^T \tag{7.10}$$

$$G^T = (\Lambda_1)^T (\Lambda_2)^T \cdots (\Lambda_n)^T \begin{pmatrix} 1/\alpha & \mathbf{0} \\ \mathbf{0} & 1/\beta \end{pmatrix}. \tag{7.11}$$

Since H^T and G^T are the kernel transformations of the dual IDWT/DWT, these formulas give us lifting-based recipes for computing the DWT, IDWT, dual IDWT, and the dual DWT, respectively.

It is desirable to have a lifting factorization where the lifting steps are as simple as possible. In the case of symmetric filters, we obtained a factorization in terms of elementary lifting matrices for both piecewise linear wavelets. It turns out that this is the case also for many other wavelets with symmetric filters. Symmetric filters mean that (see Exercise 7.4)

- $S^{(0,0)}$ and $S^{(1,1)}$ are symmetric,
- that $S^{(1,0)}$ is symmetric about $-1/2$, and
- that $S^{(0,1)}$ is symmetric about $1/2$.

Assume that we in the proof of Theorem 5 add an elementary lifting of even type. At this step we then compute $S^{(0,0)} + S_i S^{(1,0)}$ in the first entry of the first column. If $S^{(0,0)}$ is assumed symmetric, $S_i S^{(1,0)}$ must also be symmetric in order for the length to be reduced. And since the filter coefficients of $S^{(1,0)}$ are assumed symmetric about $-1/2$, S_i must be chosen with symmetry around $1/2$. It is also clear that any elementary lifting reduces the highest degree in the first column by at least one, and increases the lowest degree with at least one, while preserving symmetry.

For most our wavelets, it turns out that the filters in the first column differ in the number of filter coefficients by 1, and that this is also the case after any number of elementary lifting steps. It is then not too hard to see that, for some numbers λ_i,

- all even lifting steps have $S_i = \lambda_i\{\underline{1}, 1\}$ (i.e. the step is on the form A_{λ_i}),
- all odd lifting steps have $S_i = \lambda_i\{1, \underline{1}\}$ (i.e. the step is on the form B_{λ_i}).

Such wavelets can thus be implemented in terms of elementary lifting steps only. Since any elementary lifting can be computed in-place, this means that the entire DWT and IDWT can be computed in-place. Lifting thus provides us with a complete implementation strategy for the DWT and IDWT, where the λ_i can be pre-computed.

Clearly an elementary lifting matrix is also an MRA-matrix with symmetric filters, so that our procedure factorizes an MRA-matrix with symmetric filters into simpler MRA-matrices, also with symmetric filters.

7.2.1 Reduction in the Number of Arithmetic Operations

The number of arithmetic operations needed to apply elementary lifting matrices is easily computed. The number of multiplications is $N/2$ if symmetry is exploited as in Observation 17 (N if symmetry is not exploited). Similarly, the number of additions is N. Let K be the total number of filter coefficients in H_0, H_1. It is not too difficult to see that each lifting step reduces the number of filter coefficients in the MRA matrix by minimum 4, so that a total number of $K/4$ lifting steps are required. Thus, a total number of $KN/8$ ($KN/4$) multiplications, and $KN/4$ additions are required when a lifting factorization is used. In comparison, a direct implementation would require $KN/4$ ($KN/2$) multiplications, and $KN/2$ additions. We therefore have the following result.

Theorem 6. Reducing Arithmetic Operations.
A lifting factorization approximately halves the number of additions and multiplications needed, when compared with a direct implementation (regardless of whether symmetry is exploited or not).

The possibility of computing the DWT and IDWT in-place is perhaps even more important than the reduction in the number of arithmetic operations.

Let us now find the lifting factorizations of the wavelets we have considered, and include these in our DWT implementations. We will omit the Haar wavelet. One can easily write down a lifting factorization for this as well, but there is little to save in this factorization when comparing to the direct implementation we already have.

7.2.2 The Piecewise Linear Wavelet

For the piecewise linear wavelets we computed the lifting factorizations in Chap. 4. For the first piecewise linear wavelet, the corresponding polyphase representations of H and G are

$$\sqrt{2} \begin{pmatrix} I & 0 \\ -\frac{1}{2}\{1,\underline{1}\} & I \end{pmatrix} \text{ and } \frac{1}{\sqrt{2}} \begin{pmatrix} I & 0 \\ \frac{1}{2}\{1,\underline{1}\} & I \end{pmatrix}, \tag{7.12}$$

respectively. For the alternative piecewise linear wavelet the polyphase representation of H is

$$\sqrt{2} \begin{pmatrix} I & \frac{1}{4}\{\underline{1},1\} \\ 0 & I \end{pmatrix} \begin{pmatrix} I & 0 \\ -\frac{1}{2}\{1,\underline{1}\} & I \end{pmatrix}.$$

and that of G is

$$\frac{1}{\sqrt{2}} \begin{pmatrix} I & 0 \\ \frac{1}{2}\{1,\underline{1}\} & I \end{pmatrix} \begin{pmatrix} I & -\frac{1}{4}\{\underline{1},1\} \\ 0 & I \end{pmatrix}. \tag{7.13}$$

7.2.3 The Spline 5/3 Wavelet

Let us consider the Spline 5/3 wavelet, defined in Example 6.3.1. Recall that

$$H_0 = \left\{ -\frac{1}{4}, \frac{1}{2}, \underline{\frac{3}{2}}, \frac{1}{2}, -\frac{1}{4} \right\} \qquad H_1 = \left\{ -\frac{1}{4}, \underline{\frac{1}{2}}, -\frac{1}{4} \right\},$$

from which we see that the polyphase components of H are

$$\begin{pmatrix} H^{(0,0)} & H^{(0,1)} \\ H^{(1,0)} & H^{(1,1)} \end{pmatrix} = \begin{pmatrix} \{-\frac{1}{4}, \frac{3}{2}, -\frac{1}{4}\} & \frac{1}{2}\{\underline{1}, 1\} \\ -\frac{1}{4}\{1, \underline{1}\} & \frac{1}{2}I \end{pmatrix}$$

We see here that the upper filter has the largest length in the first column, so that we must start with an elementary lifting of even type. We need to find a filter S_1 so that $S_1\{-1/4, -1/4\} + \{-1/4, 3/2, -1/4\}$ has fewer filter coefficients than $\{-1/4, 3/2, -1/4\}$. It is clear that we can choose $S_1 = \{\underline{-1}, -1\}$, and that

$$\Lambda_1 H = \begin{pmatrix} I & \{\underline{-1}, -1\} \\ \mathbf{0} & I \end{pmatrix} \begin{pmatrix} \{-\frac{1}{4}, \frac{3}{2}, -\frac{1}{4}\} & \frac{1}{2}\{\underline{1}, 1\} \\ -\frac{1}{4}\{1, \underline{1}\} & \frac{1}{2}I \end{pmatrix} = \begin{pmatrix} 2I & 0 \\ -\frac{1}{4}\{1, \underline{1}\} & \frac{1}{2}I \end{pmatrix}$$

Now we need to apply an elementary lifting of odd type, and we need to find a filter S_2 so that $S_2 2I - \frac{1}{4}\{1, \underline{1}\} = \mathbf{0}$. Clearly we can choose $S_2 = \{1/8, \underline{1/8}\}$, and we get

$$\Lambda_2 \Lambda_1 H = \begin{pmatrix} I & \mathbf{0} \\ \frac{1}{8}\{1, \underline{1}\} & I \end{pmatrix} \begin{pmatrix} 2I & 0 \\ -\frac{1}{4}\{1, \underline{1}\} & \frac{1}{2}I \end{pmatrix} = \begin{pmatrix} 2I & 0 \\ 0 & \frac{1}{2}I \end{pmatrix}.$$

We now obtain that the polyphase representations for the Spline 5/3 wavelet are

$$H = \begin{pmatrix} I & \{\underline{1}, 1\} \\ \mathbf{0} & I \end{pmatrix} \begin{pmatrix} I & \mathbf{0} \\ -\frac{1}{8}\{1, \underline{1}\} & I \end{pmatrix} \begin{pmatrix} 2I & 0 \\ 0 & \frac{1}{2}I \end{pmatrix}$$

and

$$G = \begin{pmatrix} \frac{1}{2}I & 0 \\ 0 & 2I \end{pmatrix} \begin{pmatrix} I & \mathbf{0} \\ \frac{1}{8}\{1, \underline{1}\} & I \end{pmatrix} \begin{pmatrix} I & \{\underline{-1}, -1\} \\ \mathbf{0} & I \end{pmatrix},$$

respectively. Two lifting steps are thus required. We also see that the lifting steps involve only dyadic fractions, just as the filter coefficients themselves. This means that the lifting steps can be computed without loss of precision on a computer

7.2.4 The CDF 9/7 Wavelet

For the CDF 9/7 wavelet from Example 6.11 it is more cumbersome to compute the lifting factorization by hand. It is however, straightforward to write an algorithm which computes the lifting steps, as these are found in the proof of Theorem 5. You will be spared the details of this algorithm. Also, when we use this wavelet in implementations later they will use pre-computed values of these lifting steps, and you can take these implementations for granted too. If we run the algorithm for computing the lifting factorization we obtain that the polyphase representations of H and G are

$$\begin{pmatrix} I & 0.5861\{\underline{1},1\} \\ 0 & I \end{pmatrix} \begin{pmatrix} I & 0 \\ 0.6681\{1,\underline{1}\} & I \end{pmatrix} \begin{pmatrix} I & -0.0700\{\underline{1},1\} \\ 0 & I \end{pmatrix}$$

$$\times \begin{pmatrix} I & 0 \\ -1.2002\{1,\underline{1}\} & I \end{pmatrix} \begin{pmatrix} -1.1496 & 0 \\ 0 & -0.8699 \end{pmatrix} \text{ and}$$

$$\begin{pmatrix} -0.8699 & 0 \\ 0 & -1.1496 \end{pmatrix} \begin{pmatrix} I & 0 \\ 1.2002\{1,\underline{1}\} & I \end{pmatrix} \begin{pmatrix} I & 0.0700\{\underline{1},1\} \\ 0 & I \end{pmatrix}$$

$$\times \begin{pmatrix} I & 0 \\ -0.6681\{1,\underline{1}\} & I \end{pmatrix} \begin{pmatrix} I & -0.5861\{\underline{1},1\} \\ 0 & I \end{pmatrix},$$

respectively. In this case four lifting steps were required.

7.2.5 Orthonormal Wavelets

Finally we will find a lifting factorization for orthonormal wavelets. Note that here the filters H_0 and H_1 are not symmetric, and each of them has an even number of filter coefficients. There are thus a different number of filter coefficients with positive and negative indices, and in Sect. 6.5 we defined the filters so that they were as symmetric as possible regarding the number of nonzero filter coefficients with positive and negative indices.

We will attempt to construct a lifting factorization where the following property is preserved after each lifting step:

P1: $H^{(0,0)}$, $H^{(1,0)}$ have a minimum possible overweight of filter coefficients with negative indices.

This property stems from the assumption in Sect. 6.5 that H_0 is assumed to have a minimum possible overweight of filter coefficients with negative indices. To see that P1 holds at the start, assume as before that all the filters have $2L$ nonzero filter coefficients, so that H_0 and H_1 are on the form given by Eq. (6.27). Assume first that L is even it is clear that

$$H^{(0,0)} = \{t_{-L}, \ldots, t_{-2}, \underline{t_0}, t_2, \ldots, t_{L-2}\}$$
$$H^{(0,1)} = \{t_{-L+1}, \ldots, t_{-3}, \underline{t_{-1}}, t_1, \ldots, t_{L-1}\}$$
$$H^{(1,0)} = \{s_{-L+1}, \ldots, s_{-1}, \underline{s_1}, s_3, \ldots, s_{L-1}\}$$
$$H^{(1,1)} = \{s_{-L+2}, \ldots, s_{-2}, \underline{s_0}, s_2, \ldots, s_L\}.$$

from which clearly P1 holds. Assuming that L is odd it is clear that

$$H^{(0,0)} = \{t_{-L+1}, \ldots, t_{-2}, \underline{t_0}, t_2, \ldots, t_{L-1}\}$$
$$H^{(0,1)} = \{t_{-L}, \ldots, t_{-3}, \underline{t_{-1}}, t_1, \ldots, t_{L-2}\}$$
$$H^{(1,0)} = \{s_{-L+2}, \ldots, s_{-1}, \underline{s_1}, s_3, \ldots, s_L\}$$
$$H^{(1,1)} = \{s_{-L+1}, \ldots, s_{-2}, \underline{s_0}, s_2, \ldots, s_{L-1}\}.$$

Thus all filters have equally many filter coefficients with positive and negative indices, so that P1 holds also here.

Now let us turn to the first lifting step. We will choose it so that the number of filter coefficients in the first column is reduced with 1, and so that $H^{(0,0)}$ obtains an odd number of coefficients. We split this into the case when L is even, or odd.

If L is even, $H^{(0,0)}$ and $H^{(1,0)}$ have an even number of coefficients, so that the first lifting step must be even. To preserve P1, we must cancel t_{-L}, so that the first lifting step is

$$\Lambda_1 = \begin{pmatrix} I & -t_{-L}/s_{-L+1} \\ 0 & I \end{pmatrix}.$$

If L is odd, we saw that $H^{(0,0)}$ and $H^{(1,0)}$ had an odd number of coefficients, so that the first lifting step must be odd. To preserve P1, we must cancel s_L, so that the first lifting step is

$$\Lambda_1 = \begin{pmatrix} I & 0 \\ -s_L/t_{L-1} & I \end{pmatrix}.$$

After this first lifting step we have a difference of one filter coefficient between the entries in the first column, and we want reduce the entry with the most filter coefficients by two with a lifting step, until we have $H^{(0,0)} = \{\underline{K}\}$, $H^{(1,0)} = 0$ in the first column. We also split this into two cases, depending on whether When $H^{(0,0)}$ or $H^{(1,0)}$ has the most filter coefficients.

Assume first that $H^{(0,0)}$ has the most filter coefficients. We then need to apply an even lifting step. Before an even step, the first column has the form

$$\begin{pmatrix} \{t_{-k}, \ldots, t_{-1}, \underline{t_0}, t_1, \ldots, t_k\} \\ \{s_{-k}, \ldots, s_{-1}, \underline{s_0}, s_1, \ldots, s_{k-1}\} \end{pmatrix},$$

so that we can choose $\Lambda_i = \begin{pmatrix} I & \{\underline{-t_{-k}/s_{-k}}, -t_k/s_{k-1}\} \\ 0 & I \end{pmatrix}$ as a lifting step.

Assume then that $H^{(1,0)}$ has the most filter coefficients. We then need to apply an odd lifting step. Before an odd step, the first column has the form

$$\begin{pmatrix} \{t_{-k}, \ldots, t_{-1}, \underline{t_0}, t_1, \ldots, t_k\} \\ \{s_{-k-1}, \ldots, s_{-1}, \underline{s_0}, s_1, \ldots, s_k\} \end{pmatrix},$$

so that we can choose $\Lambda_i = \begin{pmatrix} I & 0 \\ \{-s_{-k-1}/t_{-k}, \underline{-s_k/t_k}\} & I \end{pmatrix}$ as a lifting step.

If L is even we end up with a matrix on the form $\begin{pmatrix} \alpha & \{\underline{0}, K\} \\ 0 & \beta \end{pmatrix}$, and we can choose the final lifting step as $\Lambda_n = \begin{pmatrix} I & \{\underline{0}, -K/\beta\} \\ 0 & I \end{pmatrix}$.

If L is odd we end up with a matrix on the form

$$\begin{pmatrix} \alpha & K \\ 0 & \beta \end{pmatrix},$$

and we can choose the final lifting step as $\Lambda_n = \begin{pmatrix} I & -K/\beta \\ 0 & I \end{pmatrix}$. Again using Eqs. (7.8)–(7.9), this gives us the lifting factorizations.

In summary we see that all even and odd lifting steps take the form $\begin{pmatrix} I & \{\underline{\lambda_1}, \lambda_2\} \\ 0 & I \end{pmatrix}$ and $\begin{pmatrix} I & 0 \\ \{\lambda_1, \underline{\lambda_2}\} & I \end{pmatrix}$. Symmetric lifting steps correspond to the special case when $\lambda_1 = \lambda_2$. The even and odd lifting matrices now used are

$$\begin{pmatrix} 1 & \lambda_1 & 0 & 0 & \cdots & 0 & 0 & \lambda_2 \\ 0 & 1 & 0 & 0 & \cdots & 0 & 0 & 0 \\ 0 & \lambda_2 & 1 & \lambda_1 & \cdots & 0 & 0 & 0 \\ \vdots & \vdots & \vdots & \vdots & \ddots & \vdots & \vdots & \vdots \\ 0 & 0 & 0 & 0 & \cdots & \lambda_2 & 1 & \lambda_1 \\ 0 & 0 & 0 & 0 & \cdots & 0 & 0 & 1 \end{pmatrix} \quad \text{and} \quad \begin{pmatrix} 1 & 0 & 0 & 0 & \cdots & 0 & 0 & 0 \\ \lambda_2 & 1 & \lambda_1 & 0 & \cdots & 0 & 0 & 0 \\ 0 & 0 & 1 & 0 & \cdots & 0 & 0 & 0 \\ \vdots & \vdots & \vdots & \vdots & \ddots & \vdots & \vdots & \vdots \\ 0 & 0 & 0 & 0 & \cdots & 0 & 1 & 0 \\ \lambda_1 & 0 & 0 & 0 & \cdots & 0 & \lambda_2 & 1 \end{pmatrix}, \qquad (7.14)$$

respectively. We note that when we reduce elements to the left and right in the upper and lower part of the first column, the same type of reductions must occur in the second column, since the determinant $H^{(0,0)}H^{(1,1)} - H(0,1)H^{(1,0)}$ is a constant after any number of lifting steps. In the exercises, you will be asked to implement both these non-symmetric elementary lifting steps.

This example explains the procedure for finding the lifting factorization into steps of the form given in Eq. (7.14). You will be spared the details of writing an implementation which applies this procedure.

The function `find_kernel`, which is used to extract the kernel used for orthonormal wavelets, calls a function `lifting_fact_ortho`, which follows the procedure sketched here to find the lifting factorization for an orthonormal wavelet with N vanishing moments, provided that the filter coefficients are known. The values in the lifting factorization are written to file, and read when needed.

Exercise 7.7: View of Lifting as Altering the Filters

This exercise will show that one can think of the steps in a lifting factorization as altering the low-pass/high-pass filters in alternating order. Let S be a general filter.

a) Show that

$$G \begin{pmatrix} I & 0 \\ S & I \end{pmatrix}$$

is an MRA matrix with filters $\hat{G}_0, G_1$, where

$$\lambda_{\hat{G}_0}(\omega) = \lambda_{G_0}(\omega) + \lambda_S(2\omega)e^{-i\omega}\lambda_{G_1}(\omega),$$

b) Show that

$$G \begin{pmatrix} I & S \\ 0 & I \end{pmatrix}$$

is an MRA matrix with filters $G_0, \hat{G}_1$, where

$$\lambda_{\hat{G}_1}(\omega) = \lambda_{G_1}(\omega) + \lambda_S(2\omega)e^{i\omega}\lambda_{G_0}(\omega),$$

c) Show that

$$\begin{pmatrix} I & 0 \\ S & I \end{pmatrix} H$$

is an MRA-matrix with filters $H_0, \hat{H}_1$, where

$$\lambda_{\hat{H}_1}(\omega) = \lambda_{H_1}(\omega) + \lambda_S(2\omega)e^{-i\omega}\lambda_{H_0}(\omega).$$

d)

$$\begin{pmatrix} I & S \\ 0 & I \end{pmatrix} H$$

is an MRA-matrix with filters $\hat{H}_0, H_1$, where

$$\lambda_{\hat{H}_0}(\omega) = \lambda_{H_0}(\omega) + \lambda_S(2\omega)e^{i\omega}\lambda_{H_1}(\omega).$$

Exercise 7.8: Lifting-Based Implementations of the Spline 5/3 and CDF 9/7 Wavelets

Let us use the different lifting factorizations obtained in this chapter to implement the corresponding wavelet kernels. Your functions should call the functions from Exercises 4.24 and 4.29. You will need Eqs. (7.8)–(7.11) here, in order to complete the kernels.

a) Write functions

```
dwt_kernel_53(x, bd_mode)
idwt_kernel_53(x, bd_mode)
```

which implement the DWT and IDWT kernels for the Spline 5/3 wavelet. Use the lifting factorization obtained in Example 7.2.3.

b) Write functions

```
dwt_kernel_97(x, bd_mode)
idwt_kernel_97(x, bd_mode)
```

which implement the DWT and IDWT kernels for the CDF 9/7 wavelet. Use the lifting factorization obtained in Example 7.2.4.

c) In Chap. 4, we listened to the low-resolution approximations and detail components in sound for three different wavelets. Repeat these experiments with the Spline 5/3 and the CDF 9/7 wavelet, using the new kernels you implemented in a) and b).

d) Plot all scaling functions and mother wavelets for the Spline 5/3 and the CDF 9/7 wavelets, using the cascade algorithm and the kernels you have implemented.

Exercise 7.9: Lifting-Based Implementation of Orthonormal Wavelets

a) Write functions

```
lifting_even(lambda1, lambda2, x, bd_mode)
lifting_odd(lambda1, lambda2, x, bd_mode)
```

which apply the elementary lifting matrices (7.14) to $\boldsymbol{x}$. Assume that N is even.

b) Write functions

```
dwt_kernel_ortho(x, bd_mode, wav_props)
idwt_kernel_ortho(x, bd_mode, wav_props)
```

which apply the DWT and IDWT kernels for orthonormal wavelets, using the functions `lifting_even` and `lifting_odd`. Assume that you can access lifting steps so that the lifting factorization (7.7) holds through the object `wav_props` by means of writing `wav_props.lambdas`, `wav_props.alpha`, and `wav_props.beta`, and that `wav_props.last_even` indicates whether the last lifting step is even. `wav_props.lambdas` is an $n \times 2$-matrix so that the filter coefficients $\{\underline{\lambda_1}, \lambda_2\}$ or $\{\lambda_1, \underline{\lambda_2}\}$ in the i'th lifting step are found in row i.

You can now define standard DWT and IDWT kernels in the following way, once the `wav_props` object has been computed for a given N:

```
dwt_kernel = @(x, bd_mode) dwt_kernel_ortho(x, bd_mode, wav_props);
```

c) Listen to the low-resolution approximations and detail components in sound for orthonormal wavelets for $N = 1, 2, 3, 4$.

d) Plot all scaling functions and mother wavelets for the orthonormal wavelets for $N = 1, 2, 3, 4$, using the cascade algorithm. Since the wavelets are orthonormal, we should have that $\phi = \tilde{\phi}$, and $\psi = \tilde{\psi}$. In other words, you should see that the bottom plots equal the upper plots.

Exercise 7.10: Piecewise Linear Wavelet with 4 Vanishing Moments

In Exercise 4.31 we found constants $\alpha, \beta, \gamma, \delta$ which give the coordinates of $\hat{\psi}$ in $(\boldsymbol{\phi}_1, \hat{\boldsymbol{\psi}}_1)$, where $\hat{\psi}$ was a mother wavelet in the MRA for piecewise linear functions with four vanishing moments.

a) Show that the polyphase representation of G when $\hat{\psi}$ is used as mother wavelet can be factored as

$$\frac{1}{\sqrt{2}} \begin{pmatrix} I & \mathbf{0} \\ \{1/2, \underline{1/2}\} & I \end{pmatrix} \begin{pmatrix} I & \{-\gamma, \underline{-\alpha}, -\beta, -\delta\} \\ \mathbf{0} & I \end{pmatrix}.$$

Find also a lifting factorization of H.

Exercise 7.11: Wavelet Based on Piecewise Quadratic Scaling Function

In Exercise 6.8 you should have found the filters

$$H_0 = \frac{1}{128}\{-5, 20, -1, -96, 70, \underline{280}, 70, -96, -1, 20, -5\}$$

$$H_1 = \frac{1}{16}\{1, -4, \underline{6}, -4, 1\}$$

$$G_0 = \frac{1}{16}\{1, 4, \underline{6}, 4, 1\}$$

$$G_1 = \frac{1}{128}\{5, 20, 1, -96, -70, \underline{280}, -70, -96, 1, 20, 5\}.$$

Show that

$$\begin{pmatrix} I & -\frac{1}{128}\{5, \underline{-29}, -29, 5\} \\ \mathbf{0} & I \end{pmatrix} \begin{pmatrix} I & \mathbf{0} \\ -\{1, \underline{1}\} & I \end{pmatrix} \begin{pmatrix} I & -\frac{1}{4}\{\underline{1}, 1\} \\ \mathbf{0} & I \end{pmatrix} G = \begin{pmatrix} \frac{1}{4} & 0 \\ 0 & 4 \end{pmatrix},$$

and derive a lifting factorization of G from this.

7.3 Polyphase Representations of Cosine Modulated Filter Banks and the MP3 Standard

In Sect. 5.4 we proved that the MP3 standard uses cosine modulated forward- and reverse filter bank transforms. In this section we will address two issues we have left out: How are these transforms constructed? And do they actually give perfect reconstruction? We will prove the following:

- the transforms of the MP3 standard only approximately give perfect reconstruction, and are only close to being orthogonal,
- the transforms can be modified in a very simple way so that one obtains orthogonal transforms with perfect reconstruction.

It may seem very surprising that a much used standard does not achieve perfect reconstruction, in particular when closely related transforms with perfect reconstruction exist. An explanation may be that the standard was established at about the same time as these filter banks were developed, so that one did not capture the latter perfect reconstruction filter banks.

The essential ingredient in our analysis is to compute the polyphase representations of the filter bank transforms in question, from which a 2×2 block structure will be seen. This block structure enables us to glue together $M/2$ 2-channel filter bank transforms, constructed earlier in the chapter, into one M-channel filter bank transform. The factorization we end up with is actually a much better starting point for obtaining useful cosine-modulated filter bank.

7.3.1 Polyphase Representation of the Forward Filter Bank Transform

We will start by adding $1/2$ inside the cosines in the expression we deduced from the MP3 standard in Sect. 5.4.1.

$$z_s^{(n)} = z_{32s+n}$$
$$= \sum_{m=0}^{63} \sum_{r=0}^{7} \cos\left(\frac{2\pi(n+1/2)(m+64r-16+1/2)}{2M}\right) h_{m+64r} x_{32(s+1)-(m+64r)-1},$$
$$(7.15)$$

where $z^{(n)} = \{z_{32s+n}\}_{s=0}^{\infty}$ is the n'th polyphase component of z. We will now continue to find the polyphase representations of the transforms, and we will see that, with this change of cosines, $\mathrm{DCT}_M^{(IV)}$ will be a factor in this representation (without the change in cosines, $\mathrm{DCT}_M^{(II)}$ will instead be a factor). Again we used that any $k < 512$ can be written uniquely on the form $k = m+64r$, with $0 \leq m < 64$, and $0 \leq r < 8$. Previously we used the cosine property (5.8). Due to the change inside the cosine, we will now instead adapt the properties

$$\cos\left(2\pi(n+1/2)(-m+1/2)/(2M)\right) = \cos\left(2\pi(n+1/2)(m-1+1/2)/(2M)\right)$$
$$(7.16)$$

$$\cos\left(\frac{2\pi(n+1/2)(k+2Mr+1/2)}{2M}\right) = (-1)^r \cos\left(\frac{2\pi(n+1/2)(k+1/2)}{2M}\right) \quad (7.17)$$

$$\cos\left(\frac{2\pi(n+1/2)(2M-k-1+1/2)}{2M}\right) = -\cos\left(\frac{2\pi(n+1/2)(k+1/2)}{2M}\right). \quad (7.18)$$

Equation (7.17) in particular is the new version of (5.8), and applying this first to Eq. (7.15) we obtain

$$\sum_{m=0}^{63} \cos\left(2\pi(n+1/2)(m-16+1/2)/(2M)\right) \sum_{r=0}^{7} (-1)^r h_{m+64r} x_{32(s-2r)-m+31}.$$

If we define

$$V^{(m)} = \{(-1)^0 h_m, 0, (-1)^1 h_{m+64}, 0, (-1)^2 h_{m+128}, \ldots, 0, (-1)^7 h_{m+7\cdot64}\}, \quad (7.19)$$

for $0 \leq m \leq 63$, we can rewrite this as

$$\sum_{m=0}^{63} \cos\left(2\pi(n+1/2)(m-16+1/2)/(2M)\right) (V^{(m)} x^{(31-m)})_s,$$

where we recognized $x_{32(s-r)-m+31}$ as $x_{s-r}^{(31-m)}$, and the inner sum as a convolution. Thus

$$z^{(n)} = \sum_{m=0}^{63} \cos\left(2\pi(n + 1/2)(m - 16 + 1/2)/(2M)\right) V^{(m)} x^{(31-m)}.$$

We now use properties (7.16)–(7.18) to rewrite this to an expression involving $\mathrm{DCT}_M^{(IV)}$ (see Exercise 2.34):

$$4\left(\mathrm{DCT}_M^{(IV)} \otimes^k I\right) \begin{pmatrix} \mathbf{0} & \cdots & V^{(15)} & V^{(16)} & \cdots & \mathbf{0} & \mathbf{0} & \cdots & \mathbf{0} \\ \vdots & \ddots & \vdots & \vdots & \ddots & \vdots & \vdots & \ddots & \vdots \\ V^{(0)} & \cdots & \mathbf{0} & \mathbf{0} & \cdots & \mathbf{0} & \mathbf{0} & \cdots & \mathbf{0} \\ \mathbf{0} & \cdots & \mathbf{0} & \mathbf{0} & \cdots & \mathbf{0} & \mathbf{0} & \cdots & -V^{(63)} \\ \vdots & \ddots & \vdots & \vdots & \ddots & \vdots & \vdots & \ddots & \vdots \\ \mathbf{0} & \cdots & \mathbf{0} & \mathbf{0} & \cdots & V^{(47)} & -V^{(48)} & \cdots & \mathbf{0} \end{pmatrix} \begin{pmatrix} x^{(31)} \\ \vdots \\ x^{(-32)} \end{pmatrix}$$

$$= 4\left(\mathrm{DCT}_M^{(IV)} \otimes^k I\right) \begin{pmatrix} \mathbf{0} & \cdots & V^{(15)} & V^{(16)} & \cdots & \mathbf{0} \\ \vdots & \ddots & \vdots & \vdots & \ddots & \vdots \\ V^{(0)} & \cdots & \mathbf{0} & \mathbf{0} & \cdots & V^{(31)} \\ V^{(32)} E_1 & \cdots & \mathbf{0} & \mathbf{0} & \cdots & -V^{(63)} E_1 \\ \vdots & \ddots & \vdots & \vdots & \ddots & \vdots \\ \mathbf{0} & \cdots & V^{(47)} E_1 & -V^{(48)} E_1 & \cdots & \mathbf{0} \end{pmatrix} \begin{pmatrix} x^{(31)} \\ \vdots \\ x^{(0)} \end{pmatrix}.$$

Here we substituted $x^{(i)} = E_1 x^{(i+32)}$, and used the notation $A \otimes^k B$ for the *Kronecker tensor product* of A and B, defined as the block matrix with block (i, j) being $a_{i,j} I$. The Kronecker tensor product will be examined further in Exercise 8.21. Flipping the rows and the columns we obtain the following result.

Theorem 7. Polyphase Factorization of a Forward Filter Bank Transform Based on a Prototype Filter.

Let H be a forward filter bank transform based on a prototype filter. The polyphase representation of H can be factored as

$$4\left(DCT_M^{(IV)} \otimes^k I\right) \begin{pmatrix} \mathbf{0} & \cdots & V^{(16)} & V^{(15)} & \cdots & \mathbf{0} \\ \vdots & \ddots & \vdots & \vdots & \vdots & \vdots \\ V^{(31)} & \cdots & \mathbf{0} & \mathbf{0} & \cdots & V^{(0)} \\ -V^{(63)} E_1 & \cdots & \mathbf{0} & \mathbf{0} & \cdots & V^{(32)} E_1 \\ \vdots & \ddots & \vdots & \vdots & \vdots & \vdots \\ \mathbf{0} & \cdots & -V^{(48)} E_1 & V^{(47)} E_1 & \cdots & \mathbf{0} \end{pmatrix}. \tag{7.20}$$

In this factorization the rightmost matrix corresponds to computing step 2 and 3 in the MP3 encoding procedure, while the leftmost matrix corresponds to step 4. Also, aligned elements from each polyphase component must be computed at the same time, in order to obtain the 32 output samples in each iteration as described in the standard.

7.3.2 Polyphase Representation of the Reverse Filter Bank Transform

Making the same change in the cosine modulation for the expression for the reverse transform which we deduced in Sect. 5.4.2, and applying Property (7.17) again we obtain

$$x_{32s+j} = \sum_{k=0}^{31}\sum_{i=0}^{15} g_{32i+j}\cos(2\pi(32i+j+16+1/2)(k+1/2)/(2M))z_{32(s-i)+k}$$

$$= \sum_{k=0}^{31}\sum_{i=0}^{7} g_{64i+j}\cos(2\pi(64i+j+16+1/2)(k+1/2)/(2M))z_{32(s-2i)+k}$$

$$+ \sum_{k=0}^{31}\sum_{i=0}^{7} g_{64i+32+j}\cos\left(\frac{2\pi(64i+j+32+16+1/2)(k+1/2)}{2M}\right)z_{32(s-2i-1)+k}$$

$$= \sum_{k=0}^{31}\sum_{i=0}^{7}(-1)^i g_{64i+j}\cos(2\pi(j+16+1/2)(k+1/2)/(2M))z_{32(s-2i)+k}$$

$$+ \sum_{k=0}^{31}\sum_{i=0}^{7}(-1)^i g_{64i+32+j}\cos\left(\frac{2\pi(j+48+1/2)(k+1/2)}{2M}\right)z_{32(s-2i-1)+k}.$$

If we define

$$U^{(m)} = \{(-1)^0 g_m, 0, (-1)^1 g_{m+64}, 0, (-1)^2 g_{m+128}, \ldots, 0, (-1)^7 g_{m+7\cdot64}\}, \qquad (7.21)$$

for $0 \le m \le 63$, we can rewrite this as

$$x_s^{(j)} = \sum_{k=0}^{31}\cos(2\pi(j+16+1/2)(k+1/2)/(2M))(U^{(j)}z^{(k)})_s$$

$$+ \sum_{k=0}^{31}\cos(2\pi(j+48+1/2)(k+1/2)/(2M))(U^{(32+j)}z^{(k)})_{s-1},$$

so that row j in the polyphase factorization of the reverse transform is

$$U^{(j)}\left(\cos\left(\frac{2\pi(j+16+1/2)(0+1/2)}{(2M)}\right)\cdots\cos\left(\frac{2\pi(j+16+1/2)(31+1/2)}{(2M)}\right)\right)$$

$$+ U^{(32+j)}E_1\left(\cos\left(\frac{2\pi(j+48+1/2)(0+1/2)}{(2M)}\right)\cdots\cos\left(\frac{2\pi(j+48+1/2)(31+1/2)}{(2M)}\right)\right).$$

For the full matrix we thus obtain

$$\begin{pmatrix} U^{(0)} & \cdots & \mathbf{0} \\ \mathbf{0} & \ddots & \mathbf{0} \\ \mathbf{0} & \cdots & U^{(31)} \end{pmatrix} \times$$

$$\begin{pmatrix} \cos(2\pi(16+1/2)(0+1/2)/(2M)) & \cdots & \cos(2\pi(16+1/2)(31+1/2)/(2M)) \\ \vdots & \ddots & \vdots \\ \cos(2\pi(47+1/2)(0+1/2)/(2M)) & \cdots & \cos(2\pi(47+1/2)(31+1/2)/(2M)) \end{pmatrix}$$

$$+ \begin{pmatrix} U^{(32)}E_1 & \cdots & \mathbf{0} \\ \mathbf{0} & \ddots & \mathbf{0} \\ \mathbf{0} & \cdots & U^{(63)}E_1 \end{pmatrix} \times$$

$$\begin{pmatrix} \cos(2\pi(48+1/2)(0+1/2)/(2M)) & \cdots & \cos(2\pi(48+1/2)(31+1/2)/(2M)) \\ \vdots & \ddots & \vdots \\ \cos(2\pi(79+1/2)(0+1/2)/(2M)) & \cdots & \cos(2\pi(79+1/2)(31+1/2)/(2M)) \end{pmatrix}.$$

Using properties (7.17)–(7.18) we obtain the matrix

$$4 \begin{pmatrix} \mathbf{0} \cdots \mathbf{0} & U^{(0)} & \cdots & \mathbf{0} \\ \vdots \ddots \vdots & \vdots & \ddots & \vdots \\ \mathbf{0} \cdots \mathbf{0} & \mathbf{0} & \cdots & U^{(15)} \\ \mathbf{0} \cdots \mathbf{0} & \mathbf{0} & \cdots & -U^{(16)} \\ \vdots \ddots \vdots & \vdots & \ddots & \vdots \\ \mathbf{0} \cdots \mathbf{0} & -U^{(31)} & \cdots & \mathbf{0} \end{pmatrix} \left(\text{DCT}_M^{(IV)} \otimes^k I \right)$$

$$+4 \begin{pmatrix} \mathbf{0} & \cdots & -U^{(32)}E_1 & \mathbf{0} \cdots \mathbf{0} \\ \vdots & \ddots & \vdots & \vdots \ddots \vdots \\ -U^{(47)}E_1 & \ddots & \mathbf{0} & \mathbf{0} \cdots \mathbf{0} \\ -U^{(48)}E_1 & \ddots & \mathbf{0} & \mathbf{0} \cdots \mathbf{0} \\ \vdots & \ddots & \vdots & \vdots \ddots \vdots \\ \mathbf{0} & \cdots & -U^{(63)}E_1 & \mathbf{0} \cdots \mathbf{0} \end{pmatrix} \left(\text{DCT}_M^{(IV)} \otimes^k I \right)$$

Adding together the two matrices we obtain the following result.

Theorem 8. Polyphase Factorization of a Reverse Filter Bank Transform Based on a Prototype Filter.

Let G be a reverse filter bank transform based on a prototype filter. The polyphase representation of G can be factored as

$$4 \begin{pmatrix} \mathbf{0} & \cdots & -U^{(32)}E_1 & U^{(0)} & \cdots & \mathbf{0} \\ \vdots & \ddots & \vdots & \vdots & \ddots & \vdots \\ -U^{(47)}E_1 & \ddots & \mathbf{0} & \mathbf{0} & \cdots & U^{(15)} \\ -U^{(48)}E_1 & \ddots & \mathbf{0} & \mathbf{0} & \cdots & -U^{(16)} \\ \vdots & \ddots & \vdots & \vdots & \ddots & \vdots \\ \mathbf{0} & \cdots & -U^{(63)}E_1 & -U^{(31)} & \cdots & \mathbf{0} \end{pmatrix} \left(DCT_M^{(IV)} \otimes^k I \right). \qquad (7.22)$$

7.3.3 Perfect Reconstruction

The two expressions (7.20) and (7.22) for the forward and reverse transforms are very similar. The sparse matrices in both equations can be written as sums of matrices $\sum_{i=0}^{15} B_i$ and $\sum_{i=0}^{15} C_i$, with B_i and C_i having blocks $(i, 15-i)$, $(i, 16+i)$, $(31-i, 15-i)$, and $(31 - i, 16 + i)$ as the only nonzero ones. Moreover, the only combinations of these which contribute are

$$B_i C_{15-i} = \begin{pmatrix} -U^{(47-i)} E_1 & U^{(15-i)} \\ -U^{(48+i)} E_1 & -U^{(16+i)} \end{pmatrix} \begin{pmatrix} V^{(16+i)} & V^{(15-i)} \\ -V^{(48+i)} E_1 & V^{(47-i)} E_1 \end{pmatrix}. \qquad (7.23)$$

If $U^{(i)} = V^{(i)}$ for all i (i.e. $h = g$), we see that this matches Eq. (7.2) with $\alpha = -1, d = 0$, so that Theorem 4 guarantees alias cancellation for each subblock. We will therefore assume that $h = g$ in the following.

Now, define $S_i = (V^{(16+i)} V^{(47-i)} + V^{(48+i)} V^{(15-i)}) E_1$ (i.e. $S = H^{(0,0)} H^{(1,1)} - H^{(0,1)} H^{(1,0)}$) as in the proof of Theorem 4. In order for the full H and G matrices to have alias cancellation, all the filters S_i must be equal (see Exercise 7.3). Evaluating component n in S_i we obtain

$$\sum_{r=\infty}^{-\infty} (-1)^r h_{16+i+2rM} (-1)^{n-r} h_{47-i+2(n-r)M}$$

$$+ \sum_{r=\infty}^{-\infty} (-1)^r h_{48+i+2rM} (-1)^{n-r} h_{15-i+2(n-r)M}$$

$$= (-1)^n \sum_{r=\infty}^{-\infty} h_{16+i+rM} h_{47-i-rM+2nM} = (-1)^n (h^{(16+i)} * h^{(47-i)})_{2n} \qquad (7.24)$$

where the two sums were combined, and where h was extended with zeroes in both directions. Therefore, alias cancellation means that

$$(h^{(i)} * h^{(63-i)})_{2n} = (h^{(j)} * h^{(63-j)})_{2n} \text{ for all } i, j, n.$$

Now let us also assume that $h_i = h_{511-i}$ for all i (i.e. symmetry). From this it follows

$$V^{(63-k)} = -E_{14} (V^{(k)})^T, \qquad (7.25)$$

so that in particular

$$V^{(47-i)} = -E_{14} (V^{(16+i)})^T \qquad V^{(15-i)} = -E_{14} (V^{(48+i)})^T. \qquad (7.26)$$

We also have

$$S_i = (V^{(16+i)} V^{(47-i)} + V^{(48+i)} V^{(15-i)}) E_1$$

$$= -E_{15} \left((V^{(16+i)})^T V^{(16+i)} + (V^{(48+i)})^T V^{(48+i)} \right).$$

Now, if all $(V^{(16+i)})^T V^{(16+i)} + (V^{(48+i)})^T V^{(48+i)} = I$, we have $S_i = -E_{15} = \alpha^{-1} E_{15}$, and Theorem 4 says that $B_i C_{15-i} = E_{2 \cdot 15} = E_{30}$ for each subblock. Using this for all 16 subblocks when combining Eqs. (7.20) and (7.22), and using that $(A \otimes^k B)^{-1} = (A^{-1}) \otimes^k (B^{-1})$ for any non-singular A and B (see Exercise 8.21 again), so that in particular

$$\left(\text{DCT}_M^{(IV)} \otimes^k I \right)^{-1} = \text{DCT}_M^{(IV)} \otimes^k I,$$

we get

$$GH = 16 \begin{pmatrix} E_{30} & \cdots & \mathbf{0} \\ \vdots & \ddots & \vdots \\ \mathbf{0} & \cdots & E_{30} \end{pmatrix} = 16 E_{480}$$

(see also Exercise 7.2).

Using (7.26) and setting $A = V^{(16+i)}$ and $B = V^{(48+i)}$, (7.23) becomes

$$\begin{pmatrix} A^T E_{15} & -B^T E_{14} \\ -BE_1 & -A \end{pmatrix} \begin{pmatrix} A & -B^T E_{14} \\ -BE_1 & -A^T E_{15} \end{pmatrix}$$

Since $A^T A + B^T B = I$, this is an alternative QMF filter bank (see Exercise 7.6). We can therefore use existing alternative QMF filter banks in order to construct the subblocks. The filters A and B must have 15 coefficients with alternating zeroes however (since the $V^{(m)}$ are defined like this), so let us see how we can address this. First find an alternative QMF filter bank where both A and B have 8 coefficients. The perfect reconstruction property of a QMF filter bank still holds if G_0 is delayed with any number, so delay G_0 so that its last nonzero filter coefficient has index 0. Since $G_0 = (H_0)^T$, the first nonzero filter coefficient of H_0 has index 0 as well. We can also assume that $d = 0$ (with d the delay used to relate H_1 and G_0. This can be chosen freely), so that the indices of the nonzero filter coefficients of H_1 are the same as those for G_0, and similarly for H_0 and G_1. If we change the polyphase components so that zeroes are inserted in alternating order, we will still have an alternative QMF filter bank, defined by polyphase components on the form required by $V^{(16+i)}$ and $V^{(48+i)}$.

For the case of symmetry, (7.24) can be written as

$$(S_i)_n = (-1)^n (h^{(m)} * h^{(m)})_{2n}.$$

Since $S_i = -E_{15}$ in order for perfect reconstruction we can summarize as follows.

Theorem 9. *Assume that H and G are the forward and reverse filter bank transforms for a cosine modulated filter bank, that the analysis and synthesis prototypes are equal ($h_k = g_k$), and symmetric ($h_{511-k} = h_k$). Then $GH = 16 E_{480}$ if and only if*

$$(h^{(m)} * h^{(m)})_{2n} = \delta_{n,15} \text{ for all } m, n. \tag{7.27}$$

Moreover, the $h^{(m)}$ can be constructed from alternative QMF filter banks.

As long as (7.27) holds, G and H are in fact unitary (up to a scalar), see Exercise 7.12.

For $M = 2$, (7.27) can be viewed as an already known property for alternative QMF filter banks, since their rows are 2-translates which are orthogonal to each other.

This result gives some insight into how the prototype filters can be chosen: The polyphase components can be designed separately, and for each an alternative QMF filter bank can be used. For each QMF filter bank we have some freedom in choosing the coefficients, so that numerical optimization procedures can be applied to obtain a prototype with best possible band-pass characteristics. If we drop the requirement of orthogonality, Exercise 7.16 shows that a general lifting factorization for the subblocks can be used, so that we can concentrate on optimizing the lifting variables we previously called λ_i. We will not discuss such optimization further here.

7.3.4 The MP3 Standard Does Not Give Perfect Reconstruction

The only change in this section from the MP3 standard has been the addition of $1/2$ inside the cosines in the forward and reverse transforms. It is an interesting exercise to repeat the preceding deductions without this addition (note that we can use Exercise 5.33 to simplify the deduction of the polyphase representation of the reverse transform). If you do this you will obtain polyphase factorizations very similar to (7.20) and (7.22), with $\mathrm{DCT}_M^{(IV)}$ replaced with $\mathrm{DCT}_M^{(II)}$, but one finds only 15 2×2 subblocks, for which we can invert each block in the same way as we have described. It turns out that, due to the lack of a 16th and last block, one of the diagonal entries in the polyphase representation of GH will be on the form AB, with both A and B FIR filters, and we know from Chap. 3 that AB can never be a delay if A and B both are non-trivial FIR filters. This makes perfect reconstruction impossible.

Alias cancellation is still possible, however. Assuming as before that $h = g$ is symmetric, when computing GH we still get multiplication of 15 2×2-subblocks, which give entries of the form $V_1^T V_1 + V_2^T V_2$ on the diagonal, and zero off the diagonal. The polyphase factorization of GH will thus be diagonal with 30 diagonal entries on the form $V_1^T V_1 + V_2^T V_2$, and two on the form AB. If all of these are equal we obtain alias cancellation (see Exercise 7.3), and the prototype filters of the MP3 standard are in fact chosen so that this happens. Since the entries on the form $V_1^T V_1 + V_2^T V_2$ do not equal the identity here, 2-channel perfect reconstruction filter banks are not used in the MP3 standard.

The change in cosines result in two minor changes. The first change has to do with the point of symmetry. Without adding $1/2$ the symmetry will be $h_i = h_{512-i}$, contrary to $h_i = h_{511-i}$ previously. The symmetry in h is chosen to match the symmetry in the cosine, which are different in the two cases. And when the symmetries match, all filters in the bank will have the same symmetry, so that all filters have linear phase. The second change has to do with the delay, which will differ with 1 in the two cases ($d = 480$ when $1/2$ is added as we have seen, $d = 481$ when it is not added, see Exercise 7.13).

Exercise 7.12: Connection Between Forward and Reverse Filter Bank Transforms Based on Prototypes

In Exercise 5.33 we used a direct approach to relate G and H^T. Insert instead (7.26) in Eqs. (7.20) and (7.22) to show that $G = E_{480} H^T$ and $H = E_{480} G^T$. Also, combine these with $GH = 16 E_{480}$ to show that $H^T H = G^T G = 16$. Thus, up to a scalar, the forward and reverse filter bank transforms are orthogonal.

Exercise 7.13: Run Forward and Reverse Transform

Run the forward and then the reverse transform from Exercise 5.30 on the vector $(1, 2, 3, \ldots, 8192)$. Verify that there seems to be a delay on 481 elements, as promised in Sect. 7.3.4. Do you get the exact same result back?

Exercise 7.14: Verify Statement of Filters

Use your computer to verify the following symmetries for the prototype filters:

$$C_i = \begin{cases} -C_{512-i} & i \neq 64, 128, \ldots, 448 \\ C_{512-i} & i = 64, 128, \ldots, 448. \end{cases}$$

Explain also that this implies that $h_i = h_{512-i}$ for $i = 1, \ldots, 511$. Recall that the modified version had the symmetry $h_i = h_{511-i}$. When the filters $V^{(m)}$ are defined as in this section, explain why (7.25) should be replaced by

$$V^{(64-k)} = -E_{14}(V^{(k)})^T$$

in the MP3 standard

Exercise 7.15: Verify Near Perfect Reconstruction in the MP3 Standard

In Sect. 7.3.3 we saw that the polyphase representation of GH is

$$16 \begin{pmatrix} S_0 & \cdots & \mathbf{0} \\ \vdots & \ddots & \vdots \\ \mathbf{0} & \cdots & S_{31} \end{pmatrix} E_{480},$$

where $S_i = (V^{(16+i)})^T V^{(16+i)} + (V^{(48+i)})^T V^{(48+i)}$. As explained in Sect. 7.3.4, the MP3 standard chooses the prototype filter so that the S_i are equal (so that the block matrix above is a filter), but only near perfect reconstruction is possible.

a) With $S_i = \{\ldots, t_{-1}, \underline{t_0}, t_1, \ldots\}$, explain why the (magnitude of the) frequency response of (the filter) GH equals

$$16 \left| \sum_k t_k e^{-iMk\omega} \right|.$$

b) Use the values in the table C to find the filter coefficients t_k for

$$S_1 = (V^{(17)})^T V^{(17)} + (V^{(49)})^T V^{(49)}.$$

(use (7.19), even though this filter bank does not give perfect reconstruction), and plot this frequency response. If you get an almost flat frequency response, you have verified near perfect reconstruction in the MP3 standard. It is important that you scale the vertical axis in order to see that the frequency response is close to constant.

c) Verify that the frequency response of S_1 in b) should be close to $1/(16 \cdot 32) = 1/512$. The factor 32 has to do with the relative scaling of the C and D tables.

Exercise 7.16: Lifting Factorization of M-Channel Filter Bank Transforms

The fact that step 2 and 3 in the MP3 encoding procedure can be computed in-place, corresponds in our modified scheme to multiplying with the rightmost matrix in (7.20) in-place. This exercise addresses how this can be done. We omit some details in that we make some simplifications, and don't write down any code.

a) Use a lifting factorization of each of the 16 subblocks to show that (7.20) can be factored as a product of a diagonal matrix with entries on the form $\alpha_i E_{d_i}$ on the diagonal, a permutation matrix, and matrices on the form

$$
\begin{pmatrix}
I & \cdots & 0 & 0 & \cdots & S_{15} \\
\vdots & \ddots & \vdots & \vdots & \ddots & \vdots \\
0 & \cdots & I & S_0 & \cdots & 0 \\
0 & \cdots & 0 & I & \cdots & 0 \\
\vdots & \ddots & \vdots & \vdots & \ddots & \vdots \\
0 & \cdots & 0 & 0 & \cdots & I
\end{pmatrix}
, \text{ and }
\begin{pmatrix}
I & \cdots & 0 & 0 & \cdots & 0 \\
\vdots & \ddots & \vdots & \vdots & \ddots & \vdots \\
0 & \cdots & I & 0 & \cdots & 0 \\
0 & \cdots & S_0 & I & \cdots & 0 \\
\vdots & \ddots & \vdots & \vdots & \ddots & \vdots \\
S_{15} & \cdots & 0 & 0 & \cdots & I
\end{pmatrix}.
$$

This shows that lifting factorizations can be useful in saving arithmetic operations for M-channel filter bank transforms as well.

b) Assume the 16 subblocks have a lifting factorization in terms of elementary lifting matrices. Show that the filter representation of the first matrix listed in a) looks as follows around the diagonal

$$
\left(
\begin{array}{ccc|ccc|ccc}
0 & \cdots & \lambda_{M/2-1} & 1 & \cdots & 0 & 0 & \cdots & \lambda_{M/2-1} \\
\vdots & \ddots & \vdots & \vdots & \ddots & \vdots & \vdots & \ddots & \vdots \\
\lambda_0 & \cdots & 0 & 0 & \cdots & 1 & \lambda_0 & \cdots & 0 \\
\hline
0 & \cdots & 0 & 0 & \cdots & 0 & 1 & \cdots & 0 \\
\vdots & \ddots & \vdots & \vdots & \ddots & \vdots & \vdots & \ddots & \vdots \\
0 & \cdots & 0 & 0 & \cdots & 0 & 0 & \cdots & 1
\end{array}
\right),
$$

where the shown blocks are of size $(M/2) \times (M/2)$, and the ones are on the main diagonal. Convince yourself that $M/2$ consecutive rows are unaltered, while the other $M/2$ consecutive rows are altered by combining with the $M/2$ rows above and below, so that a composition of these matrices can be computed in-place. This also gives a reduction in the number of arithmetic operations with up to 50%, since this is the case for each subblock.

c) Show also that the matrices in a) can be inverted by adding a minus sign to all the S_i-components.

This exercise can also be used in constructing useful cosine-modulated filter banks with perfect reconstruction: Simply let the λ_i in the "generalized lifting steps" in b) be parameters one needs to optimize in some way.

Exercise 7.17: DFT Modulated Filter Banks

Assume that $\{S_i\}_{i=0}^{M-1}$ are all-pass filters (see Exercise 3.44). Show that

$$(DFT_M \otimes^k I) \begin{pmatrix} S_0 & \cdots & \mathbf{0} \\ \vdots & \ddots & \vdots \\ \mathbf{0} & \cdots & S_{M-1} \end{pmatrix}$$

and

$$\begin{pmatrix} (S_0)^T & \cdots & \mathbf{0} \\ \vdots & \ddots & \vdots \\ \mathbf{0} & \cdots & (S_{M-1})^T \end{pmatrix} (IDFT_M \otimes^k I)$$

invert each other. By realizing this in terms of polyphase components, explain how one can assemble a prototype filter from all-pass filters, and forward and reverse DFT-modulated filter banks (i.e. where the filters are obtained by modulating the prototype with $e^{2\pi ikn/M}$), so that we have perfect reconstruction.

7.4 Summary

We defined the polyphase representation of a filter bank transform as a block matrix where the blocks are filters, and proved some useful properties. In particular, the filter bank transforms of wavelets are 2×2-block matrices of filters. The polyphase representation was also useful for proving a characterization of wavelets we encountered in Chap. 6.

We showed that a factorization of the polyphase representation into simpler matrices, also called a lifting factorization, was useful. For $M = 2$ in particular, lifting factors the DWT and the IDWT into simpler building blocks, and reduces the number of arithmetic operations. Lifting is used in practical implementations, and we applied it to some of the wavelets from Chap. 6. The JPEG2000 standard states the lifting steps for the Spline 5/3- and CDF 9/7 wavelets [74], as we have stated them. It is unknown to the author if other literature handles concrete lifting factorizations for orthonormal wavelets the way done here. This lifting factorization was derived for the purpose of the library.

We computed a polyphase factorization to obtain requirements on the prototype filters of cosine-modulated filter banks, in order to have perfect alias cancellation or perfect reconstruction. It turned out that the MP3 standard could not possibly give perfect reconstruction, but that a small modification could be done in order to secure this.

Much signal processing literature refer to polyphase components as polynomials in a variable z (i.e. a Z-transform), and say that such matrices with polynomial entries are non-singular if and only if the determinant is on the form αz^d. Such polynomial matrices are also called *unimodular matrices*. We have avoided this in order to stay within a pure linear algebra context, using ordinary block matrices with blocks being filters.

We focused on polyphase representations in the time-domain. It gives perfect meaning to consider the frequency domain polyphase representations as well. We will go into detail on this, only mention that Exercise 5.31 sheds some light on this matter.

The lifting scheme was first proposed by Sweldens [72], who also suggested using it for in-place computation of the DWT [71]. Daubechies and Sweldens [14] is a good tutorial for lifting in general, while [6] addresses parts of its use in M-channel filter bank transforms.

Condition (7.27) for the prototype filters of cosine-modulated filter banks comes from [43] (although expressed differently). Only a very short explanation of its proof is given, based on the expression for GH deduced in Exercise 5.32. Matrix notation, including the factorization of the polyphase matrices, is absent. The proof in [43] actually holds for filter length $2rM$ as well, and it is easy to see that the proof we have given (where $r = 8$) also holds for such filter lengths. The expression in [43] appears as a *correlation*, rather than a convolution. The reason is that the author there applied the symmetry condition. Had he not done this, he would have obtained a convolution expression (i.e. (7.27)), with validity regardless of symmetry. The corresponding forward and reverse transforms still give perfect reconstruction as long as Condition (7.27) holds, but the transforms are not orthogonal when the prototype is unsymmetric.

Lapped orthogonal transforms (LOT's) [44] is a particular class of orthogonal filter bank transforms where the number of filter coefficients is $2M$. Many such transforms were known before the cosine-modulated filter banks in [43]. As for our cosine-modulated filter bank the filters in LOT's are aligned, so that their filter representation is on the form

$$\begin{pmatrix} \ddots & \vdots & \vdots & \vdots & \ddots \\ \cdots & A_0 & A_1 & \mathbf{0} & \cdots \\ \cdots & \mathbf{0} & A_0 & A_1 & \cdots \\ \ddots & \vdots & \vdots & \vdots & \ddots \end{pmatrix},$$

where A_0 and A_1 are $M \times M$-matrices. The popular filter banks due to Princen and Bradley [57] preceded the LOT's, but can also be interpreted as such. Cosine-modulated filter banks with $r = 1$ are also particular cases of LOT, and are also called *modulated lapped transforms*. In this case the formula (7.27) can be written simply as

$$(h_i)^2 + (h_{M+i})^2 = 1 \text{ for all } i.$$

This development concludes the one-dimensional aspect of Fourier and wavelet analysis. In Chap. 8 we will define the tensor product, and use this concept to widen the scope to two dimensions and higher.

Chapter 8
Digital Images

Up to now all signals have been one-dimensional. Images, however, are two-dimensional by nature. In this chapter we will extend our theory so that it also applies in higher dimensions. The key mathematical concept in this extension is called the *tensor product*, which also will help in obtaining efficient implementations of operations on images.

First we will present some basics on images, as well as how they can be represented and manipulated with simple mathematics. In particular we will see how filter-based operations extend naturally to images. The two main examples of filter-based operations we will consider are smoothing and edge detection. We will finally consider useful coordinate changes for images. Recall that the DFT, the DCT, and the DWT were all defined as changes of coordinates for vectors or functions of one variable, and therefore cannot be directly applied to two-dimensional data like images. It turns out that the tensor product can also be used to extend changes of coordinates to a two-dimensional setting.

8.1 What Is an Image?

Before we do computations with images, it is helpful to be clear about what an image really is. Images cannot be perceived unless there is some light present, so we first review superficially what light is.

Light

Light is electromagnetic radiation with wavelengths in the range 400–700 nm (1 nm is 10^{-9} m): Violet has wavelength 400 nm and red has wavelength 700 nm. White light contains roughly equal amounts of all wave lengths.

Other examples of electromagnetic radiation are gamma radiation, ultraviolet and infrared radiation and radio waves. All electromagnetic radiation travel at the speed of light ($\approx 3 \times 10^8$ m/s). Electromagnetic radiation consists of waves and may be reflected, just like sound waves, but sound waves are not electromagnetic waves.

© Springer Nature Switzerland AG 2019
Ø. Ryan, *Linear Algebra, Signal Processing, and Wavelets - A Unified Approach*,
Springer Undergraduate Texts in Mathematics and Technology,
https://doi.org/10.1007/978-3-030-01812-2_8

We can only see objects that emit light, and there are two ways that this can happen. The object can emit light itself, like a lamp or a computer monitor, or it reflects light that falls on it. An object that reflects light usually absorbs light as well. If we perceive the object as red it means that the object absorbs all light except red, which is reflected. An object that emits light is different; if it is to be perceived as being red it must emit only red light.

Digital Output Media

A computer monitor consists of a matrix of small dots which emit light. In most technologies, each dot is really three smaller dots, and each of these smaller dots emit red, green and blue light. If the amounts of red, green and blue is varied, our brain merges the light from the three small light sources and perceives light of different colors. In this way the color at each set of three dots can be controlled, and a color image can be built from the total number of dots.

Most colors, but not all, can be obtained by mixing red, green, and blue. In addition, different monitors can use slightly different red, green and blue colors. This means that colors can look different on two different monitors, and that colors that can be displayed on one monitor may not be displayable on another.

Printers use the same principle of building an image from small dots. On most printers, however, the small dots do not consist of smaller dots of different colors. Instead as many as 7–8 different inks (or similar substances) are mixed to the right color. This makes it possible to produce a wide range of colors, but not all, and the problem of matching a color from another device like a monitor is at least as difficult as matching different colors across different monitors.

Video projectors build an image that is projected onto a wall. The final image is therefore a reflected image and it is important that the surface is white so that all colors are reflected equally.

The quality of a device is closely linked to the density of the dots. The *resolution* of a medium is the number of dots per inch (dpi). The resolution for monitors is usually in the range 70–120 dpi, while for printers it is in the range 150–4800 dpi. The horizontal and vertical resolutions may be different. On a monitor the dots are usually referred to as *pixels* (picture elements).

Digital Input Media

The two most common ways to acquire digital images are with a digital camera or a scanner. A scanner essentially takes a photo of a document in the form of a matrix of (possibly colored) dots. As for printers, an important measure of quality is the number of dots per inch. The resolution of a scanner usually varies in the range 75–9600 dpi, and the color is represented with up to 48 bits per dot.

For digital cameras it does not make sense to measure the resolution in dots per inch, as this depends on how the image is printed (its size). Instead the resolution is measured in the number of dots recorded. The number of pixels recorded by a digital camera usually varies in the range 320×240 to 6000×4000, with 24 bits of color information per pixel. The total number of pixels varies in the range 76,800–24,000,000 (i.e. 0.077 megapixels to 24 megapixels).

For scanners and cameras it is easy to think that the more dots (pixels), the better the quality. Although there is some truth to this, there are many other factors that influence the quality. The main problem is that the measured color information is very easily polluted by noise. And of course high resolution also means that the resulting files become very big; an uncompressed 6000×4000 image produces a $72\,MB$ file. The advantage of high resolution is that you can magnify the image considerably and still maintain reasonable quality.

Digital Images

So far we have not been precise about what digital images are. From a mathematical point of view a digital image P is simply a matrix of *intensity values* $\{p_{i,j}\}_{i,j=1}^{M,N}$. For grey-level images the value $p_{i,j}$ is a single number, while for color images each $p_{i,j}$ is a vector of three or more values. If the image is recorded in the rgb-model, each $p_{i,j}$ is a vector of three values,

$$p_{i,j} = (r_{i,j}, g_{i,j}, b_{i,j}),$$

that denote the amount of red, green and blue at the point (i,j).

Note that, when referring to the coordinates (i,j) in an image, i will refer to row index, j to column index, in the same was as for matrices. In particular, the top row in the image have coordinates $\{(0,j)\}_{j=0}^{N-1}$, while the left column in the image has coordinates $\{(i,0)\}_{i=0}^{M-1}$. With this notation, the dimension of the image is $M \times N$. The value $p_{i,j}$ gives the color information at the point (i,j). It is important to remember that there are many formats for this. The simplest case is plain black and white images in which case $p_{i,j}$ is either 0 or 1. For grey-level images the intensities are usually integers in the range 0–255. For color images there are many different formats, but we will just consider the rgb-format. Usually the three components are given as integers in the range 0–255 also here.

Fig. 8.1 Our test image

In Fig. 8.1 we have shown the test image we will work with, called the *Lena image* (named after the girl in the image). The image is in the rgb-format, and is used as a test image in many image processing textbooks. In Fig. 8.2 we have shown the corresponding black and white, and grey-level versions of the Lena image.

Fig. 8.2 Black and white (left), and grey-level (right) versions of the Lena image

8.2 Some Simple Operations on Images with MATLAB

Images are two-dimensional matrices of numbers, contrary to the sound signals we considered in the previous section. This means that we can manipulate an image by performing mathematical operations on the numbers. In this section we will consider some of the simpler such operations, leaving the more advanced operations to later sections (these are generalizations of operations we applied to sound). A color image can also be though of as a three-dimensional matrix, with one matrix for each color component. In the following, operations on images will be implemented in such a way that they are applied to each color component simultaneously. This is similar to the FFT and the DWT, where the operations were applied to each sound channel simultaneously.

First we need a function for reading an image from a file so that its contents are accessible as a matrix. This can be done with the function `imread`. If we write

```
X = double(imread('filename.fmt', 'fmt'))
```

the image with the given path and format is read, and stored in the matrix X. 'fmt' can be 'jpg','tif', 'gif', 'png', and so on. This parameter is optional: If it is not present, the program will attempt to determine the format from the first bytes in the file, and from the filename. After the call to `imread`, the entries in X represent the pixel values, and are of integer data type (more precisely, the data type `uint8`). To perform operations on the image, we must first convert the entries to the data type `double`, as shown above.

Similarly, the function `imwrite` can be used to write the image represented by a matrix to file. If we write

```
imwrite(uint8(X), 'filename.fmt', 'fmt')
```

the image represented by X is written to the given path, in the given format. Before the image is written to file, you see that we have converted the matrix values back to the integer data type. In other words: `imread` and `imwrite` both assume integer matrix entries, while operations on matrices assume double matrix entries.

Finally we need a function for displaying an image. If you write

```
imshow(uint8(X))
```

the matrix X will be displayed in a separate window. Also here we needed to convert the samples to integers. The following examples go through some much used operations on images.

Example 8.1: Normalizing the Intensities

We have assumed that the intensities are integer values in the range $[0, 255]$. When performing operations on images, we will assume that all values are real numbers in $[0, 1]$, however. We therefore need to map back and forth between $[0, 255]$ and $[0, 1]$, by dividing and multiplying with 255. Another thing we need to take into account is that our operations may produce output values outside $[0, 1]$ even if the input values are inside this interval. We therefore need some means of *normalizing* the intensities back to $[0, 1]$. This can be done with the simple linear function

$$g(x) = \frac{x - a}{b - a}, \quad a < b,$$

which maps the interval $[a, b]$ to $[0, 1]$. For a and b we can use the minimum and maximum intensities $p_{\min}$ and $p_{\max}$ in the input, as this will spread out the values across the entire $[0, 1]$. The following function maps to $[0, 1]$ using this approach, and will be used repeatedly in the following examples:

```
function Z=map_to_01(X)
  minval = min(min(min(X)));
  maxval = max(max(max(X)));
  Z = (X - minval)/(maxval-minval);
end
```

The functions `min` and `max` are called three times in succession here, once for each dimension. The third dimension is the color component.

Example 8.2: Extracting the Different Colors

If we have a color image $P = (r_{i,j}, g_{i,j}, b_{i,j})_{i,j=1}^{m,n}$, it is often useful to manipulate the three color components separately as the three images

$$P_r = (r_{i,j})_{i,j=1}^{m,n}, \quad P_g = (g_{i,j})_{i,j=1}^{m,n}, \quad P_b = (b_{i,j})_{i,j=1}^{m,n}.$$

As an example, let us first see how we can produce three separate images, showing the red, green, and blue color components of the Lena image. First we need to read the file:

```
img = double(imread('images/lena.png'));
```

The returned object has three dimensions, the first two representing the spatial directions (row/column), the third the color component. By using different values for the third dimension we can thus obtain the different color components as follows:

```
X1 = zeros(size(img));
X1(:,:,1) = img(:,:,1);

X2 = zeros(size(img));
X2(:,:,2) = img(:,:,2);

X3=zeros(size(img));
X3(:,:,3) = img(:,:,3);
```

The resulting images are shown in Fig. 8.3.

Fig. 8.3   The red, green, and blue components of the color image in Fig. 8.1

Example 8.3: Converting from Color to Grey-Level

A color image can easily be converted to a grey-level image: At each point in the image we have to replace the three color values (r, g, b) by a single value q that will represent the grey level. If we want the grey-level image to be a reasonable representation of the color image, the value q should somehow reflect the intensity of the image at the point. There are several ways to do this. If the image is $P = (r_{i,j}, g_{i,j}, b_{i,j})_{i,j=1}^{m,n}$, we can (for all i and j) either

- set $q_{i,j} = \max(r_{i,j}, g_{i,j}, b_{i,j})$, i.e. use the largest of the three color components,
- set $q_{i,j} = (r_{i,j} + g_{i,j} + b_{i,j})/3$, i.e. use the average of the three values,
- set $q_{i,j} = \sqrt{r_{i,j}^2 + g_{i,j}^2 + b_{i,j}^2}$, i.e. use the "length" of the color vector (r, g, b).

The three methods can be implemented as follows.

```
X1 = max(img, [], 3);

X2 = (img(:, :, 1) + img(:, :, 2) + img(:, :, 3))/3;

X3 = sqrt(img(:,:,1).^2 + img(:,:,2).^2 + img(:,:,3).^2);
X3 = 255*map_to_01(X3);
```

For the third method we had to be careful, since $\sqrt{r^2 + g^2 + b^2}$ may be outside $[0, 1]$, even if r, g and b all lie in $[0, 1]$. We thus need to normalize the values as explained above. In practice one of the last two methods are preferred, perhaps with a preference for the last method, but the actual choice depends on the application. The results of applying these three operations can be seen in Fig. 8.4.

Fig. 8.4 Alternative ways to convert the Lena color image to a grey level image

Example 8.4: Computing the Negative Image

In film-based photography a negative image was obtained when the film was developed, and then a positive image was created from the negative. We can easily simulate this and compute a negative digital image.

Suppose we have a grey-level image $P = (p_{i,j})_{i,j=1}^{m,n}$ with intensity values in the interval $[0, 1]$. Here intensity value 0 corresponds to black and 1 corresponds to white. To obtain the negative image we just have to replace an intensity p by its 'mirror value' $1 - p$. This is also easily translated to code as above. The resulting image is shown in Fig. 8.5.

Fig. 8.5 The negative versions of the corresponding images in Fig. 8.4

Example 8.5: Adjusting the Contrast

A common problem with images is that the contrast often is not good enough. This typically means that a large proportion of the grey values are concentrated in a rather small subinterval of $[0, 1]$. The obvious solution is to somehow spread out the values. This can be accomplished by applying a monotone function f which maps $[0, 1]$ onto $[0, 1]$. If we choose f so that its derivative is large in the area where many intensity values are concentrated, we obtain the desired effect. We will consider two such families of functions:

$$f_n(x) = \frac{\arctan(n(x - 1/2))}{2 \arctan(n/2)} + \frac{1}{2}$$

$$g_\epsilon(x) = \frac{\ln(x + \epsilon) - \ln \epsilon}{\ln(1 + \epsilon) - \ln \epsilon}.$$

The first type of functions have quite large derivatives near $x = 0.5$ and will therefore increase the contrast in images with a concentration of intensities near 0.5. The second type of functions have a large derivative near $x = 0$ and will therefore increase the contrast in images with a large proportion of small intensity values, i.e., very dark images. Figure 8.6 shows some examples of these functions. The three functions in the left plot in Fig. 8.6 are f_4, f_{10}, and f_{100}, the ones in the right plot are $g_{0.1}$, $g_{0.01}$, and $g_{0.001}$. In Fig. 8.7, f_{10} and $g_{0.01}$ have been applied to the image in the right part of Fig. 8.4.

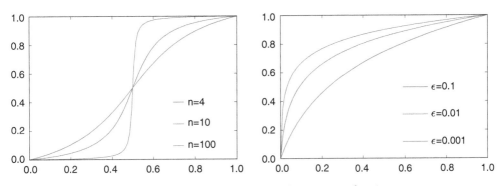

Fig. 8.6 Some functions that can be used to improve the contrast of an image

Fig. 8.7 The result after applying f_{10} and $g_{0.01}$ to the test image

Since the image was quite well balanced, f_{10} made the dark areas too dark and the bright areas too bright. $g_{0.01}$ on the other hand has made the image as a whole too bright.

Increasing the contrast is easy to implement. The following function uses the contrast adjusting function g_ϵ, with ϵ as parameter.

```
function Z=contrast_adjust(X, epsilon)
    Z = X/255; % Maps the pixel values to [0,1]
    Z = (log(Z+epsilon) - log(epsilon))/...
            (log(1+epsilon)-log(epsilon));
    Z = Z*255; % Maps the values back to [0,255]
end
```

This has been used to generate the right image in Fig. 8.7.

Exercise 8.6: Generating Black and White Images

Black and white images can be generated from grayscale images (with values between 0 and 255) by replacing each pixel value with the one of 0 and 255 which is closest. Use this strategy to generate the black and white image shown in the right part of Fig. 8.2.

Exercise 8.7: Adjusting Contrast

a) Write a function which instead uses the function f_n from Example 8.5 to adjust the contrast (rather than g_ϵ). n should be a parameter to your function.

b) Generate the left and right images in Fig. 8.7 on your own by using code which calls the function you have written, as well as `contrast_adjust`.

Exercise 8.8: Adjusting Contrast with Another Function

a) Show that the function $h_n : \mathbb{R} \to \mathbb{R}$ given by

$$h_n(x) = x^n,$$

maps the interval $[0,1] \to [0,1]$ for any n, and that $h_n'(1) \to \infty$ as $n \to \infty$.

b) The image `secret.jpg`, shown in Fig. 8.8, contains some information that is nearly invisible to the naked eye on most monitors. Use the function h_n, to reveal the secret message.

Hint

Convert to a grayscale image before you adjust the contrast with h_n.

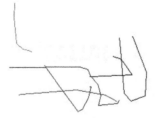

Fig. 8.8 Secret message

8.3 Filter-Based Operations on Images

The next operations on images we consider will use filters. We first need to define what is means to apply a filter to two-dimensional data. We start with the following definition of a computational molecule. This term stems from image processing.

Definition 1. *Computational Molecules.*
We say that an operation S on an image X is given by the *computational molecule*

$$A = \begin{pmatrix} & \vdots & \vdots & \vdots & \vdots & \vdots \\ \cdots & a_{-1,-1} & a_{-1,0} & a_{-1,1} & \cdots \\ \cdots & a_{0,-1} & \underline{a_{0,0}} & a_{0,1} & \cdots \\ \cdots & a_{1,-1} & a_{1,0} & a_{1,1} & \cdots \\ & \vdots & \vdots & \vdots & \vdots & \vdots \end{pmatrix}$$

if we have that

$$(SX)_{i,j} = \sum_{k_1,k_2} a_{k_1,k_2} X_{i-k_1,j-k_2}. \tag{8.1}$$

In the molecule, indices are allowed to be both positive and negative, we underline the element with index $(0,0)$ (the center of the molecule), and assume that $a_{i,j}$ with indices falling outside those listed in the molecule are zero (as for compact filter notation).

In Eq. (8.1), it is possible for the indices $i - k_1$ and $j - k_2$ to fall outside the legal range for X. We will address this as we did for filters, namely by assuming that X is extended (either periodically or symmetrically) in both directions. The interpretation of a computational molecule is that we place the center of the molecule on a pixel, multiply the pixel and its neighbors by the corresponding weights $a_{i,j}$ in reverse order, and finally sum up in order to produce the resulting value. This type of operation will turn out to be particularly useful for images. The following result expresses how computational molecules and filters are related.

Theorem 2. Filtering and Computational Molecules.
Let S_1 and S_2 be filters with compact filter notation t_1 and t_2, respectively, and consider the operation S where S_1 is first applied to the columns in the image, and then S_2 is applied to the resulting rows. Then S is an operation which can be expressed in terms of the computational molecule $a_{i,j} = (t_1)_i (t_2)_j$.

Proof. Let $X_{i,j}$ be the pixels in the image. When we apply S_1 to the columns of X we get the image Y defined by

$$Y_{i,j} = \sum_{k_1} (t_1)_{k_1} X_{i-k_1,j}.$$

When we apply S_2 to the rows of Y we get the image Z defined by

$$Z_{i,j} = \sum_{k_2} (t_2)_{k_2} Y_{i,j-k_2} = \sum_{k_2} (t_2)_{k_2} \sum_{k_1} (t_1)_{k_1} X_{i-k_1,j-k_2}$$

$$= \sum_{k_1} \sum_{k_2} (t_1)_{k_1} (t_2)_{k_2} X_{i-k_1,j-k_2}.$$

Comparing with Eq. (8.1) we see that S is given by the computational molecule with entries $a_{i,j} = (\boldsymbol{t}_1)_i (\boldsymbol{t}_2)_j$. $\square$

Note that, when we filter an image with S_1 and S_2 in this way, the order does not matter: Since applying S_1 to all columns of X is the same as computing $S_1 X$, and applying S_2 to all rows of Y is the same as computing $Y(S_2)^T$, the combined filtering operation, S, takes the form

$$S(X) = S_1 X (S_2)^T. \tag{8.2}$$

The fact that the order does not matter now follows from that matrix multiplication is associative. Applying S_1 to the columns of X is what we call a *vertical filtering operation*, while applying S_2 to the rows of X is what we call a *horizontal filtering operation*.

Most of the computational molecules we will consider can be expressed in terms of filters as in this theorem, but clearly there also exist computational molecules which are not on this form: The matrix A with entries $a_{i,j} = (\boldsymbol{t}_1)_i (\boldsymbol{t}_2)_j$ has rank one, and a general computational molecule can have any rank.

The following function computes the transformation $S(X) = S_1 X (S_2)^T$, with X, S1, and S2 the first three parameters. The function also takes a fourth parameter `bd_mode`, addressing how S1 and S2 should handle the boundaries:

```
function x=tensor2_impl(x, fx, fy, bd_mode)
    sz = 1:(length(size(x)));
    x = fx(x, bd_mode);
    x = permute(x,[2 1 sz(3:end)]);
    x = fy(x, bd_mode);
    x = permute(x,[2 1 sz(3:end)]);
end
```

Here the function `permute` was used to swap the first two dimensions of the image. In 2D this corresponds to transposing a matrix. The code above will work for both grey-color and RGB images, i.e. there may be a third dimension which should be transformed in parallel.

If a computational molecule is obtained from the filters S_1 and S_2 with filter coefficients $\boldsymbol{t}_1$ and $\boldsymbol{t}_2$, respectively, we can compute $S(X)$ with the following code:

```
S1 = @(x, bd_mode) filter_impl(t1, x, bd_mode);
S2 = @(x, bd_mode) filter_impl(t2, x, bd_mode);

Y = tensor2_impl(X, S1, S2)
```

We here called `filter_impl` with symmetric extensions, i.e. the code assumes symmetric filters. Most filters we consider will be symmetric. If the filter is non-symmetric, the last parameter should be changed to `"per"`. Note also that we assume that the filter lengths are odd and that the middle filter coefficient has index 0, since `filter_impl` assumes this.

Finally, note that the previous implementation of `filter_impl` filtered all columns in $\boldsymbol{x}$, but it may actually fail in the case when $\boldsymbol{x}$ is an object of dimension 3 or larger (such as an RGB image). The library version of `filter_impl` has addressed this. There you will see that the code `x(:, :)` "flattens" $\boldsymbol{x}$ to a two-dimensional object, for which the previous code for `filter_impl` will work.

8.3.1 Tensor Product Notation for Operations on Images

Filter-based operations on images can be written compactly using what we will call *tensor product notation*. This is part of a very general tensor product framework, and we will review parts of this framework for the sake of completeness. Let us first define the tensor product of vectors.

Definition 3. *Tensor Product of Vectors.*
If $\boldsymbol{x}, \boldsymbol{y}$ are vectors of length M and N, respectively, their *tensor product* $\boldsymbol{x} \otimes \boldsymbol{y}$ is defined as the $M \times N$-matrix with entries $(\boldsymbol{x} \otimes \boldsymbol{y})_{i,j} = x_i y_j$. In other words, $\boldsymbol{x} \otimes \boldsymbol{y} = \boldsymbol{x} \boldsymbol{y}^T$.

The tensor product $\boldsymbol{x} \boldsymbol{y}^T$ is also called the *outer product* of $\boldsymbol{x}$ and $\boldsymbol{y}$ (contrary to the inner product $\langle \boldsymbol{x}, \boldsymbol{y} \rangle = \boldsymbol{x}^T \boldsymbol{y}$). In particular $\boldsymbol{x} \otimes \boldsymbol{y}$ is a matrix of rank 1, which means that most matrices cannot be written as a tensor product of two vectors. The special case $\boldsymbol{e}_i \otimes \boldsymbol{e}_j$ is the matrix which is 1 at (i, j) and 0 elsewhere, and the set of all such matrices forms a basis for the set of $M \times N$-matrices.

Observation 4. Standard Basis for $L_{M,N}(\mathbb{R})$.
Let $\mathcal{E}_M = \{\boldsymbol{e}_i\}_{i=0}^{M-1}$ $\mathcal{E}_N = \{\boldsymbol{e}_i\}_{i=0}^{N-1}$ be the standard bases for $\mathbb{R}^M$ and $\mathbb{R}^N$. Then

$$\mathcal{E}_{M,N} = \{\boldsymbol{e}_i \otimes \boldsymbol{e}_j\}_{(i,j)=(0,0)}^{(M-1,N-1)}$$

is a basis for $L_{M,N}(\mathbb{R})$, the set of $M \times N$-matrices. This basis is often referred to as the standard basis for $L_{M,N}(\mathbb{R})$.

The standard basis thus consists of rank 1-matrices. An image can simply be thought of as a matrix in $L_{M,N}(\mathbb{R})$, and a computational molecule as a special type of linear transformation from $L_{M,N}(\mathbb{R})$ to itself. Let us also define the tensor product of matrices.

Definition 5. *Tensor Product of Matrices.*
If $S_1 : \mathbb{R}^M \to \mathbb{R}^M$ and $S_2 : \mathbb{R}^N \to \mathbb{R}^N$ are matrices, we define $S_1 \otimes S_2$ as the unique linear mapping from $L_{M,N}(\mathbb{R})$ to itself which satisfies $(S_1 \otimes S_2)(\boldsymbol{e}_i \otimes \boldsymbol{e}_j) = (S_1 \boldsymbol{e}_i) \otimes (S_2 \boldsymbol{e}_j)$ for all i and j. $S_1 \otimes S_2$ is called the *tensor product of the matrices* S_1 and S_2.

A couple of remarks are in order. First, from linear algebra we know that, when S is linear mapping from V and $S(\boldsymbol{v}_i)$ is known for a basis $\{\boldsymbol{v}_i\}_i$ of V, S is uniquely determined. In particular, since the $\{\boldsymbol{e}_i \otimes \boldsymbol{e}_j\}_{i,j}$ form a basis, there exists a unique linear transformation $S_1 \otimes S_2$ so that $(S_1 \otimes S_2)(\boldsymbol{e}_i \otimes \boldsymbol{e}_j) = (S_1 \boldsymbol{e}_i) \otimes (S_2 \boldsymbol{e}_j)$. This unique linear transformation is called the *linear extension* from the given values. Clearly, by linearity, also $(S_1 \otimes S_2)(\boldsymbol{x} \otimes \boldsymbol{y}) = (S_1 \boldsymbol{x}) \otimes (S_2 \boldsymbol{y})$ for any $\boldsymbol{x}$ and $\boldsymbol{y}$, since

$$(S_1 \otimes S_2)(\boldsymbol{x} \otimes \boldsymbol{y}) = (S_1 \otimes S_2)((\sum_i x_i \boldsymbol{e}_i) \otimes (\sum_j y_j \boldsymbol{e}_j))$$

$$= (S_1 \otimes S_2)(\sum_{i,j} x_i y_j (\boldsymbol{e}_i \otimes \boldsymbol{e}_j))$$

$$= \sum_{i,j} x_i y_j (S_1 \otimes S_2)(\boldsymbol{e}_i \otimes \boldsymbol{e}_j) = \sum_{i,j} x_i y_j (S_1 \boldsymbol{e}_i) \otimes (S_2 \boldsymbol{e}_j)$$

$$= \sum_{i,j} x_i y_j S_1 \boldsymbol{e}_i ((S_2 \boldsymbol{e}_j))^T = S_1 (\sum_i x_i \boldsymbol{e}_i)(S_2 (\sum_j y_j \boldsymbol{e}_j))^T$$

$$= S_1 \boldsymbol{x} (S_2 \boldsymbol{y})^T = (S_1 \boldsymbol{x}) \otimes (S_2 \boldsymbol{y})$$

(see also Exercise 8.16). We can now prove the following.

Theorem 6. Compact Filter Notation and Computational Molecules.

If $S_1 : \mathbb{R}^M \to \mathbb{R}^M$ and $S_2 : \mathbb{R}^N \to \mathbb{R}^N$ are matrices of linear transformations, then $(S_1 \otimes S_2)X = S_1 X (S_2)^T$ for any $X \in L_{M,N}(\mathbb{R})$. In particular $S_1 \otimes S_2$ is the operation which applies S_1 to the columns of X, and S_2 to the resulting rows. In particular, if S_1, S_2 are filters with compact filter notations $\boldsymbol{t}_1$ and $\boldsymbol{t}_2$, respectively, then $S_1 \otimes S_2$ has computational molecule $\boldsymbol{t}_1 \otimes \boldsymbol{t}_2$.

We have not formally defined the tensor product of compact filter notations, but this is a straightforward extension of the usual tensor product of vectors, where we additionally mark the element at index $(0,0)$.

Proof. We have that

$$(S_1 \otimes S_2)(\boldsymbol{e}_i \otimes \boldsymbol{e}_j) = (S_1 \boldsymbol{e}_i) \otimes (S_2 \boldsymbol{e}_j) = S_1 \boldsymbol{e}_i ((S_2 \boldsymbol{e}_j))^T$$
$$= S_1 \boldsymbol{e}_i (\boldsymbol{e}_j)^T (S_2)^T = S_1 (\boldsymbol{e}_i \otimes \boldsymbol{e}_j)(S_2)^T.$$

This means that $(S_1 \otimes S_2)X = S_1 X (S_2)^T$ for any $X \in L_{M,N}(\mathbb{R})$ also, since equality holds on the basis vectors $\boldsymbol{e}_i \otimes \boldsymbol{e}_j$. Since the matrix A with entries $a_{i,j} = (\boldsymbol{t}_1)_i (\boldsymbol{t}_2)_j$ also can be written as $\boldsymbol{t}_1 \otimes \boldsymbol{t}_2$, the result follows. $\square$

Thus, $S_1 \otimes S_2$ can be used to denote two-dimensional filtering operations. This notation also makes it easy to combine several two-dimensional filtering operations:

Corollary 7. Composing Tensor Products.

We have that $(S_1 \otimes T_1)(S_2 \otimes T_2) = (S_1 S_2) \otimes (T_1 T_2)$.

Proof. By Theorem 6 we have that

$$(S_1 \otimes T_1)(S_2 \otimes T_2)X = S_1 (S_2 X T_2^T)T_1^T = (S_1 S_2)X(T_1 T_2)^T = ((S_1 S_2) \otimes (T_1 T_2))X.$$

for any $X \in L_{M,N}(\mathbb{R})$. This proves the result. $\square$

Suppose that we want to apply the operation $S_1 \otimes S_2$ to an image. We can factorize $S_1 \otimes S_2$ as

$$S_1 \otimes S_2 = (S_1 \otimes I)(I \otimes S_2) = (I \otimes S_2)(S_1 \otimes I). \tag{8.3}$$

Moreover, since

$$(S_1 \otimes I)X = S_1 X \qquad\qquad (I \otimes S_2)X = X(S_2)^T = (S_2 X^T)^T,$$

$S_1 \otimes I$ is a vertical filtering operation, and $I \otimes S_2$ is a horizontal filtering operation in this factorization. For filters we have an even stronger result: If S_1, S_2, S_3, S_4 all are filters, we have from Corollary 7 that $(S_1 \otimes S_2)(S_3 \otimes S_4) = (S_3 \otimes S_4)(S_1 \otimes S_2)$, since all filters commute. This does not hold in general since general matrices do not commute.

The operation $X \to S_1 X (S_2)^T$ is one particular type of linear transformation from $\mathbb{R}^{N^2}$ to itself. While a general such linear transformation requires N^4 multiplications, $X \to S_1 X (S_2)^T$ can be implemented generally with only $2N^3$ multiplications (since multiplication of two $N \times N$-matrices require N^3 multiplications in general). The operation $X \to S_1 X (S_2)^T$ is thus computationally simpler than linear transformations in general. In practice the operations S_1 and S_2 are also computationally simpler, since they can be filters for instance.

We will now consider two important examples of filtering operations on images: smoothing and edge detection.

Example 8.9: Smoothing an Image

When we considered filtering of digital sound, low-pass filters dampened high frequencies. We will here similarly see that an image can be smoothed by applying a low-pass filters to the rows and the columns. Let us write down the corresponding computational molecules, and let us use low-pass filters with filter coefficients from Pascal's triangle. If we use the filter $S = \frac{1}{4}\{1,\underline{2},1\}$ (row 2 from Pascal's triangle), Theorem 2 gives the computational molecule

$$A = \frac{1}{16}\begin{pmatrix}1\,2\,1\\2\,\underline{4}\,2\\1\,2\,1\end{pmatrix}.$$

If the pixels in the image are $p_{i,j}$, this means that we compute the new pixels by

$$\hat{p}_{i,j} = \frac{1}{16}\big(4p_{i,j} + 2(p_{i,j-1} + p_{i-1,j} + p_{i+1,j} + p_{i,j+1})$$
$$+ p_{i-1,j-1} + p_{i+1,j-1} + p_{i-1,j+1} + p_{i+1,j+1}\big).$$

If we instead use the filter $S = \frac{1}{64}\{1,6,15,\underline{20},15,6,1\}$ (row 6 from Pascal's triangle), we obtain the molecule

$$\frac{1}{4096}\begin{pmatrix}1 & 6 & 15 & 20 & 15 & 6 & 1\\6 & 36 & 90 & 120 & 90 & 36 & 6\\15 & 90 & 225 & 300 & 225 & 90 & 15\\20 & 120 & 300 & \underline{400} & 300 & 120 & 20\\15 & 90 & 225 & 300 & 225 & 90 & 15\\6 & 36 & 90 & 120 & 90 & 36 & 6\\1 & 6 & 15 & 20 & 15 & 6 & 1\end{pmatrix}.$$

We anticipate that both molecules give a smoothing effect, but that the second molecule provides more smoothing. The result of applying the two molecules above to our grayscale image is shown in the two right images in Fig. 8.9.

Fig. 8.9 The two right images show the effect of smoothing the left image

To make the smoothing effect visible, we have zoomed in on the face in the image. The following function was used to obtain this portion of the image, and will be used repeatedly in the following:

```
function img=create_excerpt()
    img = double(imread('images/lena.png','png'));
    img = img(129:384,129:384,:);
end
```

The smoothing effect is best visible in the second image. Smoothing effects are perhaps more visible if we use the simple chess pattern image shown in the left part of Fig. 8.10.

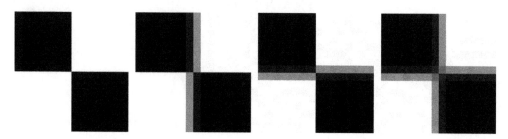

Fig. 8.10 The results of smoothing the simple image to the left in three different ways

Again we have used the filter $S = \frac{1}{4}\{1,\underline{2},1\}$. Here we also have shown what happens if we only smooth the image in one of the directions. In the right image we have smoothed in both directions. We then see the union of the two one-dimensional smoothing operations.

Example 8.10: Edge Detection

Another operation on images which can be expressed in terms of computational molecules is *edge detection*. An edge in an image is characterized by a large change in intensity values over a small distance in the image. For a continuous function this corresponds to a large derivative. We cannot compute exact derivatives for an image since it is only defined at isolated points, however, but we have a perfect situation for numerical differentiation techniques.

Partial Derivative in x-Direction

Let us first compute the partial derivative $\partial P/\partial x$ at all points in the image. Note first that it is the second coordinate in an image which refers to the x-direction you are used to from plotting functions. This means that the much used symmetric Newton quotient approximation to the partial derivative [51] takes the form

$$\frac{\partial P}{\partial x}(i,j) \approx \frac{p_{i,j+1} - p_{i,j-1}}{2},$$

where we have used the convention $h = 1$ (i.e. the derivative measures 'intensity per pixel'). This corresponds to applying the high-pass filter $S = \frac{1}{2}\{1,\underline{0},-1\}$ to all the rows, alternatively applying the tensor product $I \otimes S$ to the image, i.e. the molecule

$$\frac{1}{2} \begin{pmatrix} 0 & 0 & 0 \\ 1 & \underline{0} & -1 \\ 0 & 0 & 0 \end{pmatrix} .$$

We have included two rows of zeroes to make it clear how the computational molecule is to be applied when we place it over the pixels. If we apply this molecule to the usual excerpt of the Lena image you obtain the left image in Fig. 8.11.

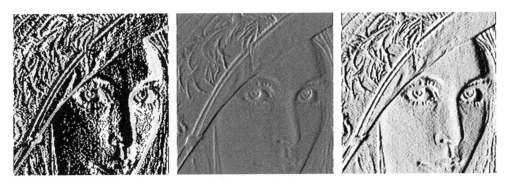

Fig. 8.11 Experimenting with the partial derivative in the x-direction for the image in Fig. 8.4. The left image has artifacts, since the pixel values are outside the legal range. We therefore normalize the intensities to lie in $[0, 255]$ (middle), before we increase the contrast (right)

This image shows many artifacts since the pixel values lie outside the legal range: many of the intensities are in fact negative. Let us therefore normalize and map all intensities to $[0, 1]$. This gives the second image. The predominant color of this image is an average grey, i.e. an intensity of about 0.5. To get more detail in the image we therefore try to increase the contrast by applying the function f_{50} from Example 8.5 to each intensity value. The result is shown in the third image, and shows indeed more detail.

It is important to understand the colors in these images. We have computed the derivative in the x-direction, and for this image it turns out that the smallest and largest value of the derivative have about the same absolute value, but with opposite signs. A value of 0 in the left image in Fig. 8.11 corresponds to no change in intensity between the two pixels, but these will be mapped to about 0.5. In the second image the edges (where the largest values of the derivative appear) have been mapped to black and white, while points with small or no changes in intensity have been mapped to a middle grey-tone. The middle image thus tells us that large parts of the image have little variation in intensity.

Partial Derivative in y-Direction

The partial derivative $\partial P / \partial y$ can be computed analogously to $\partial P / \partial x$, i.e. we apply the filter $-S = \frac{1}{2}\{-1, \underline{0}, 1\}$ to all columns of the image (alternatively, apply the tensor product $-S \otimes I$ to the image), where S is the filter we used in the x-direction. The positive direction of this axis in an image is opposite to the direction of the y-axis we use when plotting functions, and this explains the additional minus sign. We now obtain

the molecule

$$\frac{1}{2} \begin{pmatrix} 0 & 1 & 0 \\ 0 & \underline{0} & 0 \\ 0 & -1 & 0 \end{pmatrix}.$$

In Fig. 8.12 we have compared the partial derivatives in both directions. The x-derivative seems to emphasize vertical edges while the y-derivative seems to emphasize horizontal edges. This is precisely what we must expect.

Fig. 8.12 The first-order partial derivatives in the x- and y-direction, respectively. In both images values have been normalized and the contrast enhanced

The intensities have been normalized and the contrast enhanced by the function f_{50} from Example 8.5.

The Gradient

The gradient of a scalar function in two variables is defined by the vector

$$\nabla P = \left(\frac{\partial P}{\partial x}, \frac{\partial P}{\partial y} \right).$$

The length of the gradient is

$$|\nabla P| = \sqrt{\left(\frac{\partial P}{\partial x} \right)^2 + \left(\frac{\partial P}{\partial y} \right)^2}.$$

When the two first derivatives have been computed it is thus simple to compute the length of the gradient. As for the first order derivatives, it is possible for the length of the gradient to be outside the legal range of values. The computed gradient values, the gradient mapped to the legal range, and the gradient with contrast adjusted, are shown in Fig. 8.13.

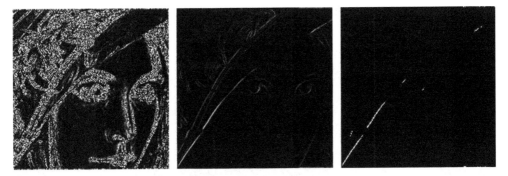

Fig. 8.13 The computed gradient (left). In the middle the intensities have been normalized to the $[0, 255]$, and to the right the contrast has been increased

The image of the gradient looks quite different from the images of the two partial derivatives. The reason is that the numbers that represent the length of the gradient are (square roots of) sums of squares of numbers. This means that the parts of the image that have virtually constant intensity (partial derivatives close to 0) are colored black. In the images of the partial derivatives these values ended up in the middle of the range of intensity values, with a final color of grey, since there were both positive and negative values. To enhance the contrast for this image we should thus do something different, like applying a one of the functions in the right plot of Fig. 8.6.

The gradient contains information about both derivatives and therefore emphasizes edges in all directions. It also gives a simpler image since the sign of the derivatives has been removed.

Example 8.11: Second-Order Derivatives

To compute the three second order derivatives we can combine the two computational molecules which we already have described. For the mixed second order derivative we get $(I \otimes S)((-S) \otimes I) = -S \otimes S$. For the last two second order derivative $\frac{\partial^2 P}{\partial x^2}$, $\frac{\partial^2 P}{\partial y^2}$, we can also use the three point approximation to the second derivative [51]

$$\frac{\partial P}{\partial x^2}(i, j) \approx p_{i,j+1} - 2p_{i,j} + p_{i,j-1}$$

(again we have set $h = 1$). This gives a smaller molecule than if we combine the two molecules for order one differentiation (i.e. $(I \otimes S)(I \otimes S) = (I \otimes S^2)$ and $((-S) \otimes I)((-S) \otimes I) = (S^2 \otimes I)$), since $S^2 = \frac{1}{2}\{1, \underline{0}, -1\}\frac{1}{2}\{1, \underline{0}, -1\} = \frac{1}{4}\{1, 0, \underline{-2}, 0, 1\}$. The second order derivatives of an image P can thus be computed by applying the computational molecules

$$\frac{\partial^2 P}{\partial x^2} : \quad \begin{pmatrix} 0 & 0 & 0 \\ 1 & \underline{-2} & 1 \\ 0 & 0 & 0 \end{pmatrix},$$

$$\frac{\partial^2 P}{\partial y \partial x} : \quad \frac{1}{4}\begin{pmatrix} -1 & 0 & 1 \\ 0 & \underline{0} & 0 \\ 1 & 0 & -1 \end{pmatrix},$$

$$\frac{\partial^2 P}{\partial y^2} : \quad \begin{pmatrix} 0 & 1 & 0 \\ 0 & -2 & 0 \\ 0 & 1 & 0 \end{pmatrix}.$$

The second-order derivatives of the Lena image are shown in Fig. 8.14.

Fig. 8.14 The second-order partial derivatives in the xx-, xy-, and yy-directions, respectively. In all images, the computed numbers have been normalized and the contrast enhanced

The computed derivatives were first normalized and then the contrast enhanced with the function f_{100} from Example 8.5.

As for the first derivatives, the xx-derivative seems to emphasize vertical edges and the yy-derivative horizontal edges. However, we also see that the second derivatives are more sensitive to noise in the image (the areas of grey are less uniform). The mixed derivative behaves a bit differently from the other two, and not surprisingly it seems to pick up both horizontal and vertical edges.

This procedure can be generalized to higher order derivatives also. To apply $\frac{\partial^{k+l}P}{\partial x^k \partial y^l}$ to an image we can compute $S_l \otimes S_k$ where S_r corresponds to any point method for computing the r'th order derivative. We can also compute $(S^l) \otimes (S^k)$, where we iterate the filter $S = \frac{1}{2}\{1, \underline{0}, -1\}$ for the first derivative. As pointed out, this gives longer filters.

Example 8.12: A Chess Pattern Image

Let us also apply the molecules for differentiation to the chess pattern test image produced by the code

```
N = 128;
img=zeros(N);
for x=0:(N-1)
    for y=0:(N-1)
        img(x+1,y+1) = 255*( (mod(x-1,64)>=32) == (mod(y-1,64)>=32) );
    end
end
```

and shown in the upper left part of Fig. 8.15. In the other parts of this figure we have applied $S \otimes I$, $I \otimes S$, and $S \otimes S$, $I \otimes S^2$, and $S^2 \otimes I$. The contrast has again been enhanced.

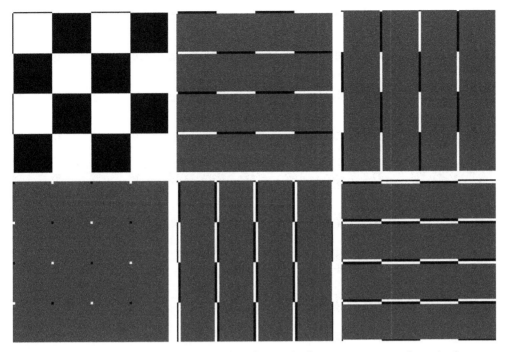

Fig. 8.15 Different tensor products applied to the simple chess pattern image shown in the upper left

These images make it even clearer that

- $S \otimes I$ detects horizontal edges,
- $I \otimes S$ detects vertical edges,
- $S \otimes S$ detects all points where abrupt changes appear in both directions.

We also see that the second order partial derivative detects exactly the same edges as the first order partial derivative, and that the edges detected with $I \otimes S^2$ are wider than the ones detected with $I \otimes S$. The reason is that the filter S^2 has more filter coefficients than S.

Edges are detected with different colors, reflecting whether the difference between the neighboring pixels is positive or negative. Note also the additional edges at the first and last rows/edges in the images. The reason is that the filter S is defined by assuming that the pixels repeat periodically (i.e. it is a circulant Toeplitz matrix).

Defining a two-dimensional filter by filtering columns and then rows is not the only way we can define a two-dimensional filter. Another possible way is to let the $MN \times MN$-matrix itself be a filter. Unfortunately, this is a bad way to define filtering, since there are some undesirable effects near the boundaries between rows: in the vector we form, the last element of one row is followed by the first element of the next row. These boundary effects are unfortunate when a filter is applied.

Exercise 8.13: Generating Images

Write code which calls the function `tensor2_impl` with appropriate filters to produce the following images:

a) The right image in Fig. 8.9.

b) The right image in Fig. 8.11.

c) The images in Fig. 8.13.

d) The images in Fig. 8.12.

e) The images in Fig. 8.14.

Exercise 8.14: Interpret Tensor Products

Let the filter S be defined by $S = \{\underline{-1}, 1\}$. Write down the 4×4-matrix $X = (1,1,1,1)\otimes (0,0,1,1)$, and compute $(S \otimes I)X$ and $(I \otimes S)X$.

Exercise 8.15: Computational Molecule of Moving Average Filter

Let S be the moving average filter of length $2L + 1$, i.e. $T = \frac{1}{L}\{\underbrace{1, \cdots, 1, \underline{1}, 1, \cdots, 1}_{2L+1 \text{ times}}\}$.

What is the computational molecule of $S \otimes S$?

Exercise 8.16: Bilinearity of the Tensor Product

Show that the mapping $F(\boldsymbol{x}, \boldsymbol{y}) = \boldsymbol{x} \otimes \boldsymbol{y}$ is bi-linear, i.e. that $F(\alpha\boldsymbol{x}_1 + \beta\boldsymbol{x}_2, \boldsymbol{y}) = \alpha F(\boldsymbol{x}_1, \boldsymbol{y}) + \beta F(\boldsymbol{x}_2, \boldsymbol{y})$, and $F(\boldsymbol{x}, \alpha\boldsymbol{y}_1 + \beta\boldsymbol{y}_2) = \alpha F(\boldsymbol{x}, \boldsymbol{y}_1) + \beta F(\boldsymbol{x}, \boldsymbol{y}_2)$.

Exercise 8.17: Attempt to Write as Tensor Product

Attempt to find matrices $S_1 : \mathbb{R}^M \to \mathbb{R}^M$ and $S_2 : \mathbb{R}^N \to \mathbb{R}^N$ so that the following mappings from $L_{M,N}(\mathbb{R})$ to $L_{M,N}(\mathbb{R})$ can be written on the form $X \to S_1 X (S_2)^T = (S_1 \otimes S_2)X$. In all the cases it may be that no such S_1, S_2 can be found. If this is the case, prove it.

a) The mapping which reverses the order of the rows in a matrix.

b) The mapping which reverses the order of the columns in a matrix.

c) The mapping which transposes a matrix.

Exercise 8.18: Computational Molecules

Let the filter S be defined by $S = \{1, \underline{2}, 1\}$.

a) Write down the computational molecule of $S \otimes S$.

b) Let us define $\boldsymbol{x} = (1, 2, 3)$, $\boldsymbol{y} = (3, 2, 1)$, $\boldsymbol{z} = (2, 2, 2)$, and $\boldsymbol{w} = (1, 4, 2)$. Compute the matrix $A = \boldsymbol{x} \otimes \boldsymbol{y} + \boldsymbol{z} \otimes \boldsymbol{w}$.

c) Compute $(S \otimes S)A$ by applying the filter S to every row and column in the matrix the way we have learned. If the matrix A was more generally an image, what can you say about how the new image will look?

Exercise 8.19: Computational Molecules 2

Let $S = \frac{1}{4}\{1, \underline{2}, 1\}$ be a filter.

a) What is the effect of applying the tensor products $S \otimes I$, $I \otimes S$, and $S \otimes S$ on an image represented by the matrix X?

b) Compute $(S \otimes S)(\boldsymbol{x} \otimes \boldsymbol{y})$, where $\boldsymbol{x} = (4, 8, 8, 4)$, $\boldsymbol{y} = (8, 4, 8, 4)$ (i.e. both $\boldsymbol{x}$ and $\boldsymbol{y}$ are column vectors).

Exercise 8.20: Eigenvectors of Tensor Products

Let $\boldsymbol{v}_A$ be an eigenvector of A with eigenvalue λ_A, and $\boldsymbol{v}_B$ an eigenvector of B with eigenvalue λ_B. Show that $\boldsymbol{v}_A \otimes \boldsymbol{v}_B$ is an eigenvector of $A \otimes B$ with eigenvalue $\lambda_A \lambda_B$.

Exercise 8.21: The Kronecker Product

The *Kronecker tensor product* of two matrices A and B, written $A \otimes^k B$, is defined as

$$A \otimes^k B = \begin{pmatrix} a_{1,1}B & a_{1,2}B & \cdots & a_{1,M}B \\ a_{2,1}B & a_{2,2}B & \cdots & a_{2,M}B \\ \vdots & \vdots & \ddots & \vdots \\ a_{p,1}B & a_{p,2}B & \cdots & a_{p,M}B \end{pmatrix},$$

where the entries of A are $a_{i,j}$. The tensor product of a $p \times M$-matrix, and a $q \times N$-matrix is thus a $(pq) \times (MN)$-matrix. Note that this tensor product in particular gives meaning for vectors: if $\boldsymbol{x} \in \mathbb{R}^M$, $\boldsymbol{y} \in \mathbb{R}^N$ are column vectors, then $\boldsymbol{x} \otimes^k \boldsymbol{y} \in \mathbb{R}^{MN}$ is also a column vector. In this exercise we will investigate how the Kronecker tensor product is related to tensor products as we have defined them in this section.

a) Explain that, if $x \in \mathbb{R}^M$, $y \in \mathbb{R}^N$ are column vectors, then $x \otimes^k y$ is the column vector where the rows of $x \otimes y$ have first been stacked into one large row vector, and this vector transposed. The linear extension of the operation defined by

$$x \otimes y \in \mathbb{R}^{M,N} \rightarrow x \otimes^k y \in \mathbb{R}^{MN}$$

thus stacks the rows of the input matrix into one large row vector, and transposes the result.

b) Show that $(A \otimes^k B)(x \otimes^k y) = (Ax) \otimes^k (By)$. We can thus use any of the defined tensor products $\otimes$, $\otimes^k$ to produce the same result, i.e. we have the commutative diagram shown in Fig. 8.16, where the vertical arrows represent stacking the rows in the matrix, and transposing, and the horizontal arrows represent the two tensor product linear transformations we have defined. In particular, we can compute the tensor product in terms of vectors, or in terms of matrices, and it is clear that the Kronecker tensor product gives the matrix of tensor product operations.

$$
\begin{array}{ccc}
x \otimes y & \xrightarrow{\;A \otimes B\;} & (Ax) \otimes (By) \\
\big\downarrow & & \big\downarrow \\
x \otimes^k y & \xrightarrow{\;A \otimes^k B\;} & (Ax) \otimes^k (By),
\end{array}
$$

Fig. 8.16 Tensor products

c) Using the Euclidean inner product on $L(M,N) = \mathbb{R}^{MN}$, i.e.

$$\langle X, Y \rangle = \sum_{i=0}^{M-1} \sum_{j=0}^{N-1} X_{i,j} \overline{Y_{i,j}}.$$

and the correspondence in a) we can define the inner product of $x_1 \otimes y_1$ and $x_2 \otimes y_2$ by

$$\langle x_1 \otimes y_1, x_2 \otimes y_2 \rangle = \langle x_1 \otimes^k y_1, x_2 \otimes^k y_2 \rangle.$$

Show that

$$\langle x_1 \otimes y_1, x_2 \otimes y_2 \rangle = \langle x_1, x_2 \rangle \langle y_1, y_2 \rangle.$$

Clearly this extends linearly to an inner product on $L_{M,N}$.

d) Show that the FFT factorization can be written as

$$\begin{pmatrix} F_{N/2} & D_{N/2} F_{N/2} \\ F_{N/2} & -D_{N/2} F_{N/2} \end{pmatrix} = \begin{pmatrix} I_{N/2} & D_{N/2} \\ I_{N/2} & -D_{N/2} \end{pmatrix} (I_2 \otimes^k F_{N/2}).$$

Also rewrite the sparse matrix factorization for the FFT from Eq. (2.16) in terms of tensor products.

8.4 Change of Coordinates in Tensor Products

Filter-based operations were not the only operations we considered for sound. We also considered the DFT, the DCT, and the wavelet transform, which were changes of coordinates which gave us useful frequency- or time-frequency information. We would like to define similar changes of coordinates for images. Tensor product notation will also be useful in this respect, and we start with the following result.

Theorem 8. The Basis $\mathcal{B}_1 \otimes \mathcal{B}_2$.
 If $\mathcal{B}_1 = \{v_i\}_{i=0}^{M-1}$ is a basis for $\mathbb{R}^M$, and $\mathcal{B}_2 = \{w_j\}_{j=0}^{N-1}$ is a basis for $\mathbb{R}^N$, then $\{v_i \otimes w_j\}_{(i,j)=(0,0)}^{(M-1,N-1)}$ is a basis for $L_{M,N}(\mathbb{R})$. We denote this basis by $\mathcal{B}_1 \otimes \mathcal{B}_2$.

Proof. Suppose that $\sum_{(i,j)=(0,0)}^{(M-1,N-1)} \alpha_{i,j}(v_i \otimes w_j) = 0$. Setting $h_i = \sum_{j=0}^{N-1} \alpha_{i,j} w_j$ we get

$$\sum_{j=0}^{N-1} \alpha_{i,j}(v_i \otimes w_j) = v_i \otimes \left(\sum_{j=0}^{N-1} \alpha_{i,j} w_j\right) = v_i \otimes h_i.$$

where we have used the bi-linearity of the tensor product mapping $(x, y) \to x \otimes y$ (Exercise 8.16). This means that

$$0 = \sum_{(i,j)=(0,0)}^{(M-1,N-1)} \alpha_{i,j}(v_i \otimes w_j) = \sum_{i=0}^{M-1} v_i \otimes h_i = \sum_{i=0}^{M-1} v_i h_i^T.$$

Column k in this matrix equation says $0 = \sum_{i=0}^{M-1} h_{i,k} v_i$, where $h_{i,k}$ are the components in h_i. By linear independence of the v_i we must have that $h_{0,k} = h_{1,k} = \cdots = h_{M-1,k} = 0$. Since this applies for all k, we must have that all $h_i = 0$. This means that $\sum_{j=0}^{N-1} \alpha_{i,j} w_j = 0$ for all i, from which it follows by linear independence of the w_j that $\alpha_{i,j} = 0$ for all j, and for all i. This means that $\mathcal{B}_1 \otimes \mathcal{B}_2$ is a basis. $\square$

In particular, as we have already seen, the standard basis for $L_{M,N}(\mathbb{R})$ can be written $\mathcal{E}_{M,N} = \mathcal{E}_M \otimes \mathcal{E}_N$. This is the basis for a useful convention: For a tensor product the bases are most naturally indexed in two dimensions, rather than the usual sequential indexing. This suggest also that we should use coordinate matrices, rather than coordinate vectors:

Definition 9. *Coordinate Matrix.*
 Let $\mathcal{B} = \{b_i\}_{i=0}^{M-1}$, $\mathcal{C} = \{c_j\}_{j=0}^{N-1}$ be bases for $\mathbb{R}^M$ and $\mathbb{R}^N$, and let $A \in L_{M,N}(\mathbb{R})$. By the coordinate matrix of A in $\mathcal{B} \otimes \mathcal{C}$ we mean the $M \times N$-matrix X (with components X_{kl}) such that $A = \sum_{k,l} X_{k,l}(b_k \otimes c_l)$.

We will have use for the following theorem, which shows how change of coordinates in $\mathbb{R}^M$ and $\mathbb{R}^N$ translate to a change of coordinates in the tensor product:

Theorem 10. Change of Coordinates in Tensor Products.
 Assume that

- $\mathcal{B}_1, \mathcal{C}_1$ are bases for $\mathbb{R}^M$, and that S_1 is the change of coordinates matrix from $\mathcal{B}_1$ to $\mathcal{C}_1$,
- $\mathcal{B}_2, \mathcal{C}_2$ are bases for $\mathbb{R}^N$, and that S_2 is the change of coordinates matrix from $\mathcal{B}_2$ to $\mathcal{C}_2$.

Both $\mathcal{B}_1 \otimes \mathcal{B}_2$ and $\mathcal{C}_1 \otimes \mathcal{C}_2$ are bases for $L_{M,N}(\mathbb{R})$, and if X is the coordinate matrix in $\mathcal{B}_1 \otimes \mathcal{B}_2$, and Y the coordinate matrix in $\mathcal{C}_1 \otimes \mathcal{C}_2$, then the change of coordinates from $\mathcal{B}_1 \otimes \mathcal{B}_2$ to $\mathcal{C}_1 \otimes \mathcal{C}_2$ can be computed as

$$Y = S_1 X (S_2)^T. \tag{8.4}$$

Proof. Denote the change of coordinates from $\mathcal{B}_1 \otimes \mathcal{B}_2$ to $\mathcal{C}_1 \otimes \mathcal{C}_2$ by S. Since any change of coordinates is linear, it is enough to show that $S(e_i \otimes e_j) = S_1(e_i \otimes e_j)(S_2)^T$ for any i, j. We can write

$$\boldsymbol{b}_{1i} \otimes \boldsymbol{b}_{2j} = \left(\sum_k (S_1)_{k,i} \boldsymbol{c}_{1k} \right) \otimes \left(\sum_l (S_2)_{l,j} \boldsymbol{c}_{2l} \right) = \sum_{k,l} (S_1)_{k,i} (S_2)_{l,j} (\boldsymbol{c}_{1k} \otimes \boldsymbol{c}_{2l})$$

$$= \sum_{k,l} (S_1)_{k,i} ((S_2)^T)_{j,l} (\boldsymbol{c}_{1k} \otimes \boldsymbol{c}_{2l}) = \sum_{k,l} (S_1 e_i (e_j)^T (S_2)^T)_{k,l} (\boldsymbol{c}_{1k} \otimes \boldsymbol{c}_{2l})$$

$$= \sum_{k,l} (S_1 (e_i \otimes e_j)(S_2)^T)_{k,l} (\boldsymbol{c}_{1k} \otimes \boldsymbol{c}_{2l})$$

This shows that the coordinate matrix of $\boldsymbol{b}_{1i} \otimes \boldsymbol{b}_{2j}$ in $\mathcal{C}_1 \otimes \mathcal{C}_2$ is $S_1(e_i \otimes e_j)(S_2)^T$. Since the coordinate matrix of $\boldsymbol{b}_{1i} \otimes \boldsymbol{b}_{2j}$ in $\mathcal{B}_1 \otimes \mathcal{B}_2$ is $e_i \otimes e_j$, this shows that $S(e_i \otimes e_j) = S_1(e_i \otimes e_j)(S_2)^T$. The result follows. $\square$

In both cases of filtering and change of coordinates in tensor products, we see that we need to compute the mapping $X \to S_1 X (S_2)^T$. As we have seen, this amounts to a row/column-wise operation, which we restate as follows:

Observation 11. Change of Coordinates in Tensor Products.
   The change of coordinates from $\mathcal{B}_1 \otimes \mathcal{B}_2$ to $\mathcal{C}_1 \otimes \mathcal{C}_2$ can be implemented as follows:

- *For every column in the coordinate matrix in $\mathcal{B}_1 \otimes \mathcal{B}_2$, perform a change of coordinates from $\mathcal{B}_1$ to $\mathcal{C}_1$.*
- *For every row in the resulting matrix, perform a change of coordinates from $\mathcal{B}_2$ to $\mathcal{C}_2$.*

We can again use the function `tensor2_impl` in order to implement change of coordinates for a tensor product. We just need to replace the filters `S1` and `S2` with the corresponding changes of coordinates. In the following examples, we will interpret the pixel values in an image as coordinates in the standard basis, and perform a change of coordinates.

Example 8.22: Change of Coordinates with the DFT

The DFT was defined as a change of coordinates from the standard basis to the Fourier basis. Let us substitute the DFT and the IDFT for S_1, S_2.

Modern image standards do typically not apply a change of coordinates to the entire image. Rather the image is split into smaller squares of appropriate size, called blocks, and a change of coordinates is performed independently on each block. In this example we will split the image into blocks of size 8×8.

Recall that the DFT values express frequency components. The same applies for the 2D DFT and thus for images, but frequencies are now represented in two different directions. Let us make a parallel to Example 2.5 for the Lena image, i.e. we will view the image after a 2D DFT, followed by discarding DFT coefficients below a given threshold, followed by a 2D IDFT. As for sound this should have little effect on the human perception of the image, if the threshold is chosen with care. DFT-coefficients in a matrix X below a threshold can be discarded with the following code:

```
X = X.*(abs(X) >= threshold);
```

`abs(X)>=threshold` returns a *threshold matrix* with 1 and 0 of the same size as X.

In Fig. 8.17 we can see the resulting images for different values of the threshold. When increasing the threshold, the image becomes more and more unclear, but the image is quite clear in the first case, where as much as more than 76.6% of the samples have been zeroed out. A blocking effect at the block boundaries is clearly visible.

Fig. 8.17 The effect on an image when it is transformed with the DFT, and the DFT-coefficients below a certain threshold are zeroed out. The threshold has been increased from left to right, from 100, to 200, and 400. The percentage of pixel values that were zeroed out are 76.6, 89.3, and 95.3, respectively

Example 8.23: Change of Coordinates with the DCT

Similarly to the DFT, the DCT was the change of coordinates from the standard basis to what we called the DCT basis. Let us substitute the DCT and the IDCT for S_1, S_2.

The DCT is used more than the DFT in image processing. In particular, the JPEG standard applies a two-dimensional DCT, rather than a two-dimensional DFT. With the JPEG standard, the blocks are always 8×8, as in the previous example. It is of course not a coincidence that a power of 2 is chosen here: the DCT, as the DFT, has an efficient implementation for powers of 2.

If we follow the same strategy for the DCT as above, i.e. zero out DCT-coefficients below a given threshold,[1] we get the images shown in Fig. 8.18. Similar effects as with the DFT can be seen. The same block sizes were used.

---

[1] The JPEG standard does not do exactly the kind of thresholding described here. Rather it performs what is called a quantization.

Fig. 8.18 The effect on an image when it is transformed with the DCT, and the DCT-coefficients below a certain threshold are zeroed out. The threshold has been increased from left to right, from 30, to 50, and 100. The percentage of pixel values that were zeroed out are 93.2, 95.8, and 97.7, respectively

It is also interesting to see what happens if we don't split the image into blocks. Of course, when we discard many of the DCT-coefficients, we should see some artifacts, but there is no reason to believe that these occur at the old block boundaries. The new artifacts can be seen in Fig. 8.19, and take a completely different shape.

Fig. 8.19 The effect on an image when it is transformed with the DCT, and the DCT-coefficients below a certain threshold are zeroed out. The image has not been split into blocks here, and the same thresholds as in Fig. 8.18 were used. The percentage of pixel values that were zeroed out are 93.2, 96.6, and 98.8, respectively

In the exercises you will be asked to write code which generates the images from these examples.

Exercise 8.24: Implementing DFT and DCT on Blocks

In this section we have used functions which apply the DCT and the DFT either to subblocks of size 8 × 8, or to the full image.

a) Implement functions dft_impl8, idft_impl8, dct_impl8, and idct_impl8 which apply the DFT, IDFT, DCT, and IDCT, to consecutive blocks of length 8.

b) Implement the two-dimensional FFT, IFFT, DCT, and IDCT on images, with the help of their one-dimensional counterparts, as well as the function `tensor2_impl`.

c) The function `forw_comp_rev_2d` in the library applies given transforms to the rows and columns of an input image, sets the coefficients below a certain threshold to zero, and transforms back. Run `forw_comp_rev_2d` for different threshold parameters on the sample image, and with the functions `dct_impl8`, `idct_impl8`, as well as DCT and IDCT applied to the entire input, as parameters. Check that this reproduces the DCT test images of this section, and that the correct numbers of values which have been discarded (i.e. which are below the threshold) are printed on screen.

Exercise 8.25: Code Example

Suppose that we have given an image by the matrix X. Consider the following code:

```
threshold = 30;
X = fft_impl(X, @fft_kernel_standard);
X = X';
X = fft_impl(X, @fft_kernel_standard);
X = X';

X = X.*(abs(X) >= threshold);

X = fft_impl(X, @fft_kernel_standard, 0);
X = X';
X = fft_impl(X, @fft_kernel_standard, 0);
X = X';
```

Comment what the code does. Comment in particular on the meaning of the parameter `threshold`, and its effect on the image.

8.5 Summary

We started by discussing what an image is, and continued with digital images and operations on them. Many of these operations could be described in terms of a row/column-wise application of filters, and more generally what we called computational molecules. We also saw how our operations could be expressed within the framework of the tensor product. Tensor products also applied to changes of coordinates for images, so that we could define two-dimensional versions of the DFT and the DCT for images. Filtering and change of coordinates on images boiled down to applying the one-dimensional counterparts to the rows and the columns in the image.

Introductory image processing textbooks also apply many other operations to images. We have limited to the techniques presented here, since our interest in images is mainly due to transformations useful for compression. Many excellent introductory books on image processing exist, but many of them are not adapted to mathematics as done here. Many are adapted to a particular programming language. An excellent textbook which uses MATLAB is [24]. This contains important topics such as image restoration and reconstruction, geometric transformations, morphology, and object recognition. The book [17] is an example of an introductory book which uses Java.

Many international standards exist for compression of images. Two of the most used are JPEG and JPEG2000. JPEG is short for *Joint Photographic Experts Group*, and was approved as an international standard in 1994. A more detailed description of the standard can be found in [56]. There are many steps an image coder and decoder must go through in these standards, and in this book we are primarily interested in the step which transforms the image to something more suitable for compression. For this JPEG uses a DCT, while JPEG2000 uses a DWT. JPEG2000 was developed to address some of the shortcomings of JPEG, such as the blocking artifacts we have seen. The standard document for JPEG2000 [27] does not focus on explaining the theory behind the standard. As the MP3 standard document, it rather states step-by-step procedures for implementing it.

There are many other important parts in data compression systems which we do not cover here (see [65, 64] for a comprehensive treatment). Two such are quantization and coding. The coding step is what actually achieves compression of the data. Different standards use different lossless coding techniques for this: JPEG2000 uses an advances type of arithmetic coding, while JPEG can use either arithmetic coding or Huffman coding.

JPEG2000 leads to as much as 20% improvement in compression ratios for medium compression rates compared to JPEG, possibly more for high or low compression rates. Although a number of components in JPEG2000 are patented, the patent holders have agreed that the core software should be available free of charge, and JPEG2000 is part of most Linux distributions. However, there appear to be some further, rather obscure, patents that have not been licensed, and this may be the reason why JPEG still has a dominant market share.

What You Should Have Learned in This Chapter

- How to read, write, and display images on a computer.
- How to extract different color components, convert from color to grey-level images, and adjust the contrast in images.
- The operation $X \to S_1 X (S_2)^T$ can be used for filter-based operations and change of coordinates on images, where S_1 and S_2 are one-dimensional counterparts.
- How tensor products can be used to conveniently express these operations.

Chapter 9
Using Tensor Products to Apply Wavelets to Images

Previously we have used wavelets to analyze sound. We would also like to use them in a similar way to analyze images. In Chap. 8 we used the tensor product to construct two dimensional objects (i.e. matrices) from one-dimensional objects (i.e. vectors). Since the spaces in wavelet contexts are function spaces, we need to extend the strategy from Chap. 8 to such spaces. In this chapter we will start with this extension, then specialize to the resolution spaces V_m, and extend the DWT to images. Finally we will look at several examples.

9.1 Tensor Product of Function Spaces

It turns out that the tensor product of two functions in on variable can be most intuitively defined as a function in two variables. This seems somewhat different from the strategy of Chap. 8, but we will see that the results we obtain will be very similar.

Definition 1. *Tensor Product of Function Spaces.*

Let U_1 and U_2 be vector spaces of functions, defined on the intervals $[0, M)$ and $[0, N)$, respectively, and suppose that $f_1 \in U_1$ and $f_2 \in U_2$. The tensor product of f_1 and f_2, denoted $f_1 \otimes f_2$, is the function in two variables defined on $[0, M) \times [0, N)$ by

$$(f_1 \otimes f_2)(t_1, t_2) = f_1(t_1)f_2(t_2).$$

$f_1 \otimes f_2$ is also called the *separable extension* of f_1 and f_2. The tensor product $U_1 \otimes U_2$ is the vector space spanned by $\{f_1 \otimes f_2\}_{f_1 \in U_1, f_2 \in U_2}$.

We will always assume that the functions in U_1 and U_2 are square integrable. In this case $U_1 \otimes U_2$ is also an inner product space, with the inner product given by a double integral,

$$\langle f, g \rangle = \int_0^N \int_0^M f(t_1, t_2)g(t_1, t_2)dt_1 dt_2. \tag{9.1}$$

© Springer Nature Switzerland AG 2019
Ø. Ryan, *Linear Algebra, Signal Processing, and Wavelets - A Unified Approach,*
Springer Undergraduate Texts in Mathematics and Technology,
https://doi.org/10.1007/978-3-030-01812-2_9

In particular, this says that

$$\langle f_1 \otimes f_2, g_1 \otimes g_2 \rangle = \int_0^N \int_0^M f_1(t_1) f_2(t_2) g_1(t_1) g_2(t_2) dt_1 dt_2$$

$$= \int_0^M f_1(t_1) g_1(t_1) dt_1 \int_0^N f_2(t_2) g_2(t_2) dt_2 = \langle f_1, g_1 \rangle \langle f_2, g_2 \rangle. \quad (9.2)$$

This is similar to what we found for the inner product of vector tensor products, as defined in Exercise 8.21. Thus, for tensor products, an inner product can be computed as the product of two inner products.

Tensor products of function spaces are useful for approximation of functions of two variables, as long as each component space has good approximation properties. We will not state a result on this, but rather consider some important examples.

Tensor Products of Polynomials

Let $U_1 = U_2$ be the space of all polynomials of finite degree. We know that U_1 can be used for approximating many kinds of functions, such as continuous functions, for example by Taylor series. The tensor product $U_1 \otimes U_1$ consists of all functions on the form $\sum_{i,j} \alpha_{i,j} t_1^i t_2^j$. It turns out that polynomials in several variables have approximation properties similar to those of polynomials in one variable.

Tensor Products of Fourier Spaces

Let $U_1 = U_2 = V_{N,T}$ be the Nth order Fourier space which is spanned by the functions

$$e^{-2\pi i N t/T}, \ldots, e^{-2\pi i t/T}, 1, e^{2\pi i t/T}, \ldots, e^{2\pi i N t/T}$$

The tensor product space $U_1 \otimes U_1$ now consists of all functions on the form

$$\sum_{k,l=-N}^{N} \alpha_{k,l} e^{2\pi i k t_1/T} e^{2\pi i l t_2/T}.$$

One can show that this space has approximation properties similar to those of $V_{N,T}$. This is the basis for the theory of Fourier series in several variables.

In the following we think of $U_1 \otimes U_2$ as a space which can be used to approximate a general class of functions. By associating a function with the vector of coordinates relative to some basis, and a matrix with a function in two variables, we have the following parallel to Theorem 8:

Theorem 2. Bases for Tensor Products of Function Spaces.

If $\{f_i\}_{i=0}^{M-1}$ is a basis for U_1 and $\{g_j\}_{j=0}^{N-1}$ is a basis for U_2, then $\{f_i \otimes g_j\}_{(i,j)=(0,0)}^{(M-1,N-1)}$ is a basis for $U_1 \otimes U_2$. Moreover, if the bases for U_1 and U_2 are orthogonal/orthonormal, then the basis for $U_1 \otimes U_2$ is orthogonal/orthonormal.

Proof. The proof is similar to that of Theorem 8: if

$$\sum_{(i,j)=(0,0)}^{(M-1,N-1)} \alpha_{i,j}(f_i \otimes g_j) = 0,$$

we define $h_i(t_2) = \sum_{j=0}^{N-1} \alpha_{i,j} g_j(t_2)$. It follows as before that $\sum_{i=0}^{M-1} h_i(t_2) f_i = 0$ for any t_2, so that $h_i(t_2) = 0$ for any t_2 due to linear independence of the f_i. But then $\alpha_{i,j} = 0$ also, due to linear independence of the g_j. The statement about orthogonality follows from Eq. (9.2). $\square$

We can now define the tensor product of two bases, and coordinate matrices, as before:

Definition 3. *Coordinate Matrix.*
  if $\mathcal{B} = \{f_i\}_{i=0}^{M-1}$ and $\mathcal{C} = \{g_j\}_{j=0}^{N-1}$, we define $\mathcal{B} \otimes \mathcal{C}$ as the basis $\{f_i \otimes g_j\}_{(i,j)=(0,0)}^{(M-1,N-1)}$ for $U_1 \otimes U_2$. We say that X is the coordinate matrix of f if

$$f(t_1,t_2) = \sum_{i,j} X_{i,j}(f_i \otimes g_j)(t_1,t_2),$$

where $X_{i,j}$ are the entries of X.

Theorem 10 can also be proved in the same way in the context of function spaces. We state this as follows:

Theorem 4. Change of Coordinates in Tensor Products of Function Spaces.
  Assume that U_1 and U_2 are function spaces, and that

- $\mathcal{B}_1, \mathcal{C}_1$ *are bases for* U_1, *and* S_1 *is the change of coordinates matrix from* $\mathcal{B}_1$ *to* $\mathcal{C}_1$,
- $\mathcal{B}_2, \mathcal{C}_2$ *are bases for* U_2, *and* S_2 *is the change of coordinates matrix from* $\mathcal{B}_2$ *to* $\mathcal{C}_2$.

Both $\mathcal{B}_1 \otimes \mathcal{B}_2$ *and* $\mathcal{C}_1 \otimes \mathcal{C}_2$ *are bases for* $U_1 \otimes U_2$, *and if* X *is the coordinate matrix in* $\mathcal{B}_1 \otimes \mathcal{B}_2$, Y *the coordinate matrix in* $\mathcal{C}_1 \otimes \mathcal{C}_2$, *then the change of coordinates from* $\mathcal{B}_1 \otimes \mathcal{B}_2$ *to* $\mathcal{C}_1 \otimes \mathcal{C}_2$ *can be computed as*

$$Y = S_1 X (S_2)^T. \qquad (9.3)$$

9.2 Tensor Product of Function Spaces in a Wavelet Setting

We will now specialize the spaces U_1, U_2 from Definition 1 to the resolution- and detail spaces V_m and W_m for a given wavelet. We can in particular form the tensor products $\phi_{0,n_1} \otimes \phi_{0,n_2}$. We will assume that

- the first component ϕ_{0,n_1} has period M (so that $\{\phi_{0,n_1}\}_{n_1=0}^{M-1}$ is a basis for the first component space),
- the second component ϕ_{0,n_2} has period N (so that $\{\phi_{0,n_2}\}_{n_2=0}^{N-1}$ is a basis for the second component space).

When we speak of $V_0 \otimes V_0$ we thus mean an MN-dimensional space with basis $\{\phi_{0,n_1} \otimes \phi_{0,n_2}\}_{(n_1,n_2)=(0,0)}^{(M-1,N-1)}$, where the coordinate matrices are $M \times N$. This difference in the dimension of the two components is done to allow for images where the number of rows and columns may be different. In the following we will implicitly assume that the component spaces have dimension M and N, to ease notation. If we use that $(\boldsymbol{\phi}_{m-1}, \boldsymbol{\psi}_{m-1})$ also is a basis for V_m, we get the following corollary to Theorem 2:

Corollary 5. Bases for Tensor Products.
Let ϕ be a scaling function and ψ a mother wavelet. Then the two sets of tensor products given by

$$\boldsymbol{\phi}_m \otimes \boldsymbol{\phi}_m = \{\phi_{m,n_1} \otimes \phi_{m,n_2}\}_{n_1,n_2}$$

and

$$(\boldsymbol{\phi}_{m-1}, \boldsymbol{\psi}_{m-1}) \otimes (\boldsymbol{\phi}_{m-1}, \boldsymbol{\psi}_{m-1})$$
$$= \{\phi_{m-1,n_1} \otimes \phi_{m-1,n_2},$$
$$\phi_{m-1,n_1} \otimes \psi_{m-1,n_2},$$
$$\psi_{m-1,n_1} \otimes \phi_{m-1,n_2},$$
$$\psi_{m-1,n_1} \otimes \psi_{m-1,n_2}\}_{n_1,n_2}$$

are both bases for $V_m \otimes V_m$. This second basis is orthogonal/orthonormal whenever the first is.

From this we observe that, while the one-dimensional wavelet decomposition splits V_m into a direct sum of the two vector spaces V_{m-1} and W_{m-1}, the corresponding two-dimensional decomposition splits $V_m \otimes V_m$ into a direct sum of four vector spaces. Let us assign names to these spaces:

Definition 6. *Tensor Product Spaces.*
We define the following subspaces of $V_m \otimes V_m$:

- The space $W_{m-1}^{(0,1)}$ spanned by $\{\phi_{m-1,n_1} \otimes \psi_{m-1,n_2}\}_{n_1,n_2}$,
- The space $W_{m-1}^{(1,0)}$ spanned by $\{\psi_{m-1,n_1} \otimes \phi_{m-1,n_2}\}_{n_1,n_2}$,
- The space $W_{m-1}^{(1,1)}$ spanned by $\{\psi_{m-1,n_1} \otimes \psi_{m-1,n_2}\}_{n_1,n_2}$.

Since these spaces are linearly independent, we can write

$$V_m \otimes V_m = (V_{m-1} \otimes V_{m-1}) \oplus W_{m-1}^{(0,1)} \oplus W_{m-1}^{(1,0)} \oplus W_{m-1}^{(1,1)}. \tag{9.4}$$

In the setting of tensor products we refer to $V_{m-1} \otimes V_{m-1}$ as a resolution space, and $W_{m-1}^{(0,1)}$, $W_{m-1}^{(1,0)}$, and $W_{m-1}^{(1,1)}$ as detail spaces. The coordinate matrix of

$$\sum_{n_1,n_2=0}^{2^{m-1}N} (c_{m-1,n_1,n_2}(\phi_{m-1,n_1} \otimes \phi_{m-1,n_2}) + w_{m-1,n_1,n_2}^{(0,1)}(\phi_{m-1,n_1} \otimes \psi_{m-1,n_2}) +$$
$$w_{m-1,n_1,n_2}^{(1,0)}(\psi_{m-1,n_1} \otimes \phi_{m-1,n_2}) + w_{m-1,n_1,n_2}^{(1,1)}(\psi_{m-1,n_1} \otimes \psi_{m-1,n_2})) \tag{9.5}$$

in the basis $(\boldsymbol{\phi}_{m-1}, \boldsymbol{\psi}_{m-1}) \otimes (\boldsymbol{\phi}_{m-1}, \boldsymbol{\psi}_{m-1})$ is

$$
\left(
\begin{array}{cc|cc}
c_{m-1,0,0} & \cdots & w^{(0,1)}_{m-1,0,0} & \cdots \\
\vdots & \vdots & \vdots & \vdots \\
\hline
w^{(1,0)}_{m-1,0,0} & \cdots & w^{(1,1)}_{m-1,0,0} & \cdots \\
\vdots & \vdots & \vdots & \vdots
\end{array}
\right).
\tag{9.6}
$$

The coordinate matrix is thus split into four submatrices:

- The c_{m-1}-values, i.e. the coordinates for $V_{m-1} \oplus V_{m-1}$. This is the upper left corner in Eq. (9.6).
- The $w^{(0,1)}_{m-1}$-values, i.e. the coordinates for $W^{(0,1)}_{m-1}$. This is the upper right corner.
- The $w^{(1,0)}_{m-1}$-values, i.e. the coordinates for $W^{(1,0)}_{m-1}$. This is the lower left corner.
- The $w^{(1,1)}_{m-1}$-values, i.e. the coordinates for $W^{(1,1)}_{m-1}$. This is the lower right corner.

The $w^{(i,j)}_{m-1}$-values are as in the one-dimensional situation often referred to as wavelet coefficients. Let us consider the Haar wavelet as an example.

Example 9.1: Piecewise Constant Functions

If V_m is a resolution space of piecewise constant functions, the resolution space $V_m \otimes V_m$ consists of functions in two variables which are constant on any square of the form $[k_1 2^{-m}, (k_1 + 1)2^{-m}) \times [k_2 2^{-m}, (k_2 + 1)2^{-m})$: Clearly $\phi_{m,k_1} \otimes \phi_{m,k_2}$ is constant on such a square and 0 elsewhere, and these functions form a basis for $V_m \otimes V_m$.

Let us compute the orthogonal projection of $\phi_{1,k_1} \otimes \phi_{1,k_2}$ onto $V_0 \otimes V_0$. Since the Haar wavelet is orthonormal, the basis functions in (9.4) are orthonormal, so that we can use the orthogonal decomposition formula to find this projection. Clearly $\phi_{1,k_1} \otimes \phi_{1,k_2}$ has different support from all except one of $\phi_{0,n_1} \otimes \phi_{0,n_2}$. Since

$$
\langle \phi_{1,k_1} \otimes \phi_{1,k_2}, \phi_{0,n_1} \otimes \phi_{0,n_2} \rangle = \langle \phi_{1,k_1}, \phi_{0,n_1} \rangle \langle \phi_{1,k_2}, \phi_{0,n_2} \rangle = \frac{\sqrt{2}}{2} \frac{\sqrt{2}}{2} = \frac{1}{2}
$$

when the supports intersect, we obtain

$$
\mathrm{proj}_{V_0 \otimes V_0}(\phi_{1,k_1} \otimes \phi_{1,k_2}) =
\begin{cases}
\frac{1}{2}(\phi_{0,k_1/2} \otimes \phi_{0,k_2/2}) & k_1, k_2 \text{ even} \\
\frac{1}{2}(\phi_{0,k_1/2} \otimes \phi_{0,(k_2-1)/2}) & k_1 \text{ even}, k_2 \text{ odd} \\
\frac{1}{2}(\phi_{0,(k_1-1)/2} \otimes \phi_{0,k_2/2}) & k_1 \text{ odd}, k_2 \text{ even} \\
\frac{1}{2}(\phi_{0,(k_1-1)/2} \otimes \phi_{0,(k_2-1)/2}) & k_1, k_2 \text{ odd}
\end{cases}
$$

So, in this case there are four different formulas. Let us also compute the projection onto the orthogonal complement of $V_0 \otimes V_0$ in $V_1 \otimes V_1$. Also here there are four different formulas. When k_1, k_2 are both even we obtain

$$\phi_{1,k_1} \otimes \phi_{1,k_2} - \text{proj}_{V_0 \otimes V_0}(\phi_{1,k_1} \otimes \phi_{1,k_2})$$

$$= \phi_{1,k_1} \otimes \phi_{1,k_2} - \frac{1}{2}(\phi_{0,k_1/2} \otimes \phi_{0,k_2/2})$$

$$= \left(\frac{1}{\sqrt{2}}(\phi_{0,k_1/2} + \psi_{0,k_1/2})\right) \otimes \left(\frac{1}{\sqrt{2}}(\phi_{0,k_2/2} + \psi_{0,k_2/2})\right) - \frac{1}{2}(\phi_{0,k_1/2} \otimes \phi_{0,k_2/2})$$

$$= \frac{1}{2}(\phi_{0,k_1/2} \otimes \phi_{0,k_2/2}) + \frac{1}{2}(\phi_{0,k_1/2} \otimes \psi_{0,k_2/2})$$

$$+ \frac{1}{2}(\psi_{0,k_1/2} \otimes \phi_{0,k_2/2}) + \frac{1}{2}(\psi_{0,k_1/2} \otimes \psi_{0,k_2/2}) - \frac{1}{2}(\phi_{0,k_1/2} \otimes \phi_{0,k_2/2})$$

$$= \frac{1}{2}(\phi_{0,k_1/2} \otimes \psi_{0,k_2/2}) + \frac{1}{2}(\psi_{0,k_1/2} \otimes \phi_{0,k_2/2}) + \frac{1}{2}(\psi_{0,k_1/2} \otimes \psi_{0,k_2/2}).$$

Here we have used the relation $\phi_{1,k_i} = \frac{1}{\sqrt{2}}(\phi_{0,k_i/2} + \psi_{0,k_i/2})$, which we have previously obtained. When either k_1 or k_2 is odd, similar formulas for the projection onto the orthogonal complement can be found. In all cases, the formulas use the basis functions for $W_0^{(0,1)}$, $W_0^{(1,0)}$, $W_0^{(1,1)}$. These functions are shown in Fig. 9.1, together with the function $\phi \otimes \phi \in V_0 \otimes V_0$.

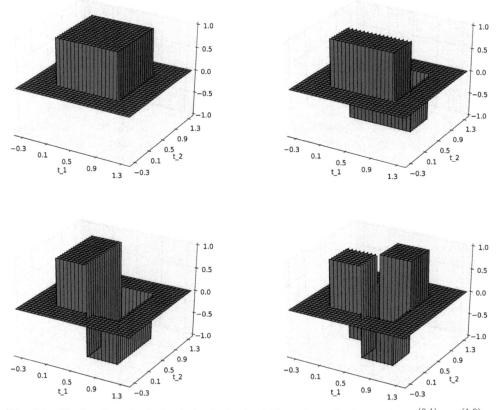

Fig. 9.1 The functions $\phi \otimes \phi$, $\phi \otimes \psi$, $\psi \otimes \phi$, $\psi \otimes \psi$, which are bases for $(V_0 \otimes V_0) \oplus W_0^{(0,1)} \oplus W_0^{(1,0)} \oplus W_0^{(1,1)}$ for the Haar wavelet

Example 9.2: Piecewise Linear Functions

If we instead use any of the wavelets for piecewise linear functions, the wavelet basis functions are not orthogonal anymore, just as in the one-dimensional case. The new basis functions are shown in Fig. 9.2 for the alternative piecewise linear wavelet.

9.2.1 Interpretation

An immediate corollary of Theorem 4 is the following:

Corollary 7. Two-Dimensional DWT.
Let

$$A_m = P_{(\phi_{m-1}, \psi_{m-1}) \leftarrow \phi_m}$$
$$B_m = P_{\phi_m \leftarrow (\phi_{m-1}, \psi_{m-1})}$$

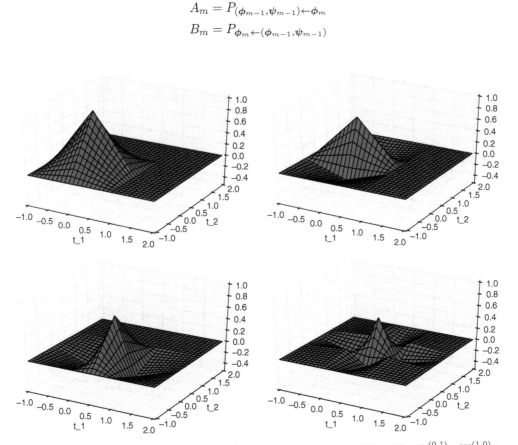

Fig. 9.2 The functions $\phi \otimes \phi$, $\phi \otimes \psi$, $\psi \otimes \phi$, $\psi \otimes \psi$, which are bases for $(V_0 \otimes V_0) \oplus W_0^{(0,1)} \oplus W_0^{(1,0)} \oplus W_0^{(1,1)}$ for the alternative piecewise linear wavelet

be the levels in the DWT and the IDWT, and let

$$X = (c_{m,i,j})_{i,j} \qquad\qquad Y = \begin{pmatrix} (c_{m-1,i,j})_{i,j} & (w^{(0,1)}_{m-1,i,j})_{i,j} \\ (w^{(1,0)}_{m-1,i,j})_{i,j} & (w^{(1,1)}_{m-1,i,j})_{i,j} \end{pmatrix} \qquad (9.7)$$

be the coordinate matrices in $\phi_m \otimes \phi_m$, and $(\phi_{m-1}, \psi_{m-1}) \otimes (\phi_{m-1}, \psi_{m-1})$, respectively. Then

$$Y = A_m X A_m^T \qquad\qquad (9.8)$$

$$X = B_m Y B_m^T \qquad\qquad (9.9)$$

By the m-level two-dimensional DWT/IDWT (or 2D DWT/2D IDWT) we mean the change of coordinates where this is repeated for $m, m-1, \ldots, 1$.

This definition of the m-level two-dimensional DWT says that we should iterate a two-dimensional coordinate change m times. The resulting basis functions are $\phi_{0,n_1} \otimes \phi_{0,n_2}$, as well as

$$\{\phi_{k,n_1} \otimes \psi_{k,n_2}, \psi_{k,n_1} \otimes \phi_{k,n_2}, \psi_{k,n_1} \otimes \psi_{k,n_2}\}_{n_1,n_2,0 \le k < m}. \qquad (9.10)$$

This is how the two-dimensional DWT is defined in the literature, and how it is applied in the JPEG2000 standard (there the Spline 5/3 and the CDF 9/7 wavelets are used).

It is straightforward to implement the 2D DWT and the 2D IDWT. Note that `dwt_impl` and `idwt_impl` are implemented as to support any dimensions, and simply forwards the computation to functions optimized for the number of dimensions in question. In the two-dimensional case these are `dwt2_impl_internal` and `idwt2_impl_internal`. They differ from `dwt1_impl_internal` and `idwt1_impl_internal` in that each dimension can use different kernels. For our purposes, however, the kernel function will always be the same for both dimensions. This means that a 2D DWT with the Haar wavelet can be implemented as follows.

```
x = dwt2_impl_internal(x, @dwt_kernel_haar, @dwt_kernel_haar);
```

Inside this method the function `tensor2_impl` from Chap. 8 is called in order to compute the tensor product.

When using many levels of the 2D DWT, the next stage is applied only to the upper left corner of the matrix (just as the DWT is applied only to the first half of the coordinates), splitting it into four equally big parts. This is illustrated in Fig. 9.3, where the different types of coordinates which appear in the first two stages are indicated.

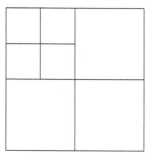

Fig. 9.3 Illustration of the different coordinates in a two level 2D DWT before the first stage, and after the first and second stages

It is instructive to see what information the different types of coordinates in an image represent. In the following examples we will discard some types of coordinates, and view the resulting image. Discarding a type of coordinates will be illustrated by coloring the corresponding regions from Fig. 9.3 black. As an example, if we perform a one-level 2D DWT, Fig. 9.4 illustrates first the set of all coordinates, and then the resulting set after discarding the different coordinate types at the first level successively. Figure 9.5 illustrates in the same way discarding the different coordinate types at the second level.

Fig. 9.4 Discarding wavelet coefficients at the first level. After the 2D DWT, the coefficients are split into four parts (left). In the following figures we have removed coefficients from $W_1^{(1,1)}$, $W_1^{(1,0)}$, and $W_1^{(0,1)}$, respectively

Fig. 9.5 Discarding wavelet coefficients at the second level. After the second stage in the 2D DWT, the coefficients from the upper left corner are also split into four parts (left). In the following figures we have removed coefficients from $W_2^{(1,1)}$, $W_2^{(1,0)}$, and $W_2^{(0,1)}$, respectively

We would also like to make another interpretation of the different coordinates $(c_{m-1,i,j})_{i,j}$, $(w^{(0,1)})_{m-1,i,j}$, $(w^{(1,0)})_{m-1,i,j}$, and $(w^{(1,1)})_{m-1,i,j}$ in terms of filters, as stated in Chap. 5. As we have stated, in a 2D DWT the DWT is first applied to the columns in the image, then to the resulting rows. When applying the DWT to the columns, we have learned that the first half of the coordinates is output from a low-pass filter H_0, the second half output from a high-pass filter H_1 (Theorem 2). As a result, the upper half of the matrix is output from a low-pass filter, the lower half output from a high-pass filter. Applying the DWT similarly to the resulting rows, we get a left part which is output from a low-pass filter, and a right part which is output from a high-pass filter. The four corners resulting from a one-level DWT can thus be characterized in terms of filters. Using the shorthand notation L for low-pass and H for high-pass, we can assign the following names to these corners:

- The upper left corner is called the LL-subband,
- The upper right corner is called the LH-subband,
- The lower left corner is called the HL-subband,
- The lower right corner is called the HH-subband.

The two letters thus indicate the type of filters which have been applied, the first indicating the filter applied to the columns, the second the one applied to the rows. The order is therefore important. The name *subband* comes from the interpretation of these filters as being selective on a certain frequency band. Since low-pass and high-pass filters extract slow variations and abrupt changes, respectively, the following hold:

Observation 8. Visual Interpretation of the 2D DWT.
After the 2D DWT has been applied to an image, we expect to see the following:

- *In the upper left corner, slow variations in both the vertical and horizontal directions are captured, i.e. this is a low-resolution version of the image.*
- *In the upper right corner, slow variations in the vertical direction are captured, together with abrupt changes in the horizontal direction.*
- *In the lower left corner, slow variations in the horizontal direction are captured, together with abrupt changes in the vertical direction.*
- *In the lower right corner, abrupt changes in both directions are captured.*

These effects will be studied through examples in the next section.

Exercise 9.3: Alternative Definition of the 2D DWT

In terms of tensor products, the way we defined the two-dimensional DWT may not seem very natural. Why not define it as the tensor product of two m-level DWT's, instead of iterated tensor products of 1-level DWT's)?

a) Explain that this alternative definition would alter the vectors listed in (9.10) to

$$\{\phi_{0,n_1} \otimes \psi_{k_2,n_2}, \psi_{k_1,n_1} \otimes \phi_{0,n_2}, \psi_{k_1,n_1} \otimes \psi_{k_2,n_2}\}_{n_1,n_2,0\leq k_1,k_2<m}.$$

b) Explain that, with this alternative definition, `dwt2_impl_internal` and `idwt2_impl_internal` could be implemented simply as

```
f = @(x, bd_mode) dwt_impl(x, 'cdf97', 4, bd_mode, 'none', 1);
img = tensor2_impl(img, f, f, 'symm');
```

and

```
invf = @(x, bd_mode) idwt_impl(x, 'cdf97', 4, bd_mode, 'none', 1);
img = tensor2_impl(img, invf, invf, 'symm');
```

This is much simpler than the implementation we have presented.
    There seems to be few comments in the literature on why the two-dimensional DWT is not defined as a tensor product of two m-level DWT's.

Exercise 9.4: Higher Order Tensor Products

Tensor products can easily be generalized to higher dimensions. For 3D, and for $x \in \mathbb{R}^M$, $y \in \mathbb{R}^N$, $z \in \mathbb{R}^K$, we define $x \otimes y \otimes z$ as the "3D matrix" with entries

$$(x \otimes y \otimes z)_{i,j,k} = x_i y_j z_k,$$

and if S_1, S_2 and S_3 are linear transformations on $\mathbb{R}^M$, $\mathbb{R}^N$, and $\mathbb{R}^K$, respectively, their tensor product is the unique linear transformation $S_1 \otimes S_2 \otimes S_3$ defined on 3D matrices by

$$(S_1 \otimes S_2 \otimes S_3)(\boldsymbol{x} \otimes \boldsymbol{y} \otimes \boldsymbol{z}) = (S_1\boldsymbol{x}) \otimes (S_2\boldsymbol{y}) \otimes (S_3\boldsymbol{z}),$$

for all $\boldsymbol{x}$, $\boldsymbol{y}$, and $\boldsymbol{z}$.

a) Explain that $S_1 \otimes S_2 \otimes S_3$ can be implemented by

- replacing $x_{0:(M-1),j,k}$ by $S_1(x_{0:(M-1),j,k})$ for all j, k,
- replacing $x_{i,0:(N-1),k}$ by $S_2(x_{i,0:(N-1),k})$ for all i, k,
- replacing $x_{i,j,0:(K-1)}$ by $S_3(x_{i,j,0:(K-1)})$ for all j, k,

and that the order of these three operations does not matter. In other words, higher order tensor products can also be split into a series of one-dimensional operations.

b) In particular we can extend the tensor product to wavelets in 3D. How many detail components will there be at each level in a 3D DWT, following the setup we did for the 2D DWT? How would you label the different components?

`dwt3_impl_internal` and `idwt3_impl_internal` are the functions in the library which compute the 3D DWT and the 3D IDWT. You are in particular encouraged to review the function `tensor3_impl`, which is used in the functions, and is a straightforward generalization of `tensor2_impl`. Also in the 3D case there is a simpler alternative definition for the DWT, following the setup from the previous exercise.

9.3 Experiments with Images Using Wavelets

In this section we will make some experiments with images using the wavelets we have considered.[1] We will apply wavelets to images by visualizing the pixel values as co-ordinates in the basis $\phi_m \otimes \phi_m$ (so that the image has size $(2^m M) \times (2^m N)$). As in the case for sound this means that we commit the wavelet crime. We then perform a change of coordinates with the 2D DWT, and set either the detail components or the low-resolution coordinates to zero, depending on what we want to inspect. Finally we apply the 2D IDWT to end up with coordinates in $\phi_m \otimes \phi_m$ again, and display the new image with pixel values equal to these coordinates.

Example 9.5: Applying the Haar Wavelet to a Chess Pattern Image

Let us apply the Haar wavelet to the sample chess pattern example image from Fig. 8.15, which was produced by the code

---

[1] Note also that MATLAB has a wavelet toolbox which could be used for these purposes. We will however not go into the usage of this, since we implement the DWT from scratch.

```
N = 128;
img=zeros(N);
for x=0:(N-1)
    for y=0:(N-1)
        img(x+1,y+1) = 255*( (mod(x-1,64)>=32) == (mod(y-1,64)>=32) );
    end
end
```

The low-pass filter of the Haar wavelet was essentially a low-pass filter with two coefficients. Also, as we have seen, the high-pass filter essentially computes an approximation to the partial derivative. Clearly, abrupt changes in the vertical and horizontal directions appear here only at the edges in the chess pattern, and abrupt changes in both directions appear only at the grid points in the chess pattern. After a 2D DWT Observation 8 thus states that we should see vertical edges from the chess pattern in one corner, horizontal edges in another, and the grid pattern in yet another. This can be verified with the following code,

```
img = dwt_impl(img, 'Haar', 1, 'symm', 'none', 2);
img = map_to_01(img);
img = img*255;
imshow(uint8(img))
```

which produces Fig. 9.6. The first lines in the code produces the simple chess pattern. We had to map the result back to $[0, 255]$, as the DWT coefficients typically take values outside the range of the image samples. Note also that we explicitly stated the dimension of the transform (2). This is again due to the assumption made by the library that, for data of more than one dimension, the DWT should be parallelized on the last dimension. For data from grey-color images this is clearly undesirable, so that the dimension needs to be passed as explicitly as parameter.

Fig. 9.6 The chess pattern example image after application of the 2D DWT. The Haar wavelet was used

Example 9.6: Creating Thumbnail Images

Let us apply the Haar wavelet to the Lena image. The following code computes the low-resolution approximation for $m = 1$:

```
X = dwt_impl(X, 'Haar', 1);
X = X(1:(size(X,1)/2), 1:(size(X,2)/2),:);
X = map_to_01(X); X = X*255;
```

Also here it was necessary to map the result back to $[0, 255]$. We also have omitted the third parameter here, as $m = 1$ is the default number of resolutions in `dwt_impl`. Also, we have not entered the number of dimensions (2) in the parameter list in this case. The reason is that this image, contrary to the previous one, is an RGB image: Since RGB images have data of dimension 3, the library correctly assumes that the DWT should be parallelized on the third dimension (the color component), which is the desirable behavior.

Repeating this up to four resolutions we obtain the images in Fig. 9.7. In Fig. 9.8 we have also shown the entire result after a 1- and 2-stage 2D DWT on the image. The first two thumbnail images can be seen as the upper left corners of the first two images. The other corners represent detail.

Fig. 9.7 The corresponding thumbnail images for the Image of Lena, obtained with a DWT of 1, 2, 3, and 4 levels

Fig. 9.8 The corresponding image resulting from a wavelet transform with the Haar-wavelet for $m = 1$ and $m = 2$

Example 9.7: Detail and Low-Resolution Approximations for Different Wavelets

Let us take a closer look at the images generated when we use different wavelets, setting the detail coefficients to zero, and viewing the result as an image of the same size. In particular, let us discard the coefficients as pictured in Figs. 9.4 and 9.5. We should expect that the lower order resolution approximations from V_0 are worse as m increases.

Figure 9.9 confirms this for the lower order resolution approximations for the Haar wavelet. Alternatively, we should see that the higher order detail spaces contain more information as m increases. Figure 9.10 confirms this.

Figures 9.11 and 9.12 confirm the same for the CDF 9/7 wavelet, Here some improvement in the low resolution approximations can be seen. Since black indicates values which are 0, most of the coefficients must be small.

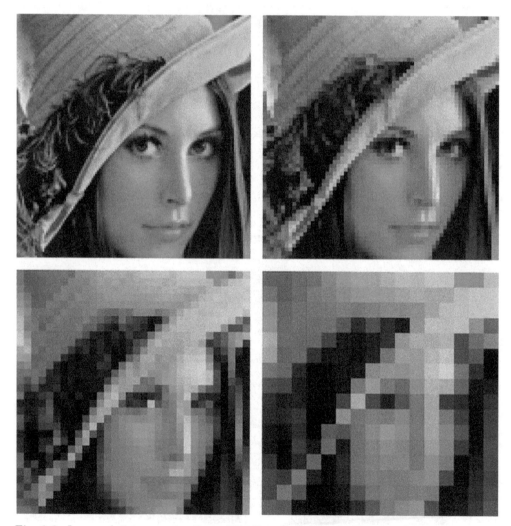

Fig. 9.9 Low resolution approximations of the Lena image, for the Haar wavelet

Example 9.8: The Spline 5/3 Wavelet and Removing Bands in the Detail Spaces

The detail components in images are split into three bands, so let us see what happens when we discard some of them (i.e. some of $W_m^{(1,1)}$, $W_m^{(1,0)}$, $W_m^{(0,1)}$). Let us use the Spline 5/3 wavelet.

The resulting images when the first level coefficients indicated in Fig. 9.4 are discarded are shown in Fig. 9.13. The corresponding plot for the second level is shown in Fig. 9.14.

The image is seen still to resemble the original one, even after two levels of wavelet coefficients have been discarded. This in itself is good for compression purposes, since we may achieve compression simply by dropping the given coefficients. However, if we continue to discard more levels, the result will look poorer.

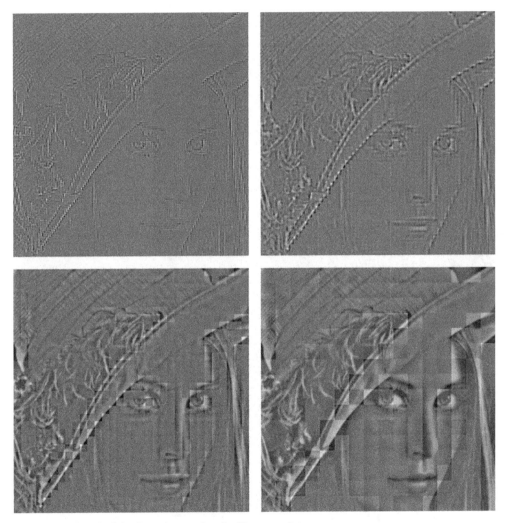

Fig. 9.10 Detail of the Lena image, for the Haar wavelet

In Fig. 9.15 we have also shown the resulting image after the third and fourth levels of detail have been discarded. Although we still can see details in the image, the quality is definitely poorer.

In Fig. 9.16, we have shown the corresponding detail for DWT's with 1, 2, 3, and 4 levels. Clearly, more detail can be seen in the image when more of the detail is included.

Exercise 9.9: Code Example

Assume that we have an image represented by the $M \times N$-matrix X, and consider the following code:

```
c = (X(1:2:M, :) + X(2:2:M, :))/sqrt(2);
w = (X(1:2:M, :) - X(2:2:M, :))/sqrt(2);
X = [c; w];
```

Fig. 9.11 Low resolution approximations of the Lena image, for the CDF 9/7 wavelet

```
c = (X(:, 1:2:N) + X(:, 2:2:N))/sqrt(2);
w = (X(:, 1:2:N) - X(:, 2:2:N))/sqrt(2);
X = [c w];
```

a) Comment what the code does, and explain what you will see if you display X as an image after the code has run.

b) The code above has an inverse transform, which reproduces the original image from the transformed values. Assume that you zero out the values in the lower left and the upper right corner of the matrix X after the code above has run, and that you reproduce the image by applying this inverse transform. What changes do you expect in the image?

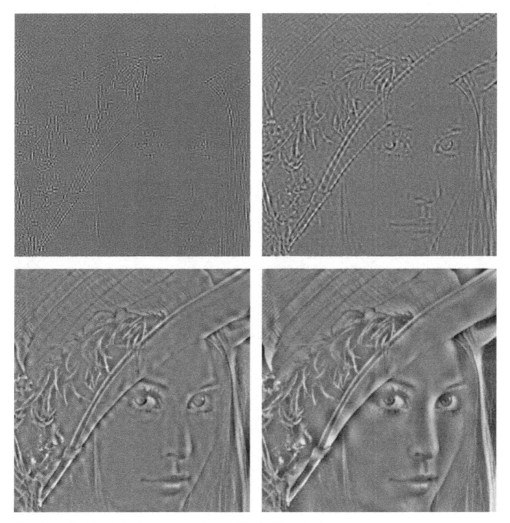

Fig. 9.12 Detail of the Lena image, for the CDF 9/7 wavelet

Fig. 9.13 Image of Lena, with various bands of detail at the first level zeroed out. From left to right, the detail at $W_1^{(1,1)}$, $W_1^{(1,0)}$, $W_1^{(0,1)}$, as illustrated in Fig. 9.4. The Spline 5/3 wavelet was used

Exercise 9.10: Applying the Haar Wavelet to Another Chess Pattern Image

The following code produces a chess pattern type image almost identical to that from Example 9.5.

Fig. 9.14 Image of Lena, with various bands of detail at the second level zeroed out. From left to right, the detail at $W_2^{(1,1)}$, $W_2^{(1,0)}$, $W_2^{(0,1)}$, as illustrated in Fig. 9.5. The Spline 5/3 wavelet was used

Fig. 9.15 Image of Lena, with detail including level 3 and 4 zeroed out. The Spline 5/3 wavelet was used

```
N = 128;
img=zeros(N);
for x=0:(N-1)
    for y=0:(N-1)
        img(x+1,y+1)=255*( (mod(x,64)>=32) == (mod(y,64)>=32) );
    end
end
imshow(uint8(img))
```

Let us now apply a 2D DWT to this image as well with the Haar wavelet:

```
img2 = img;
img2 = dwt_impl(img2, 'Haar', 1, 'per', 'none', 2);
img2 = map_to_01(img2);
img2 = img2*255;
imshow(uint8(img2))
```

The resulting images are shown in Fig. 9.17

There seem to be no detail components here, which is very different from what you saw in Example 9.5, even though the images are very similar. Attempt to explain what causes this to happen.

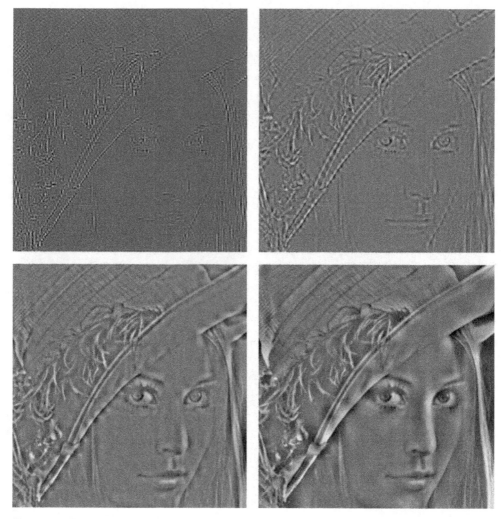

Fig. 9.16 The corresponding detail for the image of Lena. The Spline 5/3 wavelet was used

Hint

Compare with Exercise 4.14.

Fig. 9.17 A simple image before and after one level of the 2D DWT. The Haar wavelet was used

9.4 An Application to the FBI Standard for Compression of Fingerprint Images

In the beginning of the 1990s, the FBI had a major size problem with their fingerprint image archive. With more than 200 million fingerprint records, they needed to employ some compression strategy. Several strategies were tried, for instance the widely adopted JPEG standard. The problem with JPEG had to do with the blocking artifacts. Among other strategies, FBI chose a wavelet-based strategy due to its nice properties. The particular way wavelets are applied in this strategy is called *Wavelet transform/scalar quantization* (WSQ).

Fig. 9.18 A typical fingerprint image

Fingerprint images are a very specific type of images, as seen in Fig. 9.18. They differ from natural images by having a large number of abrupt changes. One may ask whether other wavelets than the ones we have considered are suitable for compressing such images. After all, wavelets with vanishing moments are most suitable for regular images. Extensive tests were undertaken to compare different wavelets, and the CDF 9/7 wavelet used by JPEG2000 turned out to perform very well also for fingerprint images.

Besides the choice of wavelet, one can also ask other questions regarding how to compress fingerprint images: What number of levels is optimal? And, is the subband decomposition obtained by splitting the upper left corner in four repeatedly (as is done in most literature) the best one? As mentioned, one obtains an alternative subband decomposition by computing the tensor product of an m-level DWT with itself. The FBI standard actually applies a subband decomposition different from these two. The first steps in this are illustrated in Fig. 9.19, and can be summarized as follows:

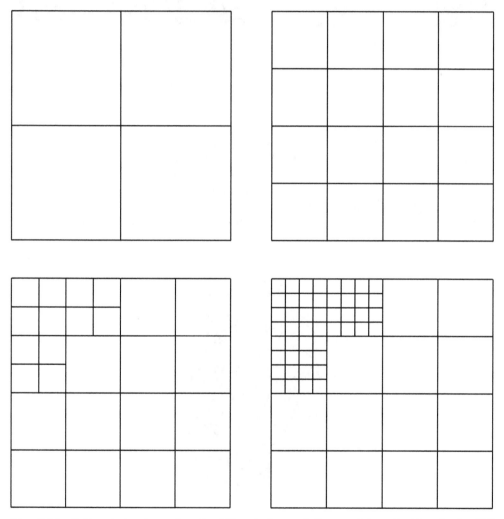

Fig. 9.19 Subband structure after the different stages of the wavelet applications in the FBI fingerprint compression scheme

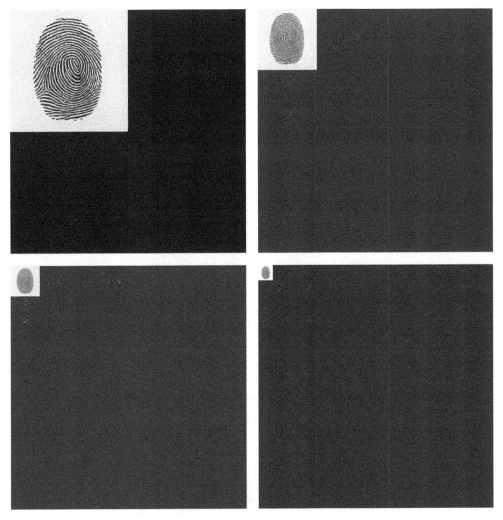

Fig. 9.20 The fingerprint image after several DWT's

1. First split the image in four corners (i.e. a one-level DWT) (Fig. 9.20). This gives the upper left part of Fig. 9.19.
2. Split each of the four corners in four (this is different from the 2D DWT, which only splits the upper left corner further). This gives the upper right part of Fig. 9.19.
3. Split three of the four corners in the upper left corner. This gives the lower left part of Fig. 9.19.
4. Split the smallest squares one more time. This gives the lower right part of Fig. 9.19.

Finally, the upper left corner is again split into four. The final decomposition is illustrated in Fig. 9.21. In Figs. 9.20 and 9.22 the resulting images after each of these steps are shown. In Fig. 9.23 we also show the corresponding low resolution approximation and detail.

In the original JPEG2000 standard it was not possible to adapt the subband decomposition like this. This has been added to a later extension of the standard, which makes the two standards more compatible.

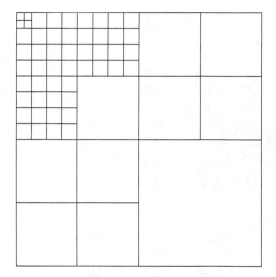

Fig. 9.21 Subband structure after all stages

Fig. 9.22 The resulting image obtained with the subband decomposition employed by the FBI

Exercise 9.11: Implementing the Fingerprint Compression Scheme

Write code which generates the images shown in Figs. 9.20, 9.22, and 9.23. Use the functions `dwt_impl` and `idwt_impl` with the CDF 9/7 wavelet.

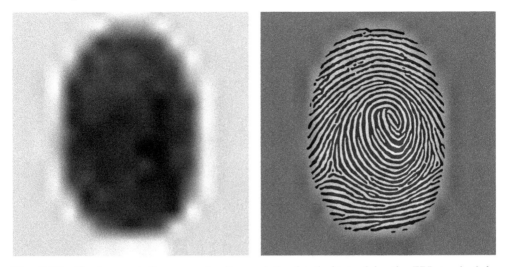

Fig. 9.23 The low-resolution approximation and the detail obtained by the FBI standard for compression of fingerprint images, when applied to our sample fingerprint image

9.5 Summary

We extended tensor products to function spaces, so that they can also be used for wavelets, and extended the wavelet transform to the tensor product setting, so that it could be applied to images. We also performed several experiments on our test images, such as creating low-resolution images and discarding wavelet coefficients, and using the Haar-, the Spline 5/3-, and the CDF 9/7 wavelet.

The specification of the JPEG2000 standard can be found in [27]. In [74], most details of this theory is covered, in particular details on how the wavelet coefficients are coded.

The FBI standard which describes how to compress fingerprint images can be found in [19]. The theory is also described in [5]. The book [21] also uses compression of fingerprint images as an application of wavelets.

Through examples we have seen that the 2D DWT is able to pick up changes in the vertical and horizontal directions. Many changes in images occur in other directions, however, and the DWT may not be have a good "directionality" in this respect. Many concepts similar to wavelets have been developed to address this, such as steerable wavelets, contourlets, surfacelets, shearlets, bandlets, and curvelets. The paper [38] gives a survey on curvelets, and also gives a short overview of the others.

What You Should Have Learned in This Chapter

- The special interpretation of the 2D DWT applied to an image as splitting into four types of coordinates (each being one corner of the image), which represent low-pass/high-pass combinations in the horizontal/vertical directions.
- How to call functions which perform different wavelet transforms on an image.
- Be able to interpret the detail components and low-resolution approximations in what you see.

Appendix A
Basic Linear Algebra

This book assumes that the student has taken a beginning course in linear algebra at university level. In this appendix we summarize the most important concepts one needs to know from linear algebra. Note that what is included here is very compact, has no exercises, and includes only short proofs, leaving more involved results to other textbooks. The material here should not be considered as a substitute for a linear algebra course: It is important for the student to go through a full such course and do many exercises, in order to get good intuition for these concepts. Recommend books.

Vectors are always written in lowercase boldface ($\boldsymbol{x}$, $\boldsymbol{y}$, etc.), and are always assumed to be column vectors, unless otherwise stated. We will also write column vectors as $\boldsymbol{x} = (x_0, x_1, \ldots, x_n)$, i.e. as a comma-separated list of values, with x_i the components of $\boldsymbol{x}$ (i.e. the components are not in boldface, to distinguish scalars and vectors).

A.1 Matrices

An $m \times n$-matrix is simply a set of mn numbers, stored in m rows and n columns. Matrices are usually written in uppercase (A, B, etc.). We write a_{kn} for the entry in row k and column n of the matrix A. Superscripts are also used to differ between vectors/matrices with the same base name (i.e. $\boldsymbol{x}^{(1)}$, $\boldsymbol{x}^{(2)}$, and $A^{(1)}$, $A^{(2)}$, etc.), so that this does not interfere with the indices of the components.

The zero matrix, denoted $\boldsymbol{0}$ is the matrix where all entries are zero. A square matrix (i.e. where $m = n$) is said to be diagonal if $a_{kn} = 0$ whenever $k \neq n$. The identity matrix, denoted I, or I_n to make the dimension of the matrix clear, is the diagonal matrix where the entries on the diagonal are 1. If A is a matrix we will denote the transpose of A by A^T. Since vectors are column vectors per default, a row vector will usually be written as $\boldsymbol{x}^T$, with $\boldsymbol{x}$ a column vector. For complex matrices we also define $A^H = (\overline{A})^T$, i.e. as the *conjugate transpose* of A. A real matrix A is said to be *symmetric* if $A = A^T$, and a complex matrix A is said to be *hermitian* if $A = A^H$.

© Springer Nature Switzerland AG 2019
Ø. Ryan, *Linear Algebra, Signal Processing, and Wavelets - A Unified Approach*,
Springer Undergraduate Texts in Mathematics and Technology,
https://doi.org/10.1007/978-3-030-01812-2

If A is an $m \times n$- matrix and $\boldsymbol{x}$ a (column) vector in $\mathbb{R}^n$, then $A\boldsymbol{x}$ is defined as the vector in $\boldsymbol{y} \in \mathbb{R}^m$ with components $y_i = \sum_{j=1}^{n} a_{ij}x_j$. If the rows of A are denoted $\boldsymbol{a}_1$, $\boldsymbol{a}_2, \ldots, \boldsymbol{a}_n$, we can also write

$$A\boldsymbol{x} = x_1\boldsymbol{a}_1 + x_2\boldsymbol{a}_2 + \cdots + x_n\boldsymbol{a}_n,$$

i.e. $A\boldsymbol{x}$ can be written as a linear combination of the columns of A. If B is a matrix with n rows,

$$B = \begin{pmatrix} \boldsymbol{b}_1 & \boldsymbol{b}_2 & \cdots & \boldsymbol{b}_k \end{pmatrix},$$

then the product of A and B is defined as

$$AB = \begin{pmatrix} A\boldsymbol{b}_1 & A\boldsymbol{b}_2 & \cdots & A\boldsymbol{b}_k \end{pmatrix}.$$

A square matrix A is said to be *invertible* if there exists a matrix B so that $AB = BA = I$. It is straightforward to show such a B is unique if it exists. B is called the *inverse* of A, and denoted by A^{-1} (so that $A^{-1}A = AA^{-1} = I$). There are many other equivalent conditions for a square matrix to be invertible. One is that $A\boldsymbol{x} = \boldsymbol{0}$ is equivalent to $\boldsymbol{x} = 0$. If the latter is fulfilled we also say that A is *non-singular*. If not, A is said to be *singular*. The terms invertible and non-singular can thus be used interchangeably.

If A is a real non-singular matrix, its inverse is also real. A real matrix A is *orthogonal* if $A^{-1} = A^T$, and a complex matrix A is said to be unitary is $A^{-1} = A^H$. Many matrices constructed in this book are unitary, such as the DFT, the DCT, and some DWT's. A very simple form of unitary matrices are *permutation matrices*, which simply reorders the components in a vector.

A matrix is called sparse if most of the entries in the matrix are zero. Linear systems where the coefficient matrix is sparse can be solved efficiently, and sparse matrices can be multiplied efficiently as well. In this book there are several examples where a matrix can be factored as

$$A = A_1 A_2 \cdots A_n,$$

with the $A_1, A_2, \ldots, A_n$ being sparse. This means that sparse matrix optimizations can be applied for A. This factorization can also be useful when A is a sparse matrix at the beginning, in order to factor a sparse matrix into a product of sparser matrices.

A.2 Block Matrices

If $m_0, \ldots, m_{r-1}, n_0, \ldots, n_{s-1}$ are integers, and $A^{(i,j)}$ is an $m_i \times n_j$-matrix for $i = 0, \ldots, r-1$ and $j = 0, \ldots, s-1$, then

$$A = \begin{pmatrix} A^{(0,0)} & A^{(0,1)} & \cdots & A^{(0,s-1)} \\ A^{(1,0)} & A^{(1,1)} & \cdots & A^{(1,s-1)} \\ \vdots & \vdots & \ddots & \vdots \\ A^{(r-1,0)} & A^{(r-1,1)} & \cdots & A^{(r-1,s-1)} \end{pmatrix}$$

will denote the $(m_0 + m_1 + \ldots + m_{r-1}) \times (n_0 + n_1 + \ldots + n_{s-1})$-matrix where the $A^{(i,j)}$ are stacked horizontally and vertically. When A is written in this way it is referred to as an $(r \times s)$ *block matrix*, and the $A^{(i,j)}$ are called the *blocks* of A.

A block matrix is called *block diagonal* if the off-diagonal blocks are zero, i.e.

$$
A = \begin{pmatrix} A_0 & 0 & \cdots & 0 \\ 0 & A_1 & \cdots & 0 \\ \vdots & \vdots & \ddots & \vdots \\ 0 & 0 & \cdots & A_{r-1} \end{pmatrix}.
$$

We will also use the notation $\mathrm{diag}(A_0, A_1, \cdots, A_{n-1})$ for block diagonal matrices. The identity matrix is clearly block diagonal, with all diagonal blocks being the identity. The (conjugate) transpose of a block matrix is also a block matrix, and we have that

$$
A^T = \begin{pmatrix} (A^{(0,0)})^T & (A^{(1,0)})^T & \cdots & (A^{(r-1,0)})^T \\ (A^{(0,1)})^T & (A^{(1,1)})^T & \cdots & (A^{(r-1,1)})^T \\ \vdots & \vdots & \ddots & \vdots \\ (A^{(0,s-1)})^T & (A^{(1,s-1)})^T & \cdots & (A^{(r-1,s-1)})^T \end{pmatrix}.
$$

If A and B are block matrices with blocks $A^{(i,j)}$, $B^{(i,j)}$, respectively, then $C = AB$ is also a block matrix, with blocks

$$
C^{(i,j)} = \sum_k A^{(i,k)} B^{(k,j)},
$$

as long as A has the same number of horizontal blocks as B has vertically, and as long as each $A^{(i,k)}$ has the same number of columns as $B^{(k,j)}$ has rows.

A.3 Vector Spaces

Given set of objects V (objects are also called *vectors*). V is called a *vector space* if it has an operation for addition, $+$, and an operation for multiplying with scalars, which satisfy certain properties. First of all $u + v$ must be in V whenever $u \in V$ and $V \in V$, and

- $+$ is commutative (i.e. $u + v = v + u$),
- $+$ is associative (i.e. $(u + v) + w = u + (v + w)$),
- there exists a zero vector in V (i.e. a vector 0 so that $u + 0 = u$ for all $u \in V$).
- For each u in V there is a vector $-u \in V$ so that $u + (-u) = 0$.

Multiplication by scalars, written cu for c a scalar and $u \in V$, is required to satisfy similar properties. Scalars may be required to be either real or complex, in which we talk about *real* and *complex vector spaces*.

We say that vectors $\{v_0, v_1, \ldots, v_{n-1}\}$ in a vector space are *linearly independent* if, whenever $\sum_{i=0}^{n-1} c_i v_i = 0$, we must have that all $c_i = 0$. We will say that a set of vectors $\mathcal{B} = \{v_0, v_1, \ldots, v_{n-1}\}$ from V is a *basis* for V if the vectors are linearly independent, and if any vector in V can be expressed as a linear combination of vectors from $\mathcal{B}$ (we say that $\mathcal{B}$ span V). Any vector in V can then be expressed as a linear combination from vectors in $\mathcal{B}$ in a unique way. Any basis for V has the same number of vectors. If the basis $\mathcal{B}$ for V has n vectors, V is said to have *dimension n*. We also write $\dim(V)$ for the dimension of V. The basis $\{e_0, e_1, \ldots, e_{n-1}\}$ for $\mathbb{R}^n$ is also called the *standard basis* of $\mathbb{R}^n$, and is denoted $\mathcal{E}_n$.

If the $n \times n$-matrix S is non-singular and $\boldsymbol{x}_1, \boldsymbol{x}_2, \ldots, \boldsymbol{x}_k$ are linearly independent vectors in $\mathbb{R}^n$, then $S\boldsymbol{x}_1, S\boldsymbol{x}_2, \ldots, S\boldsymbol{x}_k$ are also linearly independent vectors in $\mathbb{R}^n$. If not there would exist $c_1, \ldots, c_k$, not all zero, so that $\sum_{i=1}^{k} c_i S\boldsymbol{x}_i = \boldsymbol{0}$. This implies that $S(\sum_{i=1}^{k} c_i \boldsymbol{x}_i) = \boldsymbol{0}$, so that $\sum_{i=1}^{k} c_i \boldsymbol{x}_i = \boldsymbol{0}$ since S is non-singular. This contradicts that the $\boldsymbol{x}_i$ are linearly independent, so that the $S\boldsymbol{x}_i$ must be linearly independent as well. In other words, if X is the matrix with $\boldsymbol{x}_i$ as columns, X has linearly independent columns if and only if SX has linearly independent columns.

A subset U of a vector space V is called a *subspace* if U also is a vector space. Most of the vector spaces we consider are either subspaces of $\mathbb{R}^N$, matrix spaces, or function spaces. Examples of often encountered subspaces of $\mathbb{R}^N$ are

- the column space $\text{col}(A)$ of the matrix A (i.e. the space spanned by the columns of A),
- the row space $\text{row}(A)$ of the matrix A (i.e. the space spanned by the rows of A), and
- the null space $\text{null}(A)$ of the matrix A (i.e. the space of all vectors $\boldsymbol{x}$ so that $A\boldsymbol{x} = \boldsymbol{0}$).

It turns out that $\dim(\text{col}(A)) = \dim(\text{row}(A))$. To see why, recall that a general matrix can be brought to row-echelon form through a series of *elementary row operations*. Each such operation does not change the row space of a matrix. Also, each such operation is represented by an non-singular matrix, so that linear independence relations are unchanged after these operations (although the column space itself changes). It follows that $\dim(\text{col}(A)) = \dim(\text{row}(A))$ if and only if this holds for any A in row-echelon form. For a matrix in row-echelon form, however, the dimension of the row- and the column space clearly equals the number of pivot elements, and this proves the result. This common dimension of $\text{col}(A)$ and $\text{row}(A)$ is called the *rank* of A, denoted $\text{rank}(A)$. From what we showed above, the rank of any matrix is unchanged if we multiply with an non-singular matrix to the left or to the right.

Elementary row operations, as mentioned above, can be split into three types: Swapping two rows, multiplying one row by a constant, and adding a multiple of one row to another, i.e. multiplying row j with a constant λ, and add this to row i. Clearly this is the same as computing $E_{i,j,\lambda} A$ where $E_{i,j,\lambda}$ is the matrix $I_m + \lambda \boldsymbol{e}_i \boldsymbol{e}_j^T$, i.e. the matrix which equals the identity matrix, except for an additional entry λ at indices (i, j). The elementary lifting matrices of Chap. 4 combining many such operations into one. As an example it is straightforward to verify that

$$A_\lambda = \prod_{i=0}^{N/2-1} \left(E_{2i,2i-1,\lambda} E_{2i,2i+1,\lambda} \right).$$

As noted in the book, $(A_\lambda)^{-1} = A_{-\lambda}$, which can also be viewed in light of the fact that $(E_{i,j,\lambda})^{-1} = E_{i,j,-\lambda}$.

A.4 Inner Products and Orthogonality

Most vector spaces in this book are inner product spaces. A (real) *inner product* on a vector space is a binary operation, written as $\langle \boldsymbol{u}, \boldsymbol{v} \rangle$, which fulfills the following properties for any vectors $\boldsymbol{u}$, $\boldsymbol{v}$, and $\boldsymbol{w}$:

- $\langle \boldsymbol{u}, \boldsymbol{v} \rangle = \langle \boldsymbol{v}, \boldsymbol{u} \rangle$
- $\langle \boldsymbol{u} + \boldsymbol{v}, \boldsymbol{w} \rangle = \langle \boldsymbol{u}, \boldsymbol{w} \rangle + \langle \boldsymbol{v}, \boldsymbol{w} \rangle$

- $\langle c\boldsymbol{u}, \boldsymbol{v}\rangle = c\langle \boldsymbol{u}, \boldsymbol{v}\rangle$ for any scalar c
- $\langle \boldsymbol{u}, \boldsymbol{u}\rangle \geq 0$, and $\langle \boldsymbol{u}, \boldsymbol{u}\rangle = 0$ if and only if $\boldsymbol{u} = 0$.

$\boldsymbol{u}$ and $\boldsymbol{v}$ are said to be *orthogonal* if $\langle \boldsymbol{u}, \boldsymbol{v}\rangle = \boldsymbol{0}$. In this book we have seen two important examples of inner product spaces. First of all the Euclidean inner product, which is defined by

$$\langle \boldsymbol{u}, \boldsymbol{v}\rangle = \boldsymbol{v}^T \boldsymbol{u} = \sum_{i=0}^{n-1} u_i v_i \qquad (A.1)$$

for any $\boldsymbol{u}$, $\boldsymbol{v}$ in $\mathbb{R}^n$. For functions we have seen examples which are variants of the following form:

$$\langle f, g\rangle = \int f(t)g(t)dt. \qquad (A.2)$$

Functions are usually not denoted in boldface, to distinguish them from vectors in $\mathbb{R}^n$. These inner products are real, meaning that it is assumed that the underlying vector space is real, and that $\langle \cdot, \cdots \rangle$ is real-valued. We have also use for *complex inner products*, i.e. complex-valued binary operations $\langle \cdot, \cdots \rangle$ defined on complex vector spaces. A complex inner product is required to satisfy the same four axioms above for real inner products, but the first axiom is replaced by

- $\langle \boldsymbol{u}, \boldsymbol{v}\rangle = \overline{\langle \boldsymbol{v}, \boldsymbol{u}\rangle}.$

This new axiom can be used to prove the property $\langle f, cg\rangle = \bar{c}\langle f, g\rangle$, which is not one of the properties for real inner product spaces. This follows by writing

$$\langle f, cg\rangle = \overline{\langle cg, f\rangle} = \overline{c\langle g, f\rangle} = \bar{c}\overline{\langle g, f\rangle} = \bar{c}\langle f, g\rangle.$$

The inner products above can be generalized to complex inner products by defining

$$\langle \boldsymbol{u}, \boldsymbol{v}\rangle = \boldsymbol{v}^H \boldsymbol{u} = \sum_{i=0}^{n-1} u_i \overline{v_i}, \qquad (A.3)$$

and

$$\langle f, g\rangle = \int f(t)\overline{g(t)}dt. \qquad (A.4)$$

Any set of mutually orthogonal vectors are also linearly independent. A basis where all basis vectors are mutually orthogonal is called an *orthogonal basis*. If additionally the vectors all have length 1, we say that the basis is *orthonormal*. Regarding the definition of the Euclidean inner product, it is clear that a real/complex square $n \times n$-matrix is orthogonal/unitary if and only if its rows are an orthonormal basis for $\mathbb{R}^n/\mathbb{C}^n$. The same applies for the columns. Also, any unitary matrix preserves inner products, since if $A^H A = I$, and

$$\langle A\boldsymbol{x}, A\boldsymbol{y}\rangle = (A\boldsymbol{y})^H(A\boldsymbol{x}) = \boldsymbol{y}^H A^H A\boldsymbol{x} = \boldsymbol{y}^H \boldsymbol{x} = \langle \boldsymbol{x}, \boldsymbol{y}\rangle.$$

Setting $\boldsymbol{x} = \boldsymbol{y}$ here it follows that unitary matrices preserve norm, i.e. $\|A\boldsymbol{x}\| = \|\boldsymbol{x}\|$.

If $\boldsymbol{x}$ is in a vector space with an orthogonal basis $\mathcal{B} = \{\boldsymbol{v}_k\}_{k=0}^{n-1}$, we can express $\boldsymbol{x}$ as

$$\frac{\langle \boldsymbol{x}, \boldsymbol{v}_0\rangle}{\langle \boldsymbol{v}_0, \boldsymbol{v}_0\rangle}\boldsymbol{v}_0 + \frac{\langle \boldsymbol{x}, \boldsymbol{v}_1\rangle}{\langle \boldsymbol{v}_1, \boldsymbol{v}_1\rangle}\boldsymbol{v}_1 + \cdots + \frac{\langle \boldsymbol{x}, \boldsymbol{v}_{n-1}\rangle}{\langle \boldsymbol{v}_{n-1}, \boldsymbol{v}_{n-1}\rangle}\boldsymbol{v}_{n-1}. \qquad (A.5)$$

In other words, the weights in linear combinations are easily found when the basis is orthogonal. This is also called the *orthogonal decomposition theorem*.

By the *projection* of a vector $\boldsymbol{x}$ onto a subspace U we mean the vector $\boldsymbol{y} = \text{proj}_U \boldsymbol{x}$ which minimizes the distance $\|\boldsymbol{y} - \boldsymbol{x}\|$. If $\boldsymbol{v}_i$ is an orthogonal basis for U, we have that $\text{proj}_U \boldsymbol{x}$ can be written by Eq. (A.5).

A.5 Coordinates and Change of Coordinates

If $\mathcal{B} = \{\boldsymbol{b}_0, \boldsymbol{b}_1, \ldots, \boldsymbol{b}_{n-1}\}$ is a basis for a vector space, and $\boldsymbol{x} = \sum_{i=0}^{n-1} x_i \boldsymbol{b}_i$, we say that $(x_0, x_1, \ldots, x_{n-1})$ is the *coordinate vector* of $\boldsymbol{x}$ w.r.t. the basis $\mathcal{B}$. We also write $[\boldsymbol{x}]_\mathcal{B}$ for this coordinate vector.

If $\mathcal{B}$ and $\mathcal{C}$ are two different bases for the same vector space, we can write down the two coordinate vectors $[\boldsymbol{x}]_\mathcal{B}$ and $[\boldsymbol{x}]_\mathcal{C}$. A useful operation is to transform the coordinates in $\mathcal{B}$ to those in $\mathcal{C}$, i.e. apply the transformation which sends $[\boldsymbol{x}]_\mathcal{B}$ to $[\boldsymbol{x}]_\mathcal{C}$. This is a linear transformation, and we will denote the $n \times n$-matrix of this linear transformation by $P_{\mathcal{C}\leftarrow\mathcal{B}}$, and call this the *change of coordinate matrix* from $\mathcal{B}$ to $\mathcal{C}$. In other words, it is required that

$$[\boldsymbol{x}]_\mathcal{C} = P_{\mathcal{C}\leftarrow\mathcal{B}}[\boldsymbol{x}]_\mathcal{B}. \tag{A.6}$$

It is straightforward to show that $P_{\mathcal{C}\leftarrow\mathcal{B}} = (P_{\mathcal{B}\leftarrow\mathcal{C}})^{-1}$, so that matrix inversion can be used to compute the change of coordinate matrix the opposite way. It is also straightforward to show that the columns in the change of coordinate matrix from $\mathcal{B}$ to $\mathcal{C}$ can be obtained by expressing the old basis vectors in terms of the new basis vectors, i.e.

$$P_{\mathcal{C}\leftarrow\mathcal{B}} = \left([\boldsymbol{b}_0]_\mathcal{C} \; [\boldsymbol{b}_1]_\mathcal{C} \cdots [\boldsymbol{b}_{n-1}]_\mathcal{C}\right).$$

In particular, the change of coordinate matrix from $\mathcal{B}$ to the standard basis is

$$P_{\mathcal{E}\leftarrow\mathcal{B}} = \left(\boldsymbol{b}_0 \; \boldsymbol{b}_1 \cdots \boldsymbol{b}_{n-1}\right).$$

If the vectors in $\mathcal{B}$ are orthonormal this matrix is unitary, and we obtain that

$$P_{\mathcal{B}\leftarrow\mathcal{E}} = (P_{\mathcal{E}\leftarrow\mathcal{B}})^{-1} = (P_{\mathcal{E}\leftarrow\mathcal{B}})^H = \begin{pmatrix} (\boldsymbol{b}_0)^H \\ (\boldsymbol{b}_1)^H \\ \cdots \\ (\boldsymbol{b}_{n-1})^H \end{pmatrix}.$$

The DFT and the DCT are such coordinates changes.

The *Gramm matrix* of the basis $\mathcal{B}$ is the matrix with entries being $\langle \boldsymbol{b}_i, \boldsymbol{b}_j \rangle$. We will also write $(\langle \mathcal{B}, \mathcal{B} \rangle)$ for this matrix. It is useful to see how the Gramm matrix changes under a change of coordinates. Let S be the change of coordinates from $\mathcal{B}$ to $\mathcal{C}$. Another useful form for this is

$$\begin{pmatrix} \boldsymbol{b}_0 \\ \boldsymbol{b}_1 \\ \vdots \\ \boldsymbol{b}_{n-1} \end{pmatrix} = S^T \begin{pmatrix} \boldsymbol{c}_0 \\ \boldsymbol{c}_1 \\ \vdots \\ \boldsymbol{c}_{n-1} \end{pmatrix},$$

which follows directly since the columns in the change of coordinate matrix is simple the old basis expressed in the new basis. From this it follows that

$$\langle \boldsymbol{b}_i, \boldsymbol{b}_j \rangle = \left\langle \sum_{k=0}^{n-1} (S^T)_{ik} \boldsymbol{c}_k, \sum_{l=0}^{n-1} (S^T)_{jl} \boldsymbol{c}_l \right\rangle$$

$$= \sum_{k=0}^{n-1} \sum_{l=0}^{n-1} (S^T)_{ik} \langle \boldsymbol{c_k}, \boldsymbol{c}_l \rangle S_{lj} = (S^T(\langle \mathcal{C}, \mathcal{C} \rangle)S)_{ij}.$$

It follows that

$$(\langle \mathcal{B}, \mathcal{B} \rangle) = S^T(\langle \mathcal{C}, \mathcal{C} \rangle)S.$$

More generally when $\mathcal{B}$ and $\mathcal{C}$ are two different bases for the same space, we can define $(\langle \mathcal{B}, \mathcal{C} \rangle)$ as the matrix with entries $\langle \boldsymbol{b}_i, \boldsymbol{c}_j \rangle$. If $S_1 = P_{\mathcal{C}_1 \leftarrow \mathcal{B}_1}$ and $S_2 = P_{\mathcal{C}_2 \leftarrow \mathcal{B}_2}$ are coordinate changes, it is straightforward to generalize the calculation above to show that

$$(\langle \mathcal{B}_1, \mathcal{B}_2 \rangle) = (S_1)^T (\langle \mathcal{C}_1, \mathcal{C}_2 \rangle) S_2.$$

If T is a linear transformation between the spaces V and W, and $\mathcal{B}$ is a basis for V, $\mathcal{C}$ a basis for W, we can consider the operation which sends the coordinates of $\boldsymbol{x} \in V$ in $\mathcal{B}$ to the coordinates of $T\boldsymbol{x} \in W$ in $\mathcal{C}$. This is represented by a matrix, called *the matrix of T relative to the bases $\mathcal{B}$ and $\mathcal{C}$*, and denoted $[T]_{\mathcal{C} \leftarrow \mathcal{B}}$. Thus it is required that

$$[T(\boldsymbol{x})]_{\mathcal{C}} = [T]_{\mathcal{C} \leftarrow \mathcal{B}} [\boldsymbol{x}]_{\mathcal{B}},$$

and clearly

$$[T]_{\mathcal{C} \leftarrow \mathcal{B}} = \left([T(\boldsymbol{b}_0)]_{\mathcal{C}} \; [T(\boldsymbol{b}_1)]_{\mathcal{C}} \; \cdots \; [T(\boldsymbol{b}_{n-1})]_{\mathcal{C}} \right).$$

A.6 Eigenvectors and Eigenvalues

If A is a square matrix, a vector $\boldsymbol{v}$ is called an *eigenvector* if there exists a scalar λ so that $A\boldsymbol{v} = \lambda\boldsymbol{v}$. λ is called the corresponding *eigenvalue*. Clearly a matrix is non-singular if and only if 0 is not an eigenvalue. The concept of eigenvalues and eigenvectors also gives meaning for a linear transformation from a vector space to itself: If T is such a transformation, $\boldsymbol{v}$ is an eigenvector with eigenvalue λ if $T(\boldsymbol{v}) = \lambda\boldsymbol{v}$.

By the *eigenspace* of A corresponding to eigenvalue λ we mean the set of all vectors $\boldsymbol{v}$ so that $A\boldsymbol{v} = \lambda\boldsymbol{v}$. This is a vector space, and may have dimension larger than 1. In basic linear algebra textbooks one often shows that, if A is a real, symmetric matrix,

- the eigenvalues of A are real (this also implies that it has real eigenvectors),
- the eigenspaces of A are orthonormal and together span $\mathbb{R}^n$, so that any vector can be decomposed as a sum of eigenvectors from A.

If we let P be the matrix consisting of the eigenvectors of A, then clearly $AP = DP$, where D is the diagonal matrix with the eigenvalues on the diagonal. Since the eigenvectors are orthonormal, we have that $P^{-1} = P^T$. It follows that, for any symmetric matrix A, we can write

$$A = PDP^T,$$

where P is orthogonal and D is diagonal. We say that A is *orthogonally diagonalizable*. It turns out that a real matrix is symmetric if and only if it is orthogonally diagonalizable.

If A instead is a complex matrix, we say that it is *unitarily diagonalizable* if it can be written as

$$A = PDP^H,$$

For complex matrices there are different results for when a matrix is unitarily diagonalizable. It turns out that a complex matrix A is unitarily diagonalizable if and only if $AA^H = A^H A$. We say then that A is *normal*.

Digital filters as defined in Chap. 3, were shown to be unitarily diagonalizable (having the orthonormal Fourier basis vectors as eigenvectors). It follows that these matrices are normal. This also follows from the fact that filters commute, and since A^H is a filter when A is. Since real filters usually are not symmetric, they are not orthogonally diagonalizable, however. Symmetric filters are of course orthogonally diagonalizable, just as their symmetric restrictions (denoted by S_r), see Sect. 3.5 (S_r was diagonalized by the real and orthogonal DCT matrix).

A matrix A is said to be *diagonalizable* if it can be written on the form

$$A = PDP^{-1},$$

with D diagonal. This is clearly a generalization of the notion of orthogonally/unitarily diagonalizable. Also here the diagonal matrix D contains the eigenvalues of A, and the columns of P are a basis of eigenvectors for A, but they may not be orthonormal: Matrices exist which are diagonalizable, but not orthogonally/unitarily diagonalizable. It i straightforward to show that if A is diagonalizable and it also has orthonormal eigenspaces, then it is also orthogonally/unitarily diagonalizable.

A mapping on the form $T(A) = P^{-1}AP$ with P non-singular is also called a *similarity transformation*, and the matrices A and $T(A)$ are said to be similar. Similar matrices have the same eigenvalues, since

$$P^{-1}AP - \lambda I = P^{-1}(A - \lambda I)P,$$

so that the characteristic equations for A and $T(A)$ have the same roots. The eigenvectors of A and $T(A)$ are in general different: One can see that if v is an eigenvector for A, then $P^{-1}v$ is an eigenvector for $T(A)$ with the same eigenvalue. Note that A may be orthogonally/unitarily diagonalizable (i.e. on the form $P_0 D P_0^H$), while $T(A)$ is not. If P is unitary, however, $T(A)$ will also be orthogonally/unitarily diagonalizable, since

$$T(A) = P^{-1}AP = P^H P_0 D P_0^H P = (P^H P_0)D(P^H P_0)^H$$

We say that A and $T(A)$ are *unitarily similar*. In Chap. 7 we encountered unitary similarity transformations which used permutation matrices. The permutation grouped each polyphase component together.

A.7 Positive Definite Matrices

A symmetric $n \times n$-matrix A is said to be *positive semidefinite* if $x^H A x \geq 0$ for all $x \in \mathbb{C}^n$. If this inequality is strict for all $x \neq 0$, A is said to be *positive definite*. Any matrix on the form $B^H B$ is positive semidefinite, since $x^H B^H B x = (Bx)^H(Bx) = \|Bx\|^2 \geq 0$. If B also has linearly independent columns, $B^H B$ is also positive definite.

Any positive semidefinite matrix has nonnegative eigenvalues, and any positive definite matrix has positive eigenvalues. To show this write $A = UDU^H$ with D diagonal and U unitary. We then have that

$$\boldsymbol{x}^H A \boldsymbol{x} = \boldsymbol{x}^H U D U^H \boldsymbol{x} = \boldsymbol{y} D \boldsymbol{y} = \sum_{i=1}^{n} \lambda_i y_i^2.$$

where $\boldsymbol{y} = U^H \boldsymbol{x}$ is nonzero if and only if $\boldsymbol{x}$ is nonzero. Clearly $\sum_{i=1}^{n} \lambda_i y_i^2$ is nonnegative for all $\boldsymbol{y}$ if and only if all $\lambda_i \geq 0$, and positive for all $\boldsymbol{y} \neq \boldsymbol{0}$ if and only if all $\lambda_i > 0$.

For any positive definite matrix A we have that

$$\|A\boldsymbol{x}\|^2 = \|UDU^H\boldsymbol{x}\|^2 = \|DU^H\boldsymbol{x}\|^2 = \sum_{i=1}^{n} \lambda_i^2 (U^H\boldsymbol{x})_i^2.$$

Also, if the eigenvalues of A are arranged in decreasing order,

$$\lambda_n^2 \|\boldsymbol{x}\|^2 = \lambda_n^2 \|U^H\boldsymbol{x}\|^2 = \lambda_n^2 \sum_{i=1}^{n} (U^H\boldsymbol{x})_i^2 \leq \sum_{i=1}^{n} \lambda_i^2 (U^H\boldsymbol{x})_i^2$$

$$\leq \lambda_1^2 \sum_{i=1}^{n} (U^H\boldsymbol{x})_i^2 = \lambda_1^2 \|U^H\boldsymbol{x}\|^2 = \lambda_1^2 \|\boldsymbol{x}\|^2.$$

It follows that

$$\lambda_n \|\boldsymbol{x}\| \leq \|A\boldsymbol{x}\| \leq \lambda_1 \|\boldsymbol{x}\|.$$

Thus, the eigenvalues of a positive definite matrix describe the maximum and minimum increase in the length of a vector when applying A. Clearly the upper and lower bounds are achieved by eigenvectors of A.

A.8 Singular Value Decomposition

In Chap. 6 we encountered frames. The properties of frames were proved using the singular value decomposition, which we now review. Any $m \times n$-matrix A can be written on the form $A = U \Sigma V^H$ where

- U is a unitary $m \times m$-matrix,
- Σ is a diagonal $m \times n$-matrix (in the sense that only the entries $\Sigma_{n,n}$ can be non-zero) with non-negative entries on the diagonal, and in decreasing order, and
- V is a unitary $n \times n$-matrix.

The entries on the diagonal of Σ are called *singular values*, and are denoted by σ_n. $A = U \Sigma V^H$ is called a *singular value decomposition* of A. A singular value decomposition is not unique: Many different U and V may be used in such a decomposition. The matrix Σ, and hence the singular values, are always the same in such a decomposition, however. It turns out that the singular values equal the square roots of the eigenvalues of $A^H A$, and also the square roots of the eigenvalues of AA^H (it turns out that the non-zero eigenvalues of $A^H A$ and AA^H are equal). $A^H A$ and AA^H both have only non-negative eigenvalues since they are positive semidefinite.

Since the U and V in a singular value decomposition are non-singular, we have that $\text{rank}(A) = \text{rank}(\Sigma)$. Since Σ is diagonal, its rank is the number of nonzero entries on the diagonal, so that rank(A) equals the number of positive singular values. Denoting this rank by r we thus have that

$$\sigma_1 \geq \sigma_2 \geq \cdots \sigma_r > \sigma_{r+1} = \sigma_{r+2} = \cdots = \sigma_{\min(m,n)} = 0.$$

In the following we will always denote the rank of A by r. Splitting the singular value decomposition into blocks, A can also be written on the form $A = U_1 \Sigma_1 (V_1)^H$ where

- U is an $m \times r$-matrix with orthonormal columns,
- Σ_1 is the $r \times r$-diagonal matrix $\text{diag}(\sigma_1, \sigma_2, \ldots, \sigma_r)$, and
- V is an $n \times r$-matrix with orthonormal columns.

$A = U_1 \Sigma_1 (V_1)^H$ is called a *singular value factorization* of A. The matrix Σ_1 is non-singular by definition. The matrix with singular value factorization $V_1 (\Sigma_1)^{-1} (U_1)^H$ is called the *generalized inverse* of A. One can prove that this matrix is unique. It is denoted by $A^\dagger$. If A is a square, non-singular matrix, it is straightforward to prove that $A^\dagger = A^{-1}$.

Now, if A has rank n (i.e. the columns are linearly independent, so that $m \geq n$), $A^H A$ is an $n \times n$-matrix where all n eigenvalues are positive, so that it is non-singular. Moreover $V_1 = V$ is unitary in the singular value factorization of A. It follows that

$$A^\dagger A = V_1 (\Sigma_1)^{-1} (U_1)^H = U_1 \Sigma_1 (V_1)^H = V_1 (V_1)^H = I,$$

so that $A^\dagger$ indeed van be considered as an inverse. In this case the following computation also shows that there exists a concrete expression for $A^\dagger$:

$$
\begin{aligned}
(A^H A)^{-1} A^H &= (V_1 (\Sigma_1)^T (U_1)^H U_1 \Sigma_1 (V_1)^H)^{-1} V_1 (\Sigma_1)^T (U_1)^H \\
&= (V_1 \Sigma_1^2 (V_1)^H)^{-1} V_1 \Sigma_1 (U_1)^H = V_1 \Sigma_1^{-2} (V_1)^H V_1 \Sigma_1 (U_1)^H \\
&= V_1 \Sigma_1^{-2} \Sigma_1 (U_1)^H = V_1 \Sigma_1^{-1} (U_1)^H = A^\dagger.
\end{aligned}
$$

Nomenclature

$(\langle \mathcal{B}_1, \mathcal{B}_2 \rangle)$	Gramm matrix of the bases $\mathcal{B}_1$ and $\mathcal{B}_2$
$[\boldsymbol{x}]_{\mathcal{B}}$	Coordinate vector of $\boldsymbol{x}$ relative to the basis $\mathcal{B}$
$[T]_{\mathcal{C} \leftarrow \mathcal{B}}$	The matrix of T relative to the bases $\mathcal{B}$ and $\mathcal{C}$
$\boldsymbol{x} * \boldsymbol{y}$	Convolution of vectors
$\boldsymbol{x} \circledast \boldsymbol{y}$	Circular convolution of vectors
$\boldsymbol{x}^{(e)}$	Vector of even samples
$\boldsymbol{x}^{(o)}$	Vector of odd samples
$\breve{\boldsymbol{x}}$	Symmetric extension of a vector
$\breve{f}$	Symmetric extension of the function f
$\mathcal{C}_m$	Time-ordering of $(\boldsymbol{\phi}_{m-1}, \boldsymbol{\psi}_{m-1})$
$\mathcal{D}_N$	N-point DCT basis for $\mathbb{R}^N$, i.e. $\{\boldsymbol{d}_0, \boldsymbol{d}_1, \cdots, \boldsymbol{d}_{N-1}\}$
$\mathcal{D}_{N,T}$	Order N real Fourier basis for $V_{N,T}$
$\mathcal{E}_N$	Standard basis for $\mathbb{R}^N$, i.e. $\{\boldsymbol{e}_0, \boldsymbol{e}_1, \cdots, \boldsymbol{e}_{N-1}\}$
$\mathcal{F}_N$	Fourier basis for $\mathbb{R}^N$, i.e. $\{\boldsymbol{\phi}_0, \boldsymbol{\phi}_1, \cdots, \boldsymbol{\phi}_{N-1}\}$
$\mathcal{F}_{N,T}$	Order N complex Fourier basis for $V_{N,T}$
χ_A	Characteristic function for the set A
$\hat{\boldsymbol{x}}$	DFT of the vector $\boldsymbol{x}$
$\hat{f}$	Continuous-time Fourier Transform of f
$\lambda_s(\nu)$	Frequency response of a filter
$\lambda_S(\omega)$	Continuous frequency response of a digital filter
$\langle \boldsymbol{u}, \boldsymbol{v} \rangle$	Inner product
ν	Frequency
ω	Angular frequency
$\oplus$	Direct sum
$\otimes$	Tensor product
$\overline{A}$	Conjugate of a matrix
ϕ	Scaling function
$\phi_{m,n}$	Scaled and translated version of ϕ
σ_i	Singular values of a matrix
DCT_N	$N \times N$-DCT matrix
DFT_N	$N \times N$-DFT matrix
$\text{Supp}(f)$	Support of f

© Springer Nature Switzerland AG 2019
Ø. Ryan, *Linear Algebra, Signal Processing, and Wavelets - A Unified Approach*,
Springer Undergraduate Texts in Mathematics and Technology,
https://doi.org/10.1007/978-3-030-01812-2

$\tilde{\phi}$	Dual scaling function
$\tilde{\psi}$	Dual mother wavelet
$\tilde{V}_m$	Dual resolution space
$\tilde{W}_m$	Dual detail space
$\boldsymbol{\phi}_m$	Basis for V_m
$\boldsymbol{\psi}_m$	Basis for W_m
A^H	Conjugate transpose of a matrix
A^T	Transpose of a matrix
A^{-1}	Inverse of a matrix
$A^\dagger$	Generalized inverse of A
A_λ	Elementary lifting matrix of even type
B_λ	Elementary lifting matrix of odd type
$c_{m,n}$	Coordinates in $\boldsymbol{\phi}_m$
E_d	Filter which delays with d samples
F_N	$N \times N$-Fourier matrix
f_N	N'th order Fourier series of f
f_s	Sampling frequency. Also used for the square wave
f_t	Triangle wave
G	IDWT kernel, or reverse filter bank transform
G_0, G_1	IDWT filter components
H	DWT kernel, or forward filter bank transform
H_0, H_1	DWT filter components
$l(S)$	Length of a filter
N	Number of points in a DFT/DCT
$O(N)$	Order of an algorithm
$P_{\mathcal{C} \leftarrow \mathcal{B}}$	Change of coordinate matrix from $\mathcal{B}$ to $\mathcal{C}$
$S^{\leftrightarrow}$	Matrix with the columns reversed
S_r	Symmetric restriction of S
T	Period of a function
T_s	Sampling period
V_m	Resolution space
$V_{N,T}$	N'th order Fourier space
W_m	Detail space
$W_m^{(0,1)}$	Resolution m Complementary wavelet space, LH
$W_m^{(1,0)}$	Resolution m Complementary wavelet space, HL
$W_m^{(1,1)}$	Resolution m Complementary wavelet space, HH
$w_{m,n}$	Coordinates in $\boldsymbol{\psi}_m$

References

[1] A. Ambardar, *Digital Signal Processing: A Modern Introduction* (Cengage Learning, Belmont, 2006)

[2] T. Barnwell, An experimental study of sub-band coder design incorporating recursive quadrature filters and optimum ADPCM, in *Acoustics, Speech, and Signal Processing, IEEE International Conference on ICASSP*, pp. 808–811 (1981)

[3] A. Boggess, F.J. Narcowich, *A First Course in Wavelets with Fourier Analysis.* (Prentice Hall, Upper Saddle River, 2001)

[4] O. Bratteli, P. Jorgensen, *Wavelets Through a Looking Glass* (Birkhauser, Boston, 2002)

[5] C.M. Brislawn, Fingerprints go digital. Not. AMS **42**(11), 1278–1283 (1995)

[6] A.A.M.L. Bruekens, A.W.M. van den Enden, New networks for perfect inversion and perfect reconstruction. IEEE J. Sel. Areas Commun. **10**(1), 130–137 (1992)

[7] B.A. Cipra, The best of the 20th century: Editors name top 10 algorithms. SIAM News **33**(4) (2000). http://www.uta.edu/faculty/rcli/TopTen/topten.pdf

[8] A. Cohen, I. Daubechies, Wavelets on the interval and fast wavelet transforms. Appl. Comput. Harmon. Anal. **1**, 54–81 (1993)

[9] A. Cohen, I. Daubechies, J.-C. Feauveau, Biorthogonal bases of compactly supported wavelets. Commun. Pure Appl. Math. **45**(5), 485–560 (1992)

[10] J.W. Cooley, J.W. Tukey, An algorithm for the machine calculation of complex fourier series. Math. Comput. **19**, 297–301 (1965)

[11] A. Croisier, D. Esteban, C. Galand, Perfect channel splitting by use of interpolation/decimation/tree decomposition techniques, in *International Conference on Information Sciences and Systems*, pp. 443–446 (1976)

[12] I. Daubechies, Orthonormal bases of compactly supported wavelets. Commun. Pure Appl. Math. **41**(7), 909–996 (1988)

[13] I. Daubechies, *Ten Lectures on Wavelets.* CBMS-NSF Conference Series in Applied Mathematics (SIAM, Philadelphia, 1992)

[14] I. Daubechies, W. Sweldens, Factoring wavelet transforms into lifting steps. J. Fourier Anal. Appl. **4**(3), 247–269 (1998)

[15] A. Deitmar, *A First Course in Harmonic Analysis*, 2nd edn. (Springer, New York, 2005)

© Springer Nature Switzerland AG 2019

Ø. Ryan, *Linear Algebra, Signal Processing, and Wavelets - A Unified Approach*,
Springer Undergraduate Texts in Mathematics and Technology,
https://doi.org/10.1007/978-3-030-01812-2

[16] P. Duhamel, H. Hollmann, 'split-radix' FFT-algorithm. Electron. Lett. **20**(1), 14–16 (1984)

[17] N. Efford, *Digital Image Processing. A Practical Introduction Using Java* (Addison-Wesley, Boston, 2000)

[18] Y.C. Eldar, *Sampling Theory* (Cambridge University Press, Cambridge, 2015)

[19] FBI, WSQ gray-scale fingerprint image compression specification. Technical report, IAFIS-IC (1993)

[20] S. Foucart, H. Rauhut, *A Mathematical Introduction to Compressive Sensing* (Birkhauser, Basel, 2013)

[21] M.W. Frazier, *An Introduction to Wavelets Through Linear Algebra* (Springer, New York, 1999)

[22] M. Frigo, S.G. Johnson, The design and implementation of FFTW3. Proc. IEEE **93**(2), 216–231 (2005)

[23] T.W. Gamelin, *Complex Analysis* (Springer, New York, 2001)

[24] R.C. Gonzalez, R.E. Woods, S.L. Eddins, *Digital Image Processing Using MATLAB* (Gatesmark Publishing, Knoxville, 2009)

[25] F.J. Harris, On the use of windows for harmonic analysis with the discrete Fourier transform. Proc. IEEE **66**(1), 51–83 (1978)

[26] ISO/IEC, Information technology – coding of moving pictures and associated audio for digital storage media at up to about 1.5 mbit/s. Technical report, ISO/IEC (1993)

[27] ISO/IEC, JPEG2000 part 1 final draft international standard. ISO/IEC FDIS 15444-1. Technical report, ISO/IEC (2000)

[28] S.G. Johnson, M. Frigo, A modified split-radix FFT with fewer arithmetic operations. IEEE Trans. Signal Process. **54**, 1562 (2006)

[29] G. Kaiser, *A Friendly Guide to Wavelets* (Birkhauser, Basel, 1994)

[30] Y. Katznelson, *An Introduction to Harmonic Analysis*, 3rd edn. Cambridge Mathematical Library (Cambridge University Press, Cambridge, 2002)

[31] H.P. Langtangen, *A Primer on Scientific Programming with Python*, 5th edn. (Springer, New York, 2016)

[32] D.C. Lay, *Linear Algebra and Its Applications*, 4th edn. (Addison-Wesley, Boston, 2011)

[33] S.J. Leon, *Linear Algebra with Applications*, 8th edn. (Pearson, Upper Saddle River, 2010)

[34] S. Linge, H.P. Langtangen, *Programming for Computations – MATLAB/Octave* (Springer, New York, 2016)

[35] S. Linge, H.P. Langtangen, *Programming for Computations – Python* (Springer, New York, 2016)

[36] T.D. Lookabaugh, M.G. Perkins, Application of the Princen-Bradley filter bank to speech and image compression. IEEE Trans. Acoust. Speech Signal Process. **38**(11), 1914–1926 (1990)

[37] T. Lyche, *Numerical Linear Algebra and Matrix Factorizations* (Springer, New York, 2018)

[38] J. Ma, G. Plonka, The curvelet transform. IEEE Signal Process. Mag. **27**, 118–133 (2010)

[39] S.G. Mallat, Multiresolution approximations and wavelet orthonormal bases of $L^2(\mathbb{R})$. Trans. Am. Math. Soc. **315**(1), 69–87 (1989)

[40] S. Mallat, *A Wavelet Tour of Signal Processing* (Tapir Academic Press, Boston, 1998)

[41] A. Malthe-Sørenssen, *Elementary Mechanics Using Matlab* (Springer, New York, 2015)

[42] A. Malthe-Sørenssen, *Elementary Mechanics Using Python* (Springer, New York, 2015)

[43] H.S. Malvar, Modulated QMF filter banks with perfect reconstruction. Electron. Lett. **26**(13), 906–907 (1990)

[44] H.S. Malvar, *Signal Processing with Lapped Transforms* (Artech House, Norwood, 1992)

[45] S. Martucci, Symmetric convolution and the discrete sine and cosine transforms. IEEE Trans. Signal Process. **42**, 1038–1051 (1994)

[46] Y. Meyer, Principe d'incertitude, bases hilbertiennes et algebres d'operateurs. Seminaire Bourbaki **662**, 209–223 (1985/1986)

[47] Y. Meyer, Ondelettes et functions splines, in *Seminaire EDP, Ecole Polytecnique*, Paris, France, Dec 1986

[48] Y. Meyer, *Wavelets and Operators* (Cambridge University Press, Cambridge, 1992)

[49] C.D. Meyer, *Matrix Analysis and Applied Linear Algebra* (SIAM, Philadelphia, 2000)

[50] Y. Meyer, R. Coifman, *Wavelets. Calderon-Zygmund and Multilinear Operators* (Cambridge University Press, Cambridge, 1997)

[51] K. Mørken, *Numerical Algorithms and Digital Representation* (UIO, 2013)

[52] P. Noll, MPEG digital audio coding. IEEE Signal Process. Mag. **14**, 59–81 (1997)

[53] P.J. Olver, C. Shakiban, *Applied Linear Algebra* (Pearson, Upper Saddle River, 2006)

[54] A.V. Oppenheim, R.W. Schafer, *Discrete-Time Signal Processing* (Prentice Hall, Upper Saddler River, 1989)

[55] D. Pan, A tutorial on MPEG/audio compression. IEEE Multimedia **2**, 60–74 (1995)

[56] W.B. Pennebaker, J.L. Mitchell, *JPEG Still Image Data Compression Standard* (Van Nostrand Reihnold, New York, 1993)

[57] J.P. Prince, A.B. Bradley, Analysis/synthesis filter bank design based on time domain aliasing cancellation. IEEE Trans. Acoust. Speech Signal Process. **34**(5), 1153–1161 (1986)

[58] J.P. Princen, A.W. Johnson, A.B. Bradley, Subband/transform coding using filter bank designs based on time domain aliasing cancellation, in *IEEE Proceedings of the International Conference on Acoustics, Speech, and Signal Processing*, pp. 2161–2164 (1987)

[59] J.G. Proakis, D.G. Manolakis, *Digital Signal Processing. Principles, Algorithms, and Applications*, 5th edn. (Pearson, Upper Saddle River, 2007)

[60] C.M. Rader, Discrete Fourier transforms when the number of data samples is prime. Proc. IEEE **56**, 1107–1108 (1968)

[61] T.A. Ramstad, S.O. Aase, J.H. Husøy, *Subband Compression of Images: Principles and Examples*, vol. 6 (Elsevier, Amsterdam, 1995)

[62] T.D. Rossing, *Handbook of Acoustics* (Springer, New York, 2015)

[63] J.H. Rothweiler, Polyphase quadrature filters – a new subband coding technique, in *ICASSP 83*, Boston, pp. 1280–1283 (1983)

[64] D. Salomon, *Data Compression. The Complete Reference*, 5th edn. (Springer, New York, 2007)

[65] K. Sayood, *Introduction to Data Compression*, 2nd edn. (Morgan Kaufmann, Cambridge, 2000)

[66] C.E. Shannon, Communication in the presence of noise. Proc. Inst. Radio Eng. **37**(1), 10–21 (1949)

[67] M.J.T. Smith, T.P. Barnwell, A new filter bank theory for time-frequency representation. IEEE Trans. Acoust. Speech Signal Process. **35**(3), 314–327 (1987)

[68] P. Stoica, R. Moses, *Spectral Analysis of Signals* (Prentice Hall, Upper Saddler River, 2005)

[69] G. Strang, *Linear Algebra and Its Applications*, 3rd edn. (Brooks/Cole, Belmont, 1988)

[70] G. Strang, T. Nguyen, *Wavelets and Filter Banks* (Cambridge Press, Wellesley, 1996)

[71] W. Sweldens, The lifting scheme: a new philosophy in biorthogonal wavelet constructions, in *Wavelet Applications in Signal and Image Processing III*, pp. 68–79 (1995)

[72] W. Sweldens, The lifting scheme: a custom-design construction of biorthogonal wavelets. Appl. Comput. Harmon. Anal. **3**, 186–200 (1996)

[73] T. Tao, *Analysis II*, 3rd edn. (Springer, New York, 2015)

[74] D.S. Taubman, M.W. Marcellin, *JPEG2000. Image Compression. Fundamentals, Standards and Practice* (Kluwer Academic Publishers, Boston, 2002)

[75] F. Uhlig, *Transform Linear Algebra* (Prentice-Hall, Upper Saddler River, 2002)

[76] M. Vetterli, J. Kovacevic, *Wavelets and Subband Coding* (Prentice Hall, Upper Saddler River, 1995)

[77] M. Vetterli, H.J. Nussbaumer, Simple FFT and DCT algorithms with reduced number of operations. Signal Proc. **6**, 267–278 (1984)

[78] M. Vetterli, E. Kovasevic, V.K. Goyal, *Foundations of Signal Processing* (Cambridge University Press, Cambridge, 2014)

[79] Z. Wang, Fast algorithms for the discrete W transform and for the discrete Fourier transform. IEEE Trans. Acoust. Speech Signal Process. **32**(4), 803–816 (1984)

[80] S. Winograd, On computing the discrete Fourier transform. Math. Comput. **32**, 175–199 (1978)

[81] R. Yavne, An economical method for calculating the discrete Fourier transform. Proc. AFIPS Fall Joint Comput. Conf. **33**, 115–125 (1968)

Index

© Springer Nature Switzerland AG 2019
Ø. Ryan, *Linear Algebra, Signal Processing, and Wavelets - A Unified Approach*,
Springer Undergraduate Texts in Mathematics and Technology,
https://doi.org/10.1007/978-3-030-01812-2

CPSIA information can be obtained
at www.ICGtesting.com
Printed in the USA
LVHW060953100319
610105LV00005B/12/P